WITHDRAWN
WRIGHT STATE UNIVERSITY LIBRARIES

Virtual Prototyping & Bio Manufacturing in Medical Applications

Bopaya Bidanda · Paulo J. Bártolo
Editors

Virtual Prototyping & Bio Manufacturing in Medical Applications

Editors

Bopaya Bidanda
Ernest Roth Professor & Chairman
Department of Industrial Engineering
University of Pittsburgh
Pittsburgh, PA 15261
USA

Paulo Bártolo
Centre for Rapid and Sustainable Product
Development
Institute for Polymers and Composites
Polytechnic Institute of Leiria
School of Technology and Management
Morro do Lena
Alto do Veria, Leiria
Portugal

ISBN: 978-0-387-33429-5   e-ISBN: 978-0-387-68831-2

Library of Congress Control Number: 2007928989

© 2008 Springer Science+Business Media, LLC
All rights reserved. This work may not be translated or copied in whole or in part without the written permission of the publisher (Springer Science+Business Media, LLC, 233 Spring Street, New York, NY 10013, USA), except for brief excerpts in connection with reviews or scholarly analysis. Use in connection with any form of information storage and retrieval, electronic adaptation, computer software, or by similar or dissimilar methodology now known or hereafter developed is forbidden. The use in this publication of trade names, trademarks, service marks and similar terms, even if they are not identified as such, is not to be taken as an expression of opinion as to whether or not they are subject to proprietary rights.

Printed on acid-free paper.

9 8 7 6 5 4 3 2 1

springer.com

*This book is dedicated to
our parents
Maria Alice and Francisco Bártolo
&
Neena and Monapa Bidanda*

*And our families
Helena and Pedro
Louella, Maya & Rahul
for their constant support and cheeriness
throughout this project!*

Preface

We are especially pleased to present our edited book in an area that is quickly emerging as one of the most active research areas that integrate both engineering and medicine. Preliminary research results show significant potential in effecting major breakthroughs ranging from a reduction in the number of corrective surgeries needed to the 'scientific miracle' of generating tissue growth. Over $600 million has been in invested in tissue engineering last year alone—a large and significant component of this is in the area of virtual and physical prototyping.

Virtual & Physical prototyping can broadly be divided into three categories: Modeling, Manufacturing & Materials. This book focuses on the first part and some areas of the second. The second book in this series will focus on the areas in the second and third categories. As you will see from this book, the principles utilized draw heavily from the more traditional engineering fields including mechanical engineering, industrial engineering, civil engineering (structures), electrical engineering and bio engineering.

The first chapter by Winder provides an insight into the practicalities of creating custom made implants for the skull details. The concept of cranioplasty, describing the correction of a bone defect or deformity in the cranium using a bio-compatible material, is described and explored from a computational point of view

In Chapter 2, Ming Leu and his group of researchers review the use of virtual reality technology for the development of a virtual bone surgery system, which can be used for training in orthopedic surgery and planning of bone surgery procedures. Then, they discuss the basic methods and techniques used to develop these systems.

Petros Patias compares medical imaging with photogrammetry in Chapter 3. The mathematical foundations of photogrammetry are reviewed and important issues explored like multi-sensor image registration and fusion, matching of image sequences, image segmentation and motion tracking analysis.

In Chapter 4, Naing, *et al.* present a system of CAD structures based on convex polyhedral for use with Rapid Prototyping (RP) technology in tissue engineering applications that allows the designer (given the unit cell and the required dimensions), to automatically generate a structure that is suitable for the intended tissue engineering application.

Tissue engineering scaffolds must fill complex anatomic defects, provide temporary mechanical function (which can be represented by constitutive models like elasticity, viscoelasticity and poroelasticity), and enhance tissue regeneration through biologic delivery and mass transport to support in-growing cells and tissue. Tissue engineering scaffolds must provide a porous architecture that can satisfy that can balance these often conflicting requirements. Furthermore, the porous architecture is hierarchical and must be designed over scales ranging from microns to centimeters. Computational scaffold design methods must thus be able to represent scaffold topology over many orders of scale and be able to compute effective mechanical and mass transport properties from design porous architecture. In addition, the capability to simulate scaffold degradation and tissue growth within the scaffold would help improve design of tissue engineering treatments. The effective design of scaffolds is critical in moving this field forward. Wettergreen *et al.*, in Chapter 5, also describe computer-aided tissue engineering system of unit blocks and interfaces, that serve as the basic building unit for engineered bone replacement scaffolds. The sixth Chapter by Hollister *et al.*, describes an image-based hierarchical design approach for computational design, optimization and simulation of tissue engineering scaffolds.

Scaffolds are key structures in Tissue Engineering as they provide an initial biochemical substrate for the novel tissue until cells can produce their own extracellular matrix. Therefore scaffolds not only define the 3D space for the formation of new tissues, but also serve to provide tissues with appropriate functions. Several techniques have been developed to produce scaffolds. In Chapter 7, Morsi *et al.*, present and discuss the progress of the so-called *conventional methods* that have been used to construct 3D porous scaffolds. In Chapter 8, Bártolo *et al.*, describe the current state of the art of rapid prototyping in Tissue Engineering. Different additive techniques are described and their main advantages and disadvantages analyzed.

In Chapter 9, Alexander Woez reviews the various attempts to use rapid prototyping techniques to directly or indirectly produce scaffolds with a defined architecture from various materials for bone replacement and summarize the research work carried out by his group at both Cambridge (UK) and Max Planck Institute (Germany).

The ability to print tissue and organs layer-by-layer or cell-by-cell is a topical area of research. There are many techniques capable of printing living cells including ink jet printers, electrospray deposition, and various laser-based technologies. In Chapter 10 Reigeisen *et al.*, describe cell printing via two modified laser induced forward transfer (LIFT) methods. The techniques, referred as matrix assisted pulsed laser evaporation direct write (MAPLE DW) and biological laser printing (BioLP), are the only cell printers that do not require a print-head, nozzle, or orifice to obtain micron-scale resolution, eliminating potential contamination and clogging issues common with other techniques. These technologies also have the proven ability to print 2D and 3D cell structures, and can print with resolutions ranging from a single cell per spot to hundreds of cells per spot with maximum speeds of thousands of spots per second. Specific experimental details are provided including results that demonstrate no damage to the genotype and phenotype of printed cells.

In Chapter 11, Suman Das describes recent work on selective laser sintering (SLS) for making tissue engineering scaffolds in bioresorbable polymers and polymer-ceramic composites. A general overview about the physical fundamentals of SLS is provided as well as models of heat, mass, and momentum transport phenomena associated with sintering, melting and resolidification of polymer powders. The Chapter also provides in-depth information about selective laser sintering of Nylon-6, a biocompatible polymer, polycaprolactone (PCL), a bioresorbable polymer, and composites of polycaprolactone and tricalcium phosphate for making bone tissue engineering scaffolds. Mechanical property data as a function of scaffold composition and interior architecture are presented.

The last Chapter by Hutmacher et al., the authors define scaffold properties and attempt to define some broad criteria and constraints for scaffold design. A review on the current state-of-the-art in rapid prototyping applied to scaffold fabrication within the tissue engineering context is provided. The chapter completes with the design and development of 3D scaffold family with a range of morphological and biomechanical properties using synthetic biopolymer via an in-house developed extrusion based rapid prototyping technique. It focuses on the aspects related to the process parameters (temperature, pressure and deposition speed), structural features (lay-down pattern and filament distance), and mechanical properties of the scaffolds.

The production of this book has been a most enjoyable experience. We thank the authors for their valuable and timely contributions to this volume. We also would like to thank Dr Luciano de Almeida, President of the Polytechnic Institute of Leiria and Dean Gerald Holder, US Steel Dean of Engineering, University of Pittsburgh for their support of our academic endeavors.

We would especially like to acknowledge the unlimited patience and constant support of Steve Elliot and Jennifer Mirski from Springer. Without their untiring efforts, this book would not have been possible.

Bopaya Bidanda
Paulo Bártolo

Contents

1. **Computer Assisted Cranioplasty**
   Dr John Winder 1

2. **Virtual Bone Surgery**
   Ming C. Leu, Qiang Niu and Xiaoyi Chi 21

3. **Medical Imaging Challenges Photogrammetry**
   Petros Patias 45

4. **Computer Aided Tissue Engineering Scaffold Fabrication**
   M. W. Naing, C. K. Chua and K. F. Leong 67

5. **CAD Assembly Process for Bone Replacement Scaffolds in Computer-Aided Tissue Engineering**
   M. A. Wettergreen, B. S. Bucklen, M. A. K. Liebschner and W. Sun 87

6. **Computational Design and Simulation of Tissue Engineering Scaffolds**
   Scott J. Hollister, Chia-Ying Lin, Heesuk Kang and Taiji Adachi 113

7. **Virtual Prototyping of Biomanufacturing in Medical Applications**
   Y. S. Morsi, C. S. Wong and S. S. Patel 129

8. **Advanced Processes to Fabricate Scaffolds for Tissue Engineering**
   Paulo J. Bártolo, Henrique A. Almeida, Rodrigo A. Rezende, Tahar Laoui and Bopaya Bidanda 149

9. **Rapid Prototyping to Produce Porous Scaffolds with Controlled Architecture for Possible use in Bone Tissue Engineering**
   Alexander Woesz 171

10. **Laser Printing Cells**
    Bradley R. Ringeisen, Jason A. Barron, Daniel Young, Christina M. Othon, Doug Ladoucuer, Peter K. Wu and Barry J. Spargo 207

11. **Selective Laser Sintering of Polymers and Polymer-Ceramic Composites**
    Suman Das 229

12 **Design, Fabrication and Physical Characterization of Scaffolds Made from Biodegradable Synthetic Polymers in combination with RP Systems based on Melt Extrusion**
D. W. Hutmacher, M. E. Hoque and Y. S. Wong 261

Index 293

List of Contributors

Taiji Adachi
Department of Mechanical
Engineering and Science Kyoto
University Japan

Henrique A. Almeida
Centre for Rapid and Sustainable
Product Development
Institute for Polymers and Composites
Polytechnic Institute of Leiria
School of Technology and Management
Morro do Lena
Alto do Vieiro
Leiria
Portugal

Jason A. Barron
U.S. Naval Research Laboratory
4555 Overlook Ave. SW
Washington
DC 20375, USA

Paulo J. Bártolo
Centre for Rapid and Sustainable
Product Development
Institute for Polymers and Composites
Polytechnic Institute of Leiria
School of Technology and Management
Morro do Lena
Alto do Vieiro
Leiria
Portugal

Bopaya Bidanda
Deprtment of Industrial Engineering
University of Pittsburgh
Pittsburgh
PA 15261, USA

B.S. Bucklen
Department of Bioengineering
Rice University
Houston
TX, 77098

Xiaoyi Chi
Department of Mechanical and
Aerospace Engineering
University of Missouri-Rolla
Rolla
Missouri 65409, USA

C.K. Chua
School of Mechanical and Aerospace
Engineering
Nanyang Technological University
Singapore

Scott J. Hollister
Departments of Biomedical Engineering, Mechanical Engineering, Surgery,
and Neurosurgery
The University of Michigan
USA

M. Enamul Hoque
Department of Mechanical Engineering
Faculty of Engineering
National University of Singapore
Engineering Drive 1
Singapore

Dietmar W. Hutmacher
Division of Bioengineering
Faculty of Engineering
Department of Orthopaedic Surgery
Yong Loo Lin School of Medicine
National University of Singapore
Engineering Drive 1
Singapore

Heesuk Kang
Departments of Mechanical
Engineering
The University of Michigan
USA

Doug Ladoucuer
U.S. Naval Research Laboratory
4555 Overlook Ave. SW
Washington,
DC 20375, USA

Tahar Laoui
University of Wolverhampton
RIATec, School of Engineering and
Built Environment
Wolverhampton, UK

K.F. Leong

Ming C. Leu
Department of Mechanical and
Aerospace Engineering
University of Missouri-Rolla
Rolla
Missouri 65409, USA

M.A.K. Liebschner
Department of Bioengineering
Rice University
Houston, TX, 77098

Chia-Ying Lin
Departments of Biomedical
Engineering
Neurosurgery The University
of Michigan
USA

Y.S. Morsi

M.W. Naing
School of Mechanical and Aerospace
Engineering
Nanyang Technological University
Singapore

Qiang Niu
Department of Mechanical and
Aerospace Engineering
University of Missouri-Rolla
Rolla
Missouri 65409, USA

Christina M. Othon
U.S. Naval Research Laboratory
4555 Overlook Ave. SW
Washington
DC 20375, USA

S.S. Patel

Petros Patias
The Aristotle University of Thessaloniki
Dept. of Cadastre
Photogrammetry and Cartography
University Box 473
GR-54006 Thessaloniki
Greece
patias@topo.auth.gr

Rodrigo A. Rezende
Centre for Rapid and Sustainable
Product Development
Institute for Polymers and Composites
Polytechnic Institute of Leiria
School of Technology and Management
Morro do Lena
Alto do Vieiro
Leiria
Portugal

List of Contributors

Bradley R. Ringeisen
U.S. Naval Research Laboratory
4555 Overlook Ave. SW
Washington
DC 20375, USA

Barry J. Spargo
U.S. Naval Research Laboratory
4555 Overlook Ave. SW
Washington
DC 20375, USA

Suman Das
Mechanical Engineering Department
University of Michigan
Ann Arbor 48109-2125
USA

W. Sun
Department of Mechanical Engineering
Drexel University
Philadelphia, PA

M.A. Wettergreen
Department of Bioengineering
Rice University
Houston, TX, 77098

Dr John Winder
Reader in Rehabilitation Sciences (job title)
Health and Rehabilitation Sciences Research Institute (Department)
University of Ulster (Institution)

Alexander Woesz Max Planck
Institute of Colloids and Interfaces
Potsdam
Germany and Cambridge Centre for Medical Materials
University of Cambridge
UK

C.S. Wong

Y. S. Wong
Department of Mechanical Engineering
National University of Singapore
Engineering Drive 1
Singapore

Peter K. Wu
Southern Oregon University
Ashland, OR 97520

Daniel Young
Wright State University
Dayton, OH 45435

Chapter 1
Computer Assisted Cranioplasty

Dr John Winder

1.1 Introduction to Cranioplasty and Materials for Repair

1.1.1 Introduction to Chapter

The purpose of this chapter is to provide the engineer, medical imaging specialist and surgeon an introduction and insight into the practicalities of creating custom made implants for the skull. This field of research has grown steadily over the past decade. However, we still do not have a definitive solution to the problem—"How do we create a bio-compatible, easy to fit, durable and cost effective implant for repairing holes in the skull?" We have a number of solutions to the problem, although it is not clear from the literature which brings the most benefit to the patient, to the surgeon and which is most cost-effective.

The term cranioplasty describes the correction of a bone defect or deformity in the cranium using a bio-compatible material. Historically, many substances have been used to fill holes in the skull which were caused either by injury, primitive operations or disease. For example, defects have been filled with precious metals (gold or silver), animal bone, autologous bone graft (the patient's own bone) and more recently methyl methacrylate (Sanan and Haines 1997). Skull remnants from ancient graveyards in Peru showed evidence of trephination, the creation of a hole in the skull, and repair. Nowadays, the two most common methods of creating a repair are autologous bone (harvested from the patient's skull or the body) (Sheikh 2006) or acrylic resin, commonly known as bone cement, in the form of polymethylmethacrylate (PMMA) (Park et al. 2001; Chiarini et al. 2004; Raja and Linskey 2005). Other materials have also been used to varying degrees of success. Thin titanium sheet, with a thickness of between 0.5 and 1.0 mm can be formed under pressure to fit the contour of the skull. The plate is fitted with lugs to take fixation screws used to attach the plate to the outer surface of the skull (Joffe et al. 1999a). Hydroxyapatite cement on its own or used with titanium mesh has also been employed to repair small holes in the skull (Arriaga and Chen 2002). Titanium mesh has also been moulded into shape on polyurethane skull replica (Brandt and Haug 2002) and also compared to thin titanium sheet in a limited number of cases (Schipper et al. 2004). Schipper noted that titanium mesh on its own was useful for defects smaller than 100 cm² as it had reduced stability and shock resistance compared to the sheet.

The following sections detail methods of cranioplasty, computer assisted cranioplasty and implant design and current research in the field.

1.1.2 Clinical Indication for Cranioplasty

Holes in the skull may be caused by trauma, the result of decompressive surgery or a re-implanted bone flap becoming infected after a neurosurgical procedure (Artico et al. 2003). In each of these cases the patient's own bone has been damaged or become unusable due to infection and has to be replaced with another biocompatible material. An autologous implant is one which has been derived from the same patient whereas an alloplast implant is one that has been constructed from an inert biocompatible material. There is evidence in the literature that not only does repairing a cranial defect provide protection to the brain and a good cosmetic outcome for the patient but may also relieve neurological symptoms like headache, dizziness, depression and anxiety (Rotaru et al. 2006). It has also been demonstrated that a patient will recover neurocognitive function after cranioplasty (Anger et al. 2002). Normally, if the patient has had recent trauma or surgery, the surgeon will wait around 6 months for healing to occur before embarking on the repair.

1.1.3 Materials and Methods of Repair

As mentioned earlier a range of materials may be used for repairing a cranial defect. Acrylic resin, titanium sheet/mesh and hydroxyapatite have all been employed. Each of these materials have different physical and biological properties. They may be used alone or in combination with calvarial bone, bone dust or other alloplast materials. Cranial implants are normally preformed following moulage of the defect (the creating of a plaster impression through the overlying skin) (Gronet 2003). The patient's scalp is shaved and the palpated margins of the defect are outlined with an indelible marker. This margin is transposed onto the plaster impression. Swelling, haemorrhage, oedema and overlying muscle all contribute to the imprecise definition of the defect margin. Implants created using this method often require further manipulation during theatre to accurately fit the skull defect contour and are flat in appearance (Joffe et al. 1992).

Acrylic resin is provided to surgeons in kit form and is also known as bone cement. Its principal use has traditionally been in securing hip and knee implants in orthopaedic surgery. There is extensive research on a range of resins in these applications and many reviews in the field (Gousain 2004). The acrylic resin kit requires the surgeon to mix a polymer powder with a liquid monomer component (methyl methacrylate) to create a malleable paste. The powder may also contain a colour agent and a radio-opaque substance to make it visible after implantation by X-ray imaging. Some bone cements will contain an antibiotic, for example gentamicin, which is released slowly after implantation. The mixing of the powder polymer and liquid monomer results in an exothermic reaction during which the temperature of the paste will rise dramatically. It is not uncommon in laboratory experiments for the paste to reach a temperature of >70 °C (Lara et al. 1998; Lung and Darvell 2005). In orthopaedic surgery, it has been speculated that thermal tissue damage may cause loosening of hip implants. However, necrosis observed during implantation may be due to the reaming and drilling associated with the operation (Jiranek 2005). During cranioplasty the paste is laid onto the skull defect whilst warm and malleable using damp gauze over the dura to provide protection from the heat. At this stage the paste is formed manually to fit the skull defect and allowed to set forming a hard barrier. Once hardened the resin implant is held in place using titanium mini-plates which are secured to the implant and the skull with screws. There is also concern about placing a heated material close to the surface of the brain or dura during an operation. However, protection from heat is normally provided using gauze damp with saline solution placed between the acrylic resin and dura tissue whilst it is in the hot polymerisation phase (10–15 minutes after mixing). Concern has also been expressed about patient (and staff) exposure to the monomer. There has been one death attributed to a systemic allergic reaction due to exposure to the methyl methacrylate monomer (Meel 2004). It should be noted that the use of acrylic resin for cranioplasty in paediatric patients is not recommended as the infection/rejection rate has been measured at 23% in a group of 75 children over a 15-year period (Blum et al. 1997). However, in a long-term follow up of 312 patients (449 procedures) over a 20 year period it was found that bone graft and PMMA were the best materials for a long term outcome (Moreira-Gonzalez et al. 2003).

The use of thin titanium sheet for cranioplasty has been reported by a number of authors (Joffe et al. 1999b; Winder and Fannin 1999a). The technique was originally developed in Belfast, Northern Ireland, during a period of civil strife (Gordon and Blair 1974). At this time a simple, effective technique for repairing skull defects caused by trauma was required. An impression of the skull defect was made through the overlying skin using plaster of Paris. This impression was used as a template to create a dental stone former onto which the titanium sheet was pressed under high pressure. The 3D contour of the impression and identification of the defect edges was unreliable as the overlying skin may be between 5 and 8 mm. The titanium plate was placed over the defect and held in place by flat headed screws overlapping the defect by approximately 2 mm. Titanium has been shown to have a very low rejection rate of around 2% (Hieu et al. 2002). However, a more recent study has demonstrated a lower success rate for titanium cranioplasty for 169 patients (Eufinger et al. 2005). 143 patients of the total group were assessed post-operatively, with 9.8% requiring removal of the implant due to infection. It should be noted that complications arose more frequently with larger defects (>100 cm^2) in this study, with 18% requiring explantation.

Hydroxyapatite is an alternative bone cement and has been used in craniofacial reconstruction in conjunction with titanium mesh (Ducic et al. 2002). It is mainly composed of calcium phosphate and as a powder is mixed with water to form a paste. It has excellent bio-compatibility and is osteoconductive which means it becomes infiltrated with bone cells over time. It has generally been used for small defects as "slumping" occurs when it is in the paste phase and distorts whist hardening in the defect. In one recent small study of nine patients, two cranioplasties had to be removed due to infection, 22% (Durham et al. 2003). Hydroxyapatite cranioplasties have also been implanted in a group of 63 paediatric patients, however, there was a 25% delamination (splitting along a plane) rate and removed implants demonstrated bacterial contamination (Beber et al. 2003).

All methods for cranioplasty have their advantages and disadvantages, although it is argued that a patient's own bone is cheaper than any implant material (Hayward 1999). In this situation a section of the outer table of the skull from the opposite side of the defect is removed and implanted within the defect. This requires a second skin incision to harvest the new bone flap and the patient may be at risk of further infection and morbidity associated with a second wound. It has been demonstrated that the storage of a cranial bone flap in the subcutaneous tissue of the abdominal wall was safe, efficacious and cost-effective when decompressive craniectomy was performed (Flannery and McConnell 2001). Each of the common materials used for cranioplasty have their relative strengths and weaknesses. To date we do not have a definitive bio-compatible implant material that solves all of the problems associated with repairing large skull defects.

1.1.4 Computer Assisted Cranioplasty

There has been considerable interest in the production of custom made cranial implants for the repair of skull defects which has become known as Computer Assisted Cranioplasty (CAC). Medical image processing and computer assisted design/manufacturing (CAD/CAM) software, or a combination of the two, are necessary tools for computer based design and manufacture of customised implants. CAC implants offer a range of benefits over the traditional technique of manual repair. They provide a better cosmetic outcome (Joffe et al. 1999a), offer a potential reduction in theatre time, increase surgical confidence in the implant and enable a range of materials to be used for repair. However, this may be seen as a disadvantage given that long term studies of rejection rates for cranioplasty materials are not generally available in the literature.

A review of the literature demonstrates that CAC follows a set of general principles.
Step 1—3D computed tomography (CT) scan

> CT is the imaging modality of choice as it demonstrates high inherent image contrast between bone and soft tissue. This enables relatively easy segmentation of the bone from the rest of the soft tissue data.

Step 2—Image processing

> The CT data volume may require some pre-processing before the model is generated. The pre-processing steps may include conversion to an isotropic data set, image thresholding, voxel connectivity, volume data mirroring and 3D image editing (this allows the removal of metal artifacts or implants).

Step 3—STL file generation

After segmentation the bone surface contours are mathematically modeled using triangles or polygons to create a mesh and converted in a suitable Computer Aided Design (CAD) file format. The STL file format (an abbreviation for stereolithography), is the most commonly used as it is widely transferable between a variety of software systems.

Step 4—Creation of physical model

The STL data is transferred to one of many rapid prototyping systems available and a physical model is created. Stereolithography and fused deposition modeling are the most widely used RP systems, however CNC milling is rapid and inexpensive for surfaces that do not require undercuts in the model. This physical model is used as an anatomically accurate template on which an implant is designed.

Step 5—Implant production

The RP model is used as an anatomically accurate template from which an implant may be manually constructed. The method of manufacture will depend on the material being used as an implant and may involve manual sculpting, forming, moulding and milling.

1.2 Data Source and Image Processing

1.2.1 3D Computed Tomography

The purpose of acquiring a 3D CT scan is to produce a geometrically accurate volumetric data set of the patients head. The volumetric data set is composed of individual voxels (3D volume elements). The imaging parameters (field of view and slice thickness) are set by the CT scan operator so that the voxels are nearly isotropic. This means that they have approximately equal x, y and z dimensions. A CT image may be described by its field of view (FOV), usually quoted in centimetres and is typically 220 mm for the head. The image matrix is made of 512×512 pixels. In CT, pixel values, also known as Hounsfield units, range from -1024 (air), through 0 (water), to $+3072$. Bone, depending on its density, is represented by the CT number range from approximately $+100$ to $+2000$, where the highest numbers indicate the most dense bone. There are no other natural human tissues within this range although you may find some metal implants have a similar or greater pixel values to dense bone. Data acquisition and image reconstruction has been comprehensively described by (Kalender 2000). The kV and mAs used should be sufficient to ensure good image contrast between bone and soft tissue (typical values would be kV = 140 and mA = 250–300). Generally, the gantry tilt should be set to 0°, as software interpreting the CT data acquired with a gantry tilt, may not take account of the shear generated in CT volume data (Winder et al. 2005). A cranial defect should be scanned in its entirety, from below the lower boundaries of the defect to above its upper boundary ensuring that the full extent of the defect is imaged. This range will include normal skull bone both below and above the defect and will aid the creation of a computer generated surface when the CT data is transferred to CAD software.

After data acquisition a series of axial slices (axial is the plane perpendicular to the long axis of the body) are reconstructed, each representing a finite slice of tissue. A typical slice thickness encountered during routine scanning of the head is between 1.0 and 3.0 mm. There are typically 150 × 1.0 mm axial slices required to image the skull, not including the mandible or orbits. The physical size of the data volume is defined by the FOV, the slice thickness and the number of slices. This volume of data may be considered as a cuboid made up of smaller elements called voxels (3D pixels). A FOV of 200 mm will produce a pixel size of approximately 0.4 mm. It should be noted that the pixel size is small in CT, therefore the in-plane spatial resolution of the system is lower than conventional X-ray images and should demonstrate objects as small as 1.0 mm. However, the slice thickness also influences the spatial resolution and should be as small as is consistent with maintaining image quality.

Figure 1.1 (a) shows a typical 2.0 mm CT slice through the middle of a cranial defect. A large break in the circumference of the normal skull (represented as white in the image) is demonstrated with soft tissues (represented as mid-grey) following the contour of the defect. There is a deep depression in the overlying skin which is an external visual feature of a large skull defect. Figure 1.1 (b) shows a surface shaded 3D CT scan of the same subject. It was created with a threshold setting to remove all soft tissue (CT threshold = 150). The image shows a large lateral defect on the left side. The maximum dimensions of the defect are 13.0 cm × 12.0 cm.

1.2.2 DICOM Data Extraction

Modern CT scanners encountered in clinical departments will have a facility for exporting CT images either through the hospital network to an image processing PC or directly burned onto CD. Each CT image is written as an individual file within a directory structure. The format of the images, nearly exclusively, follows the medical industry standard, called DICOM V3.0 (**D**igital **I**maging and **CO**mmunications in **M**edicine, see www.nema.org for full details). A useful introduction to the DICOM standard is given by Flanders and Carrino 2003. Adhering to this standard ensures that vital information concerning the scaling of the image volume and also the right/left orientation is available to other software used for image editing and bone surface modeling. It is essential to confirm that data imported to software external to the CT scanner is tested for scaling and orientation as it is not unheard of for right/left swapping to occur. This has obvious consequences in the treatment of patients. A simple way to test and validate CT data transfer is to perform a scan (using the same imaging parameters that would be used for patients) on an object of well-characterised dimensions with an identifiable right/left orientation. The geometry and orientation of the data set can be validated against the original object. There is no reason why this procedure should not be repeated at regular intervals or when there is an upgrade or change in software, either on the CT scanner or

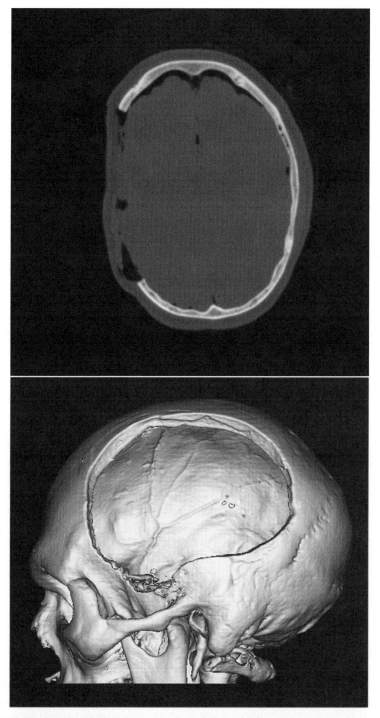

Fig. 1.1 (a) shows an axial CT image through a skull defect and (b) shows a 3D surface rendered CT volume with a large lateral defect

in image processing or CAD workstations. Individual DICOM images from CT are approximately 0.5 MB in size. If we require 150 × 1.0 mm slices to provide anatomical coverage of the skull defect then the data set will be 75 MB. One should ensure that the PC and image processing and CAD software is capable of handling these large data sizes.

1.2.3 CT Volume Data Editing

A number of basics steps are required to condition the CT data so that a good quality STL file can be produced. Firstly, the pixel size in CT will be typically 0.4 mm and the slice thickness will be 1.0 to 2.0 mm, depending on the image reconstruction settings. It is usual to create an isotropic data set where the voxel dimensions are all the same size. This is achieved by performing interpolation of the data so that the slice thickness is converted to 0.4 mm. This will inevitably add additional slices to the data volume increasing its overall size. After interpolation, structures or edges within the data which lie at an angle to the slice plane will be presented more smoothly. However, it should be noted that although the image volume will appear smoother this does not indicate higher resolution, there is an increase in aesthetics but not information. Secondly, image segmentation is used to remove any unwanted structures like soft tissues (skin, fat, brain tissue). Using CT as the source data this is generally easily performed using a simple CT number thresholding technique. A CT number is identified by analysis of the Hounsfield units representing bone, above which only bone tissue is visualised. Thresholding may create "floating objects" (these are regions of voxels which lie above the CT threshold but are not connected to the skull). These may be problematic at the rapid prototyping stage and should be removed as early as possible. To remove floating data the volume may be conditioned with an algorithm that applies a "connectivity" rule. Any bone or other object present that isn't physically connected to the main skull i.e. voxels are not adjacent or touching are deleted. Sometimes there are devices (for example drainage tubes) implanted within the skull which may interfere with the model making. Software should be available to manually edit the CT data volume either in 3D or on a slice-by-slice basis to enable the user to remove unwanted structures or to reduce the data volume size.

1.2.4 STL Surface Generation

After bone segmentation the skull surface is mathematically modeled. This is commonly known as tessellation or tiling (for a simple explanation see http://mathforum.org/sum95/suzanne/whattess.html). Tiling may be carried out using simple triangles or polygons. The most common algorithm for surface modeling 3D medical image data sets is the marching cubes algorithm which uses a set of triangle templates in various orientations to represent where the bone surface cuts through each voxel

(Karron 1992). The resulting data set is a listing of the triangle corner coordinates and the direction of the triangle surface normal. The surface normal describes the angle at which the triangle lies in space. The full data set (known as an STL file) describes the orientation and position in 3D space of all the triangles that represent the skull surface. The STL file format is the most commonly used as it is easily transferable between many CAD software packages. The quality of the STL data file depends on the software being used. A poor STL file will have holes in the surface description or regions where the triangle edges do not interface correctly. Another method of modeling the skull surface is to use a series of axial contours. The outline of the skull is identified on each CT scan slice and the contour coordinates written to a file. The whole skull is described by a series of stacked contours. There are many file formats used. Some common examples are Initial Graphics Exchange Specification IGES and Data Exchange Format DXF, although the reader should be aware that there are many more.

There are a number of software packages available to perform this function. They are Mimics by Materialise, Leuven, Belgium, Analyze AVW by AnalyzeDirect, Lenexa, Kansas and Biobuild by Anatomics, Queensland, Australia. Other centres have developed there own in house software for this purpose. Typically, the number of triangles used to represent a whole skull may range from 500,000 to a few million. The number of triangles depends on the resolution of the original CT data and the method of tiling. It should be noted that the STL file size may be anywhere between a few Mbytes and 100 MB. The number of triangles used to represent the surface may be reduced, especially in areas of low curvature where larger elements may be used to model the surface. Automated software is commercially available to perform surface modeling. This is called decimation and results in a reduced file size which improves data transfer and manipulation.

Figure 1.2 (a) shows a portion of a skull around the orbit which has been surface modeled. Note the contouring which indicates the size of the surface triangles, also known as facets. In this case the surface was modeled using the original CT scan voxel size as the basis for the facet size. Figure 1.2 (b) shows the same area modeled with software that uses interpolation. The surface is much smoother compared to 2 (a), however, it should be noted that smoothing may remove some very fine detail in the model.

1.3 CAD/CAM in Cranioplasty

Rapid prototyping (RP) is a generic name given to a range of related technologies that are used to fabricate physical objects directly from computer aided design data sources. These enable design and manufacturing processes to be performed much more quickly than conventional manual methods of prototyping. In all aspects of manufacture the speed of moving from concept to product is an important part of making a product competitive. RP technologies enable an engineer/designer to produce a working prototype of a CAD design for visualisation and testing purposes. They were developed to increase the rate of producing a physical prototype model

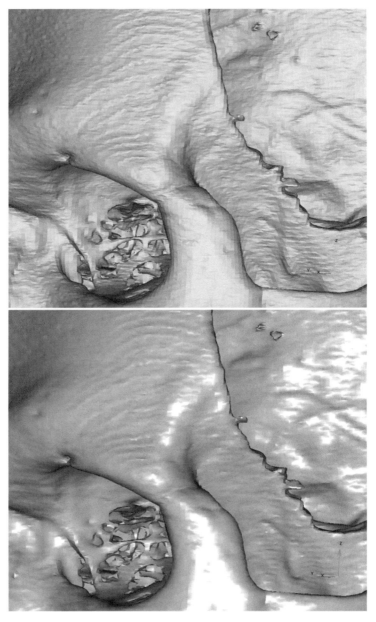

Fig. 1.2 (a) shows a surface model without interpolation and **(b)** shows a surface model with interpolation

in the design process for anything from an aeroplane engine part to a mobile phone cover. There are a number of texts describing the development of rapid prototyping technology and its applications in engineering and design (Jacobs P 1996; Kai and

Fai 1997). A description will be given in later sections on stereolithography (SL), fused deposition modeling (FDM) and computer numerically controlled milling (CNC) which are the most common rapid prototyping techniques used in medicine.

1.3.1 Medical Rapid Prototyping

Medical modeling became possible in the early 1990s due to the integration of 3D software and RP technology (Barker et al. 1994]. Two useful reviews of RP applied to medicine are given by Petzold et al. (1999) and Webb (2000). The material in which models are constructed is governed by the RP machine. However, the initial model may be used as a precursor to the final model material, acting as a design mould or template. The important issues of model cost, build material, build accuracy and the model application will be addressed.

1.3.1.1 Stereolithography

A stereolithography (SL) RP system consists of a bath of photosensitive resin, a model building platform and an ultraviolet (UV) laser for curing the resin. A mirror is used to guide a laser focus onto the surface of the resin. The resin becomes cured when exposed to UV radiation. The mirror is computer controlled and its movement is guided to cure the resin on a slice-by-slice basis. A model is initially designed with CAD software and a suitable file format (commonly STL, as described in Section 1.4) and transferred to the SL machine for building. The CAD data file is converted into individual slices of known dimensions. This slice data are then fed to the RP machine which guides the exposure path of the UV laser onto the resin surface. The layers are cured sequentially and bond together to form a solid object beginning from the bottom of the model and building upwards. As the resin is exposed to ultraviolet (UV) light a thin well-defined layer thickness becomes cured. After a layer of resin is exposed, the resin platform is lowered within the bath by a small known distance. A new layer of resin is wiped across the surface of the previous layer using a wiper blade and this second layer is subsequently exposed and cured. The process of curing and lowering the platform into the resin bath is repeated until the full model is complete. The self-adhesive property of the material causes the layers to bond to one another and eventually form a complete, robust, three-dimensional object. The model is then removed from the bath and cured for a further period of time in an UV cabinet. The build part may contain layers which significantly overhang layers below. If this is the case then a network of support structures, made of the same material, are added beneath these over-hanging layers at the design stage to support it during the curing process. These support structures, analogous to a scaffold, are removed manually after the model is fully cured. This is labour intensive and time-consuming. Generally, SL is considered to provide the greatest accuracy and best surface finish of any rapid prototyping technology. The model material is robust, slightly brittle and relatively light, although it is hydroscopic and may physically warp over time (a few weeks or months) if left in the air.

1.3.1.2 Fused Deposition Modeling

Fused deposition modeling (FDM) employs a similar principle to SL in that it builds models on a layer-by-layer basis beginning at the bottom. The main difference is that the layers are extruded as a thermoplastic from a fine nozzle. A commonly used material is acrylonitrile butadiene styrene (ABS) nylon. The physical properties of ABS nylon are, it is rigid, has dimensional stability, has thermoplastic properties and is inexpensive. The model is constructed by extruding the heated nylon material onto a foam surface in a path guided by the model data. Once a layer has been deposited the nozzle is raised by one layer thickness (0.178–0.356 mm). The next layer is deposited on top of the previous layer. This process is repeated until the model is complete. As with SL, support structures are required for FDM models as it takes time for the nylon to harden and the layers to bond together. The supports are added to the model at the design stage and built using a different coloured nylon material, extruded through a second nozzle. This enables easy identification and subsequent removal of the supports by hand after the model is complete. Modern FDM systems have support structures that are dissolved in a warm caustic bath. As with SL, the build accuracy of the FDM machine is greater than that for the image acquisition technique and was therefore very suitable for medical models generated from CT image data.

1.3.1.3 CNC Milling

Although not generally considered as part of the RP family of machines, CNC milling has proved useful for the production of cranial implants (Joffe et al. 1999a). It compares favourably in terms of model accuracy and repeatability, but has the distinct advantage of lower cost and more rapid build times. Many materials are easily milled and large diameter cutting tools (5–12 mm) may be used to decrease build time. A CNC mill, due to the method of cutting is most suited to modeling large or small cranial defects in areas of low or high curvature. However, it is limited for complex surfaces as the cut can only be performed in one direction. Complex 3D objects may be produced by rotating the model a number of times and milling from different directions. However, this negates the benefit of milling simple surfaces therefore other RP systems should be considered for these cases.

1.3.2 Accuracy of Medical Models

Of the three RP technologies used for medical modeling each are suitable from the point of view of model accuracy. It has been pointed out that models created using stereolithography are dimensionally accurate to 0.62 mm ± 0.35 mm (Choi et al 2002), the limiting factor being the CT imaging technique rather than the RP technology employed. In a report of a series of 20 linear measurements comparing three dry skulls with stereolithography models, errors were typically in the range of 0.0–2.0 mm. However, one model had a maximum difference of 7.0 mm across the maximum breadth of the skull (Wulf et al. 2001). An FDM skull model has been similarly assessed and found to have a maximum error of less than 2.0 mm and an average error of 0.46 mm (Galantucci et al. 2006). Computer controlled milling has been clinically validated and is used routinely in a number of centres for the creation of custom titanium cranioplasty (Joffe et al. 1999b; Hieu et al. 2002).

1.3.3 Computer Assisted Manufacture of Implants

There has been a range of developments in the computer assisted manufacture of cranial implants. Initially, 3D visualisation of the patient skull defect from a CT scan was used for surgical planning (Cutting et al. 1982; Lambrecht and Brix 1990; Zonnefeld 1994). The purpose of the early work was to demonstrate the feasibility of the technique and create a satisfactory computer generated contour to cover the defect. It was pointed out, at this time, that CAD and CAM software could not handle the very large data files generated from human image data. These software systems were designed to handle simple geometric shapes and not the detailed complex contours required for the head and face. Software and hardware developments are now such that 100 MB data files may be handled with relative ease. Nowadays, CAD is routinely employed to create anatomically accurate contours (based on the patient's skull symmetry) and to design a suitable implant to ensure close fitting and ease of attachment to the remaining bone.

There are a number of approaches to designing an implant for the repair of skull defects. These approaches may be sub-divided into two types, use of medical imaging software and use of CAD software. Medical imaging software enables the user to mirror a CT volume of the skull along the mid-sagittal plane and merge this new data set with the original. The data sets are merged using a 3D registration algorithm to align the skulls with the purpose of filling the skull defect with a new contour from the contra-lateral side (Studholme et al. 1997). The new contour will provide a surface based on the patient's own anatomy with good cosmetic outcome. However, it should be noted that the 3D registration method may not align the data perfectly, due to asymmetry in the patient's anatomy, and there may be steps in the contour at the defect edge (Winder et al. 2006).

CAD may be used to reverse engineer a new contour based on the local curvature of the skull around the defect. It is common for a CAD contour to be generated using non-uniform rational B-splines (NURBS). In this case contour profiles are projected across the defect using the defect edge surface as a starting point. These profiles begin on the undamaged bone around the circumference of the defect. The NURBS surface is then interpolated towards the centre of the defect from all directions and blended with the surface from the opposite edge. This NURBS surface may be constrained using the mirrored contra-lateral side of the skull to provide an anatomically accurate surface (Winder et al. 2006). Figure 1.3 shows a NURBS surface generated for a large lateral defect. The defect contour was based on the mirrored side of the skull and the contour edges are blended with the remaining skull around the defect edges. This ensures that the implant edges closely fit the defect edges. Also, this surface may be projected along a line perpendicular to the centre of the surface plane. The edge between the two surfaces is filled creating a solid 3D model.

The design of PMMA implants required further CAD work as the implant is normally positioned within the defect itself (as opposed to placement on the outer surface of the skull). The defect contour was geometrically modeled by identifying points around the defect on the skull surface. Software automatically generated a set of curves through these user-defined points. Manual interaction enabled the user to define the skull defect edges and to allow trimming of the implant. Finally a surface was generated based on these contour lines (Hieu et al. 2002). The implant design was finished by extending the outer surface thickness by the required implant thickness. This enabled a mould of the implant to be created. A refinement to this approach was described where a segment of the skull containing the defect was isolated and seed points were positioned around the edge of the defect. These points were then connected using Dijkstra's algorithm which minimised the path across the defect based on maximum local curvature (Dean et al. 2003). The final contour was produced and stored as B-splines (a smooth class of approximating curves based on defined control points). Twisting curvature and torsion of the surface could be smoothed to remove sharp changes in the implant surface. This method provides an implant design where the implant edges are adjacent to the skull defect edge, something not easily achieved in the manual implantation of PMMA. Bezier patches

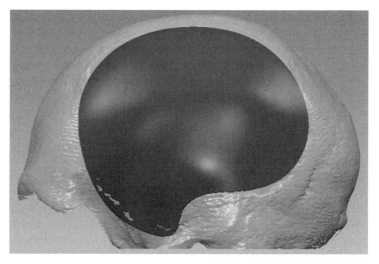

Fig. 1.3 shows a CAD generated implant contour providing excellent anatomical replication of the contra-lateral side and good fit along the defect edges

have been recently employed to generate an accurate contour to fill the hole of a cranial defect. The surface is achieved using a customized advancing-front meshing technique using surface approximations based on a Quartic Bezier patch/surface. (Chong et al. 2006).

It is useful to test the accuracy of a particular surface modeling algorithm. Quality assurance of surface contours generated using three different methods was performed on a cadaver specimen (Winder et al. 2006). A large lateral defect was created in the specimen skull and a contour modeled using two CAD methods and data volume mirroring with medical imaging software. The modeled surface was compared to the original surface. Colour coded difference maps were used to measure the success of each technique. Figure 1.4 shows a difference map (presented as grey scale) for the NURBS surface. This demonstrates that there was a maximum deviation of ±2 mm from the original skull surface and was clinically acceptable.

The full cost of manufacturing custom made implants is rarely published. Some costs have been given for stereolithography models and range from £125-00 (Winder et al. 1999b) to 4000 HK dollars (Hieu et al. 2004). These figures do not include the personnel costs of specialist engineers and technicians. Also, any savings to Health Service Institution have not been documented, therefore it is difficult to determine the overall economic impact of computer assisted cranioplasty. A novel solution to decrease the cost of CAD designed cranioplasty implants was proposed (Hieu et al. 2004). One hundred and twenty patients with cranial defects were CT scanned and the size and distribution of the defects recorded. The authors developed a standard design template from which a range of cranial implants were created from PMMA. These implant designs were mainly used in simple defects or areas of the skull

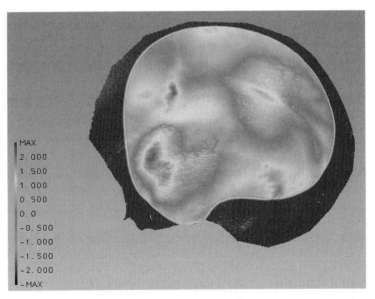

Fig. 1.4 shows the result of comparing a NURBS generated surface contour with the original skull surface

where cosmetic outcome was not deemed important. Three implant templates were designed up to a maximum size of 120×150 mm, however they recommend that for complex curvatures or very large defects a custom implant should still be used.

Finally, CAD/CAM has assisted surgeons to plan and execute a cranioplasty procedure in one step rather than the normal two. Individualised surgical templates were developed which enabled accurate resection of a skull section. At the same time a corresponding implant was produced. This enabled the surgeon to accurately resect part of the skull and insert the implant in one surgical procedure. (Eufinger et al. 1998; Winder et al. 1999b; Weihe et al. 2000). Although limited in its application this work has demonstrated the potential for surgical robotics to be used in cranioplasty.

Further research into computer assisted cranioplasty is required in the area of tissue engineered scaffolds and bioresorbable polymers. An ideal bioresorbable material would degrade slowly within the body and be osteoconductive to allow its replacement by new bone (Zein et al. 2002; Rohner et al. 2003). Polycaprolactone (PCL) is such a material, however, it rarely becomes replaced by bone but rather by fibrous tissue. Fused deposition modeling has been used to build 3D scaffolds of polycaprolatone (Woodfield et al. 2004; Hoque et al. 2005). Current research in this area is focused on the scaffold architecture and pore structure to investigate the mechanical properties and cell distribution. Customised PCL scaffolds with precisely controlled microarchitecture have been fabricated and cultured with progenitor cells and osteoblasts. Seeded scaffolds had 50% yield strength compared to calvarial bone but demonstrated increased calcification measured by CT scanning (Schantz et al. 2003).

1.4 Summary

No single method exists for repairing holes in the skull. This is surprising as modeling of complex surfaces is relatively straightforward using modern hardware and CAD software. It seems that the reverse engineering aspect of cranioplasty and implant design has been solved. Cranial implants made from a range of materials can be designed and constructed to an accuracy of ± 1 mm. This ensures that the cosmetic outcome for the patient is excellent and ease of fit of the implant may reduce surgical time. One of the main variables in cranioplasty is the range of materials available for repair. Different materials require different manufacturing strategies. However, we do not have a singular implant material that has excellent biocompatibility, the required strength and good integration with local bone tissue used on a global basis.

Computer assisted cranioplasty has evolved rapidly over recent times with the use of sophisticated CAD techniques used to produce anatomically accurate contours. There have been many studies on the clinical efficacy of new materials for cranioplasty. However, these tend to be carried out on low numbers of patients. The length of follow-up in many studies is relatively short and measured in months rather than years. We are at a point where we could begin to "over-engineer" a solution to the cranioplasty problem. Concerted effort should be made in determining the best implant material to act as a bone substitute and then developing the design and manufacturing techniques to best produce a cranioplasty implant.

Acknowledgment Data used to create Figs. 1.1–1.4 were given by kind permission of Dr Joerg Wulf and Professor Luder Busch, Department of Anatomy, Medical University of Luebeck, Germany.

References

Anger C, Dujovny M, Gaviria M (2002) Neurocognitive assessment before and after cranioplasty. Acta Neurochir 144:1033–1040
Arriaga MA, Chen DA (2002) Hydroxyapatite cement cranioplasty in translabyrinthine acoustic neuroma surgery. Otolaryngol Head Neck Surg 126(5):512–517
Artico M, Ferrante L, Pastore FS, Ramundo EO, Cantarelli D, Scopoletti D Iannetti G (2003) Bone autografting of the calvaria and craniofacial skeleton: historical background, surgical results in a series of 15 patients and review of the literature. Surg Neuro 60:71–79
Barker TM, Earwaker WJ, Frost N, Wakeley G (1994) Accuracy of stereolithographic models of human anatomy. Australas Radiol 38(2):106–111
Beber B, Philips JH, Forrest CR (2003) Hydroxyapatite cement cranioplasty in the pediatric population. Craniofacial Surgery 10:227–230
Blum KS, Schneider SJ, Rosenthal AD (1997) Methyl methacrylate cranioplasty in children: Long-term results. Pediatr Neurosurg 26(1):33–35
Brandt MT, Haug RH (2002) The use of a polyurethane skull replica as a template for contouring titanium mesh. J Oral and Maxillofac Surg 60:337–338
Chiarini L, Figurelli S, Pollastri G, Torcia E, Ferarri F, Albanese M, Nocine PF (2004) Cranioplasty using acrylic material: a new technical procedure. J Cranio-Maxillofacial Surgery 32:5–9
Choi JY, Choi JH, Kim NK, Kim Y, Lee JK, Kim MK, Lee JH, Kim MJ (2002) Analysis of errors in medical rapid prototyping models. Int J Oral Maxillofac Surg 31(1):23–32
Chong CS, Lee H, Kumar AS (2006) Automatic hole repairing for cranioplasty using Bezier surface approximation. J Craniofac Surg 17(2):344–352
Cutting C, Bookstein FL, Grayson B (1982) Three dimensional computer assisted design of craniofacial surgical procedures: Optimisation and interaction with cephalometric and computer tomographic based models. Plastic Reconstructive Surg 77:877
Dean D, Min KJ, Bond A (2003) Computer aided design of large-format prefabricated cranial plates. J Craniofacial Surg 14(6):819–832
Ducic Y (2002) Titanium mesh and hydroxyapatite cement cranioplasty: a report of 20 cases. J Oral Maxillofac Surg 60(3):272–6.
Durham SR, McComb JG, Levy ML (2003) Correction of large (>25 cm^2) cranial defects with "reinforced" hydroxyapatite cement: technique and complications. Neurosurgery 52 (4): 842–845.
Eufinger H, Wittkampf AR, Wehmoller M, Zonneveld FW (1998) Single-step fronto-orbital resection and reconstruction with individual resection template and corresponding titanium implant: a new method of computer aided surgery. J Craniomaxillofac Surg 26(6):373–378
Eufinger H, Rasche C, Wehmoller M, Schmieder K, Scholz M, Weihe S, Scherer P (2005) CAD/CAM titanium implants for cranioplasty—An evaluation of success and quality of life of 169 consecutive implants with regard to size and location International. Cong Ser 1281:827–831
Flanders AE, Carrino JA (2003) Understanding DICOM and IHE. Semin Roentgenol 38(3):270–81
Flannery T, McConnell R (2001) Cranioplasty: Why throw the bone flap out? Br J Neurosurg 15(6):518–20

Galantucci LM, Percoco G, Angelelli G, Lopez C, Introna F, Liuzzi C, De Donno A (2006) Reverse engineering techniques applied to a human skull, for CAD 3D reconstruction and physical replication by rapid prototyping. J Med Eng Techn 30(2):102–111

Gordon DS, Blair GAS (1974) Titanium cranioplasty. Br Med J 2:478–481

Gousain AK (2004) Biomaterials for reconstruction of the cranial vault. Plast Reconstr Surg 116(2):663–666

Gronet PM, Waskewicz GA, Richardson C (2003) Preformed acrylic cranial implants using fused deposition modeling: a clinical report. J Prosthet Dent 90(5):429–33.

Hayward RD (1999) Cranioplasty: Don't forget the patient's own bone is cheaper than titanium. Br J Neurosurg 13(5):490–491

Hieu LC, Bohez E, Vander Sloten J, Oris P, Phien HN, Vatcharaporn E, Binh PH (2002) Design and manufacture of cranioplasty implants by 3-axis CNC milling. Technol and Health Care 10:412–423

Hieu LC, Vander SJ, Bohez E, Phien HN, Vatcharaporn E, An PV, To NC, Binh PH (2004) A cheap technical solution for cranioplasty treatments. Technol and Health Care 12:281–292

Hoque ME, Hutmacher D, Feng W, Li S, Huang MH, Vert M, Wong YS (2005) Fabrication using a rapid prototyping system and in vitro characterization of PEG-PCL-PLA scaffolds for tissue engineering. J Biomater Sci Polymer Ed 16(12):1595–1610

Jacobs P (1996) Stereolithography and other rapid prototyping and manufacturing technologies. American Association of Engineers Press, Dearborn, MI

Jiranek W (2005) Thermal manipulation of bone cement. Orthopaedics 28(8):863–866

Joffe J, McDermott P, Linney A, Mosse C, Harris M (1992) Computer-generated titanium cranioplasty: Report of a new technique for repairing skull defects. Br J Neurosurg 6: 343–350

Joffe J, Harris M, Kahugu F, Nicoll S, Linney A, Richards R (1999a) A prospective study of computer-aided design and manufacture of titanium plate for cranioplasty and its clinical outcome. Br J Neurosurg 13(6):576–80

Joffe J, Nicoll S, Richards R, Linney, A, Harris M (1999b) Validation of computer assisted manufacture of titanium plates for cranioplasty. Int J Oral Maxillofac Surg 28:309–13

Kai CC, Fai LK (1997) Rapid prototyping: principles & applications in manufacturing. John Wiley & Sons Ltd, Singapore

Kalender WA, (2000) Computed Tomography, Publicis MCD Verlag, Munich

Karron D (1992) The "spider web" algorithm for surface construction in noisy volume data. In: SPIE Visualisation in Biomed Computing 1808:462–476

Lambrecht JT, Brix F (1990) Individual skull model fabrication for craniofacial surgery. Cleft Palate 27:382

Lara WC, Schweitzer J, Lewis RP, Odum BC, Edlich RF Gampper TJ (1998) Technical considerations in the use of polymethylmethacrylate in cranioplasty. J Long Term Effects of Med Implants 8(1):43–53

Lung CYK, Darvell BW (2005) Minimisation of the inevitable residual monomer in denture based acrylic. Den Mater 21:1119–1128

Meel BL (2004) Fatal systemic allergic reaction following acrylic cranioplasty: A case report. J Clinic Foren Med 11:205–207

Moreira-Gonzalez A, Jackson IT, Miyawaki T, Barakat K, DiNick V (2003) Clinical outcome in cranioplasty: Critical review in long term follow up. J Craniofacial Surg 14(2): 144–153

Park HK, Dujivny M, Agner C, Diaz FG (2001) Biomechanical properties of calvarium prosthesis. Neurol Res 23:267–276

Petzold R, Zeilhofer H, Kalender W (1999) Rapid prototyping technology in medicine-basics and applications. Computerised Med Imaging and Graphics, 23:277–84

Raja AI and Linskey ME (2005) In situ cranioplasty with methylmethacrylate and wire lattice. Br J Neurosurg 19(5):416–419

Rohner D, Hutmacher DW, Cheng TK, Oberholzer M, Hammer B (2003) In vivo efficacy of bone amrrow coated polycaprolactone scaffolds for the reconstruction of orbital defects in pigs. J Biomed Mater Res Part B: Appl Biomater 66B:574–580

Rotaru H, Bacuit M, Stan H, Bran S, Chezan H, Iosif A, Tomescu M, Kim SG, Rotaru A, Baciut G (2006) Silicone rubber mould cast polyethylmethacrylate-hydroxyapatite plate used for repairing a large skull defect. J Craniomaxillofac Surg. 34(4):242–246

Sanan A, Haines S (1997) Repairing holes in the head: A history of cranioplasty. Neurosurgery 40(3):588–603

Schantz JT, Hutmacher DW, Lam CX, Brinkman M, Wong KM, Lim TC, Chou N, Guldberg RE, Teoh SH (2003) Repair of calvarial defects with customized tissue engineered bone grafts II. Evaluation of cellular efficiency and efficacy in vivo. Tiss Engg Suppl 1:127–139

Schipper J, Ridder GJ, Spetzger U, Teszler CB (2004) Individual prefabricated titanium implants and titanium mesh in skull base reconstructive surgery. A report of cases. Eur Arch Otorhinolaryngol 261:282–290

Sheikh BY (2006) Simple and safe method of cranial reconstruction after posterior fossa craniectomy. Surg Neurol 65:63–66

Studholme C, Hill DL, Hawkes DJ (1997) Automated three-dimensional registration of magnetic resonance and positron emission tomography brain images by multiresolution optimization of voxel similarity measures. Med Phys 24(1):25–35

Webb PA (2000) A review of rapid prototyping (RP) techniques in the medical and biomedical sector. J Med Eng Technol 24(4):149–153

Weihe S, Wehmoller M, Schliephake H, Hassfeld S, Tschakaloff A, Raczkowsky J, Eufinger H (2000) Synthesis of CAD/CAM, robotics and biomaterial implant fabrication: Single step reconstruction in computer aided frontotemporal bone resection. Int J Oral Maxillofac Surg 29(5):284–388

Winder J, Bibb R (2005) Medical rapid prototyping technologies: state of the art and current limitations for application in oral and maxillofacial surgery. J Oral Maxillofac Surg 63(7):1006–15

Winder RJ and Fannin T (1999a) Virtual neurosurgery and medical rapid prototyping applied to the treatment of fibrous dysplasia. Proceedings of 13th international congress and exhibition of computer assisted radiology and surgery, pp 1033

Winder RJ, Cooke RS, Gray J, Fannin T, Fegan T (1999b) Medical rapid prototyping and 3D CT in the manufacture of custom made cranial titanium plates. J Med Engg Technol 23(1):26–28

Winder RJ, McKnight W, Golz T, Giese A, Busch LC, Wulf J (2006) Comparison of custom cranial implant source data: Manual, mirrored and CAD generated skull surfaces. Cranial & Maxillofacial Workshop, MICCAI, Copenhagen, 5th October 2006

Woodfield TB, Malda J, de Wijn J, Peters F, Riesle J, van Blitterswijk CA (2004) Design of porous scaffolds for cartilage tissue engineering using a three-dimensional fiber-deposition technique. Biomaterials 25(18):4149–4161

Wulf J, Vitt KD, Gehl HB, Busch LC (2001) Anatomical accuracy in medical 3D modeling. Phidias Newsletter 7:1–2 (http://www.materialise.com/medical/files/ph7.pdf) accessed 23 October 2006

Zein I, Hutmacher DW, Tan KC, Teoh SC (2002) Fused deposition modeling of novel scaffold architectures for tissue engineering applications. Biomaterials 23(4):1169–1185

Zonnefeld FW (1994) A decade of clinical three dimensional imaging: A review. Part III Image analysis and interaction, display options and physical models. Investigative Radiol 29:716

Chapter 2
Virtual Bone Surgery

Ming C. Leu, Qiang Niu and Xiaoyi Chi

2.1 Introduction

To become a skillful surgeon requires rigorous training and iterative practice. Traditional training and learning methods for surgeons are based upon the Halstedian apprenticeship model, i.e., "see one, do one, teach one" which is almost 100 years old (Haluck et al. 2000). For bone surgery, students often watch and perform operations on cadaveric or synthetic bones under the tutelage of experienced physicians before performing the procedure themselves under expert supervision. They need to learn and perform material removal operations such as drilling, burring, etc., as shown in Fig. 2.1. Mistakes can lead to irreparable defects to the bone and the surrounding soft tissue during such procedures, which can result in complications such as early loosening, mal-alignment, dislocation, altered gait, and leg length discrepancy (Conditt et al. 2003). The current system of surgery education has many challenges in terms of flexibility, efficiency, cost and safety. In addition, as new types of operations are developed rapidly, more efficient methods of surgical skill education are needed for practicing surgeons (Gorman et al. 2000).

Virtual Reality (VR) is one of the most active research areas in Computer simulation. Virtual reality systems use computers to create virtual environments to simulate real-world scenarios. Special devices such as head-mounted displays, haptic devices, and data gloves are used for interacting with virtual environments to give real-world like feedback to the user. The most important contributing factor to VR development has been the arrival of low-cost, industry-standard multimedia computers and high-performance graphic hardware. VR has been integrated into many aspects of the modern society such as engineering, architecture, entertainment, etc.

The concept of developing and integrating computer-based simulation and training aids for surgery training has begun with VR simulators. VR techniques provide a realistic, safe, controllable environment for novice surgeons to practice surgical operations, allowing them to make mistakes without serious consequences. It promises to change the world of surgical training and practice. With a VR simulator, novice surgeons can train and perfect their skills on simulated human models, and experienced surgeons will be able to use the simulator to plan surgical procedures. VR training also offers the possibility of providing a standardized performance evaluation for the trainees.

B. Bidanda and P. J. Bártolo (eds.), *Virtual Prototyping & Bio Manufacturing in Medical Applications.*
© Springer 2008

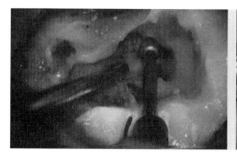

Fig. 2.1 Example orthopedic operations: (a) bone burring for mastoidectomy surgery (Agus et al. (2002); (b) drilling of the metatarsal phalanx in toe arthroplasty surgery (http://www.orthosonics.com/)

Bone surgery is one of the medical applications which can be simulated using VR technology. There exist some surgical simulation tools for orthopedic applications such as knee surgery, but most of them involve only soft tissues. Few have considered the simulation of cutting, sawing, burring, etc., which involve operating on bones as well as on ligaments and muscles. The development of a virtual bone surgery system is very desirable for training surgeons, allowing them to visualize surgical operations simulated with the added sense of touch during the process. As the Minimally Invasive Surgery (MIS) takes a foot-hold in orthopedics, VR technology will become more and more valuable for assisting actual surgery operations. As surgical techniques are developed to sequentially reduce access to the surgical site (via smaller incisions), and instruments and implants are miniaturized to accommodate for these techniques, surgical dexterity and bone preparation and implant positioning will become a less and less forgiving part of the operation. It will be necessary to integrate VR models with images obtained during the actual surgery operations, the so-called Augmented Reality (AR) technology, in order to assist the surgeons in performing the MIS process.

This book chapter reviews the bone surgery simulation systems being developed in various research laboratories and discusses the basic methods and techniques used to develop these systems.

2.2 State-of-the-Art in Surgery Simulation

2.2.1 Current State of Digital Surgery

Previous research on surgery simulation has covered a wide range of operations. Some of the simulators were developed to provide a virtual reality environment as a training tool. The VRMedLab networked facility at the University of Illinois-Chicago (VRMedLab, 2003) was designed to provide an educational resource to otolaryngology surgeons, enabling them to visualize bone-encased structures within the temporal bone using interactive 3D visualization technology. The Ohio

Supercomputer Center and the Ohio State University Hospital (Edmond et al. 1997) developed an endoscopic sinus-surgery simulator, which provided the capability of intuitive interaction with complex volume data and haptic feedback sensation. Cathsim^{TM} (Barker, 1999) was an example of a commercially available training system for venipuncture. Users of Cathsim could practice inserting a needle into a virtual vein with different scenarios available. Røtnes et al. (2002) designed a coronary anastomosis simulator called SimMentor^{TM}, which exploits a common geometric model for providing animated 3D visualization and interactive simulation during the training session.

Some of the surgery simulations involved deformable geometric models in the software system. Delp et al. (1997) developed tissue cutting and bleeding models for this purpose. Bro-Nielsen et al. (1998) described a HT Abdominal Trauma Simulator (HATS), which was developed for open surgery from the front to remove a kidney that had been shattered as the result of a blunt trauma to the body. Berkley et al. (1999) described a simulator for training in-wound suturing with real-time deformable tissues to increase the fidelity of the simulation.

Research on virtual bone surgery is still at the infancy stage. Most researchers on virtual bone surgery considered only deformation of muscles and cutting of soft tissues. Few have considered mechanical removal of bone material, which is an integral part of bone surgery. Gibson et al. (1997) presented some early results of their effort to develop an arthroscopic knee surgery simulator. This simulator used a volumetric approach to model organs. Computers were still too slow to allow realistic deformation of a volumetric representation at that time.

Agus et al. (2002) presented a haptic and visual implementation of a bone-cutting burr, which was developed as one component of a training system for temporal bone surgery, as shown in Fig. 2.2(a). They used a physics-based based model to describe the burr-bone interaction, which included the haptic force evaluation, the bone erosion process, and some secondary effects such as the obscuring of the operational site due to the accumulation of bone dust and the resulting burring debris in the surgical penetration. The implementation operates on a voxel discretization of patient-specific CT and MRI data and is efficient enough to provide real-time haptic and graphic rendering on a multi-processor PC platform. It is expected that the ability of directly using patient specific data as input will help in the accumulation of a large number of training cases. The cutting tool is simplified as a sphere.

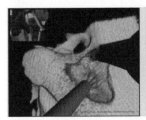

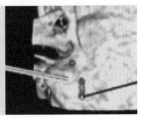

Fig. 2.2 Some examples of temporal bone surgery simulators: (a) Agus et al. (2002); (b) Morris et al. (2004); (c) Wiet et al. (2000) and Bryan et al. (2001)

However, volume modeling imposes strict requirements on the computer memory and algorithm efficiency.

Morris et al. (2004) described a framework for training-oriented simulation of temporal bone surgery as shown in Fig. 2.2(b). The system allows two users to observe and manipulate a common model in two different locations while performing a collaborative bone surgery, and allows each user to experience the effects of forces generated by the other user's contact with the bone surface. This system also permits an instructor to remotely observe a trainee and to provide real-time feedback and demonstration. It uses hybrid data representation for graphic rendering and volumetric data representation for haptic feedback.

The Ohio Supercomputer Center and the Ohio State University Medical Center (Wiet et al. 2000; Bryan et al. 2001) developed a working prototype system for the simulation of temporal bone dissection as shown in Fig. 2.2(c). It provides a customizable and more complete alternative to learning anatomy and surgical techniques as opposed to printed media or physical dissections.

In the project IERAPSI (Integrated Environment for the Rehearsal and Planning of Surgical Intervention), John et al. (2001) and Jackson et al. (2002) focused on surgery for the petrous bone. This project brought together a consortium of European clinicians and technology providers working in this field. Three independent subsystems, i.e., pre-operation planning, surgical simulation, and usage demonstration and training, were provided. It included direct volume rendering for visualization, construction of a synthetic dissection workstation, emphasis on training through a combination of identification and exposure of key structures, and integration of an intelligent tutor.

Pflesser et al. (2002) and Petersik et al. (2002) presented a system of virtual petrous bone surgery that allows realistic simulation of laterobasal surgical approaches. The system was based on a volumetric, high-resolution model of the temporal bone derived from CT scan data. Interactive volume cutting methods using a new multi-volume scheme were developed, allowing high-quality visualization of interactively generated cut surfaces.

In the above simulators, most of the researchers focused on temporal bone surgery. Only a small portion of temporal bone was used in the simulation, the data was not huge, and tool-bone interaction was limited to burring/milling. In real

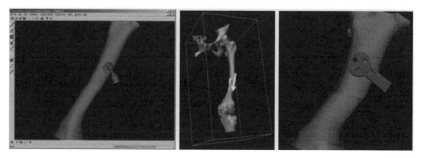

Fig. 2.3 Bone burring, free drilling and guided drilling in bone surgery simulation (Chi et al. 2005; Niu et al. 2005; Niu and Leu 2007)

orthopedic surgery, however, there are also other machining operations like drilling, broaching, sawing, reaming, and milling. These operations are often needed prior to an orthopedic operation, such as pin or screw insertion to the bone. The Virtual Reality and Rapid Prototyping Lab at University of Missouri—Rolla has been developing a bone surgery simulation system to accomplish these tasks. The various system components are integrated in a Windows GUI environment for purpose of implementation. Some of the screenshots can be seen in Fig. 2.3. The system development involves medical image processing, geometric modeling, force modeling, graphic rendering, and haptic rendering (Peng et al. 2003; Chi et al. 2004, 2005; Chi and Leu, 2006; Niu et al. 2005, Niu and Leu 2007).

The simulation of material removal for bone surgery can be achieved similar to the simulation of a virtual sculpting process for creating a 3D freeform object from a stock. It should be noted, however, that bone surgery simulation deals with inhomogeneous materials while virtual sculpting deals with homogeneous materials. The material removal can be achieved by continuously performing Boolean subtraction of the geometric model of the moving tool from the bone model. Galyean and Hughes (1991) introduced the concept of voxel-based sculpting as a method of creating free from 3D shapes by interactively editing a model represented in a voxel raster, and developed a virtual sculpting system with a simple tool. Wang and Kaufman (1995) presented a similar sculpting system with carving and sawing tools. In order to achieve real-time interaction, that system reduced the operations between the 3D tool and the 3D object to voxel-by-voxel operations. Bærentzen (1998) proposed octree-based volume sculpting and discussed its support of multi-resolution sculpting.

2.2.2 Key Technologies and System Overview

The schematic of a basic bone surgery simulation system is shown in Fig. 2.4. The user can use a personal computer based system to manipulate the interaction between the virtual bone and the virtual surgical tool, and perform virtual bone surgery

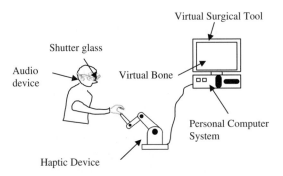

Fig. 2.4 Schematic of a basic bone surgery simulation system

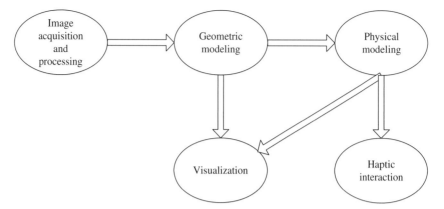

Fig. 2.5 Key elements involved in bone surgery simulation

by "seeing" bone material removal through a graphic display device, "feeling" the force via a haptic device, and "hearing" the sound of tool-bone interaction.

Generally speaking, bone surgery simulation involves several issues: image acquisition and processing, geometric modeling, physical modeling, visualization, and haptic interaction. The relationships between these key elements are illustrated in Fig. 2.5. Usually image acquisition and processing precedes the simulation and is done off-line in order to save the data processing time during on-line simulation.

A bone surgery simulation system consists of the following key elements as shown in Fig. 2.5:

1. Input the CT or MRI data of the bones and construct a geometric model attached with properties such as material and density.
2. Develop physical models to represent the tool-bone interaction, of which the interface force is updated continuously, to simulate the deformation and cutting of bones, ligaments, etc.
3. Implement real-time graphic rendering of volumetric data to obtain realistic visualization of bone surgery.
4. Provide force feedback to the user with haptic rendering.

To provide a meaningful virtual bone surgery system with realistic force feedback and visual effects, there are several requirements that must be met:

1. The medical data obtained from image acquisition must be processed to minimize noise and irrelevant data (Jackson et al. 2002; Niu et al. 2005). This data processing must be done before the bone surgery simulation.
2. In order to feed appropriate sensorial inputs to the human perceptual system, the simulation system must update the data at very different frequencies: about 30 Hz for visual rendering and above 1k Hz for haptic response (Mark et al. 1996).
3. Interactive data modification is required for both visual and force feedback, so data modification calculation should involve only local data Avila and Sobierajski, 1996; Astley and Hayward, 2000).

4. The amount of force computation time should be small for real-time haptic rendering (Avila and Sobierajski, 1996).

2.3 Medical Image Processing and Segmentation

2.3.1 Imaging Procedures

Computer imaging techniques have become an important diagnostic tool in the practice of modern medicine. Today, advanced medical scanners can provide high-quality and highly detailed images for surgeons prior to performing the actual physical procedures. Medical data obtained from imaging techniques typically represent the values of some properties at various locations (Kaufman et al. 1993). The most commonly used medical imaging techniques include CT (Computed Tomography), MRI (Magnetic Resonance Imaging), SPECT (Single-Photon Emission Computed Tomography) and PET (Positron Emission Tomography). These techniques use a data acquisition process to capture information about the internal anatomy of a patient. This information is in the form of slice-plane images, similar to conventional photographic X-rays (Schroeder et al. 2002).

CT and MRI are most commonly employed in obtaining medical images. CT provides high spatial resolution bone images while MRI provides better images for soft tissues. For most bone surgery simulators, CT scan data are used in the research because they show better contrast between bones and soft tissues. For reporting and displaying reconstructed CT values, Hounsfield Unit (HU) is usually used. This unit describes the amount of x-ray attenuation of each volume element in the three-dimensional image. There are good correlations between CT scan data and bone's material properties such as density and mechanical strength (Bentzen et al. 1987), so HU value is usually used for each data point to represent bone density.

The process of constructing a VR environment from the imaging data is a major challenge. The process can be divided into three stages: (1) spatial co-registration of data from multiple modalities; (2) identification of tissue types (segmentation); (3) definition of tissue boundaries for the VR environment (Jackson et al. 2002).

2.3.2 Image Processing

Noise and other artifacts are inherent in all methods of data acquisition. Due to noise in many signals and lots of irrelevant information in the medical data, image processing is necessary. Filtering and smoothing techniques, e.g., Gaussian filters and median filters, are usually used to reduce noise on images (Schroeder et al. 2002). Since information gained from various images acquired in a medical imaging procedure is usually complementary, proper integration of useful data obtained from the separate images is desired. Image reiteration is the process of determining the spatial transform that maps points from one image to homologous points on the second image (Luis et al. 2003). These images could have different or same format. The most common registration methods could be found from the survey of medical image registration by Maintz and Viergever (1998).

It is also necessary to identify which type of tissue is present in the data space and to identify the precise location of edges between different tissue types. Image segmentation is the process of identifying the distribution of different tissue types within the data set. Bones can be extracted by manual or partially automated segmentation methods. Usually, threshold segmentation is used to distinguish pixels or voxels within an image by their gray-scale value. An upper or lower threshold can be defined, separating the image into the structure of interest and background. This method works well for bone segmentation from CT scans since bone tissue attenuated significantly more during image acquisition and is therefore represented by much higher values on the Housfield scale compared to soft tissues. Whereas thresholding focuses on the difference of pixel intensities, segmentation looks for regions of pixels or voxels with similar intensities (Ritter et al. 2004).

Segmentation methods are usually divided into two kinds: region-based and edge-based (Kovacevic et al. 1999). Region-based methods search for connected regions of pixels/voxels with some similar features such as brightness, texture pattern, etc. After dividing the medical image into regions in some way, similarity among pixels is checked for each region, and then neighboring regions with similar features are merged into a bigger region, and a region with different features within it are split into smaller regions. These steps are repeated until there is no more splitting or merging. A main issue of this approach is how to determine exact borders of an object because regions are not necessarily split on natural borders of the object. Edge-based algorithms search for pixels with high gradient values, which are usually edge pixels, and then try to connect them to form a curve which represents a boundary of the object. A difficult problem here is how to connect high gradient pixels because in real images they might not be neighbors.

2.4 Geometric Modeling and Data Manipulation

2.4.1 Volume Modeling

The sequence of 2D slices of data obtained by CT or MRI can be represented as a 3D discrete regular grid of voxels (volume elements), as shown in Fig. 2.6. For surgical simulation, voxel-based modeling has a number of advantages over the use of surface polygons or solid primitives. First, voxel-based representation is natural for 3D digital images obtained by medical scanning techniques such as MRI or CT. Second, since no surface extraction or data reformatting is required, errors introduced by fitting surfaces or geometric primitives to the scanned images can be avoided. Third, volumetric objects can incorporate detailed information about the internal anatomical or physiological structure of organs and tissues. This information

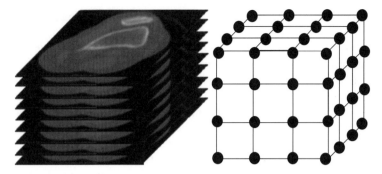

Fig. 2.6 A volume seen as a stack of images and a volume seen as a 3D lattice of voxels

is particularly important for realistic modeling and visualization of complex tissues (Gibson et al. 1997).

In volume representation, the basic elements are voxels (Bærentzen, 2001). Just as a pixel is a small rectangle, a voxel can be viewed as a small block. A voxel can be represented by the coordinates of its center point and the three orthogonal dimensions plus some attributes. If the voxels have fixed dimensions, then they can be represented by the vertices of a 3-D lattice, which are characterized by their positions and associated values of attributes. For example, it can be expressed as an array $(x, y, z, v_1, v_2, \ldots v_n)$, where (x, y, z) represents the position of each voxel and v_i represents a property. These properties can be physical properties such as density, material classification, stiffness, and viscosity as well as display properties such as color, shading, etc.

In general, the samples may be taken at random locations. Depending on how the samples are connected to form a grid structure, there are two classes of volumetric data: structured and unstructured. Structured data has a logical organization of the samples into a three-dimensional array and a mapping of each sample into the physical domain. Unstructured data is not based upon a logical organization of arrays, but instead upon a group of cells of certain shapes such as tetrahedra, hexahedra, or prisms.

An interpolation function is used to produce a continuous scalar field for each property. This is critical for producing smooth volume rendering and haptics rendering (Avila and Sobierajski, 1996). In order to meet the system requirements, it is often desirable to pre-compute and store the contents of each voxel, so there is no need to change every voxel during the surgical operation simulation. By storing the volumetric data in a space-efficient, hierarchical structure, the storage requirements can be reduced.

2.4.2 Data Manipulation

The data set for surgery simulation is usually very large. For example, for a medium resolution of 512^3, 2 bytes per voxel, the volume buffer must have 256 MB

(Kaufman et al. 1993). How to organize and manipulate such huge data is a challenging problem.

Zhu et al. (1998) used a finite element method (FEM) of analysis in their study of muscle deformation. A muscle was modeled with 8-node, 3D brick elements equivalent to the voxel structure. The simulation was achieved by solving a sparse linear system of equations which governs the behavior of the model. Gibson et al. (1997) developed a linked volume model to represent the volumetric data. The links were stretched, contracted or sheared during object deformation, and they were deleted or created when objects were cut or joined. Compared with the FEM method, the linked volume approach can be used for creating models with high geometric complexity and it can achieve interactivity with the use of low-cost mathematical modeling.

Bærentzen (1998) proposed an octree-based volume sculpting method in order to quickly separate homogeneously empty regions outside the object of interest. An octree structure as shown in Fig. 2.7 was chosen to organize the huge set of volumetric data and to improve the efficiency of data storage. A volume was subdivided until the leaf level of a prescribed size had been reached. This technique can significantly reduce the memory requirement and speed up the graphic rendering and modeling task. Basically, octrees are a hierarchical variant of spatial-occupancy enumeration that can be used to address the demanding storage requirements in volume modeling (Foley et al. 1996).

In bone surgery simulation, operation tools such as drills, mills and broaches remove voxels occupied by the cutting tool's volume during the course of operation. For static data structure, e.g. 3D arrays, voxels can only be removed in the defined size. That is, the cells representing the interaction between the cutting tool and the bone are constant in size and thus the resolution is static. Due to this limitation, voxel removal can only be done on a rough level. Octree modeling can provide a flexible data structure for performing material removal simulation dynamically. High resolution can be achieved in the region of interest, which is usually the current surgical tool location and its neighborhood. The octree nodes representing cells in the region of interest are subdivided to generate children nodes representing sub-cells. The material removal operation is then done on the children node level. The subdivision process can be repeated until the desired resolution is reached. To control the resolution automatically, a criterion to end the subdivision can be set.

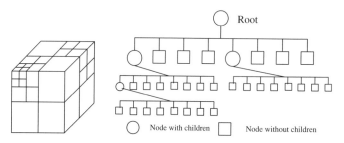

Fig. 2.7 Octree representation

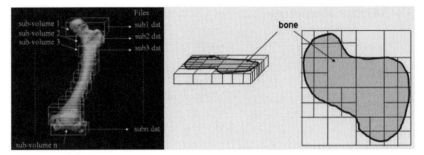

Fig. 2.8 Bounding volume and quadtree subdivision for human bone

One criterion could be that the smallest linear dimension of the voxel is equal to the radius of the drill multiplied by a sizing factor.

Another method is using Bounding Volume together with Quadtree Subdivision (Niu et al. 2005) to deal with irregular long bones. This method uses AABB (Axis Aligned Bounding Box) as the bounding volume type to find a tight bounding box for the bone. The whole bone volume after pre-processing is divided into many sub-volumes, which have certain slices/layers in Z direction (along the axis of the bone) and different dimensions in X and Y directions. All these sub-volumes should have relatively tight bounding boxes around the objects as shown in Fig. 2.8(a). Then, quadtree subdivision is obtained by successively dividing the sub-volumes from 1 to n in both x and y dimensions to form quadrants as shown in Fig. 2.8(b). Each quadrant of the sub-volumes may be full, partially filled, or empty, depending if the object of consideration intersects the area of concern. This method has been applied to remove irrelevant data and to organize the rest data, in order to make the bone surgery simulation system interactive in real time (Niu et al. 2005).

2.5 Graphic Rendering

Volume visualization is the technique used to display the information inside volumetric data using interactive graphics and imaging. Basically, the methods of graphic rendering of three-dimensional data (volumetric data) can be grouped into two: (1) Surface Rendering or indirect rendering and (2) Volume Rendering or direct rendering. To decide on which of the rendering methods is suitable for the bone surgery system, the following considerations are important: (1) real-time rendering and (2) surface quality in graphic display. Surface rendering extracts polygons from volumetric data and renders the surface interactively. Volume rendering can not give interactive real-time performance based on current graphics hardware and software unless very coarse approximation is made.

2.5.1 Surface Rendering

Marching Cube (Lorensen and Cline, 1987) is the most popular algorithm in surface rendering. The marching cube algorithm traverses all boundary cells of the entire volume and determines the triangulation within each cell based on the values of cell vertices. This method first partitions a volume data into cubes. Each cube consists of eight voxels. Then it decides the surface configuration of each cube according to 15 possible configurations, as shown in Fig. 2.9. Marching cube leads to satisfactory results for small or medium datasets. However, for simulation in the medical field, there usually exists a very large dataset which may restrict the real-time performance of interactive manipulation. Use of octrees for faster isosurface generation proposed by Wilhelms and van Gelder (1990) is an improved algorithm for extracting surfaces from volumetric data. This algorithm stores min/max voxel values at each octree node, and then traverses octree nodes that may contain an isosurface to obtain the triangles that form the surface. Other researchers (Shekhar et al. 1996; Sutton and Hansen, 1999; Velasco and Torres, 2001) also presented improved octree-based marching cube algorithms and their applications. These methods used some techniques to save data storing space and improve computational performance, but they do not support multi-resolution isosurface extraction.

Adaptive resolution surface rendering is the method mostly used in bone surgery simulation. Some researchers (Westermann et al. 1999; Boada and Navazo, 2001) presented ideas on this rendering method. The rendering algorithms are based on an extended marching cube algorithm for octree data as follows:

1. Find the focus of interest (i.e. the current surgical tool location and its neighborhood).
2. The region of interest is rendered in high resolution, meaning that the cells in the region are subdivided into sub-cells and the surface is extracted at the sub-cell level using the marching cube algorithm.
3. The rest regions are rendered in lower resolution. The cells in these regions are merged to form coarser cells.

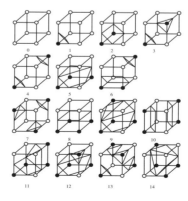

Fig. 2.9 The marching cube algorithm for surface rendering of voxel data

Trade-off exists between surface quality and interactivity. Although octree can address this problem to some extent, interactivity is still difficult to achieve for a large set of data. In order to improve performance, the initial resolution (usually not a very fine level) for the surface rendering needs to be specified. Dynamic resolution can be used depending on how the surgical tool interacts with the bone material. Parallel computing can be used to increase the computation and hence the resolution.

2.5.2 Volume Rendering

In this rendering method, the volume data are directly rendered, which means that the images are generated through the transformation, shading and projection of 3D voxels into 2D pixels. Volume rendering demands greater computational processing but produces images with greater versatility. Since all the voxels located in the line of view are used in image generation, this method allows the visualization of parts inside the surface. Although real-time rendering can hardly be achieved, this method is a good choice for some applications with special visualization requirements. Volume rendering will become more attractive in the future as computers are becoming faster and cheaper with larger memory.

The most popular algorithm of volume rendering is *Ray-Casting* (Levoy, 1988 and 1990). Traditionally, the ray-casting algorithm spans the projection plane and traces the rays into the scene. Usually parallel rays that are orthogonal to the projection plane are cast. These rays are traced from the observer position to the volume data. For each ray, sample points are calculated considering a fixed step on the path. The algorithm can calculate and accumulate both color and opacity values along the ray for obtaining the pixel color. Besides ray-casting, there are other popular algorithms in the volume rendering approach, e.g., splatting (Westover, 1990), shear-warp (Lacroute and Levoy, 1994), and 3D texture-mapping (Cabral et al. etc., 1994). Meiÿner et al. (2000) did an extensive survey on volume rendering algorithms.

Currently, most bone surgery simulation systems do not use volume rendering because of the interactivity restriction, the need for expensive, dedicated graphics hardware, and the need for large amounts of computation time and storage space. However, the merits of volume rendering along with the continuing decrease of computation costs may compel the researchers to use this method in the future.

2.6 Haptic Rendering

Haptic interface can enhance the realism of virtual surgery by giving a realistic feel of the surgical operation. Haptic rendering is the process of applying reactive forces to the user through a force-feedback device (Okamura, 1998). The rendering consists of using information about the tool-object interface to assign forces to be displayed, given the action at the operational point. The major challenge in simulating force-reflecting volume models is to achieve an optimal balance between the complexity of geometric models and the realism of the visual and haptic displays.

The following issues must be addressed in order to provide meaningful force feedback (Peng et al. 2003; Hua and Qin, 2002):

1. Force computation rate: This rate must be high enough and the latency must be low enough in order to generate a proper feel of the operation.
2. Generation of contact force: This creates the feel of the object during the simulated surgery. Interaction forces between the tool and the bone can be calculated using mathematical models.

For haptic rendering, there are several important components: force modeling, collision detection, force computation, and haptic rendering as shown in Fig. 2.10.

2.6.1 Force Modeling

Bone material removal operations are of considerable importance in orthopedic surgery (Plaskos et al. 2003). In hip and knee replacement procedures, for instance, the geometrical accuracy of the prepared bone surface is particularly relevant to achieving accurate placement and good fixation of the implant.

Bone drilling is needed prior to many orthopedic operations, such as pin or screw insertion to the bone, and it requires a high surgical skills. There were several studies on bone drilling reported in the literature. Wiggins and Malkin (1976) investigated the interrelationships between thrust pressure, feed rate, torque, and specific cutting energy (energy per unit volume required to remove material) for three types of drill bits. Jacob et al. (1976) presented research results showing that the drill point geometry was critical when attempting to minimize drilling force and that a softening effect occurred when the bone was drilled at relatively high speeds. Hobkirk and Rusiniak (1977) studied the relationships between drilling speeds, operator techniques, types of drills and applied forces in bone drilling. Through experiments they showed that the peak vertical force exerted on the drill varied between 5.98 and 24.32 N, and that the mean vertical force ranged from 4.22 to 18.93 N. Karalis and Galanos (1982) tested the drilling force against the bone hardness and triaxial

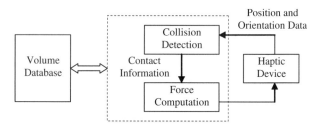

Fig. 2.10 Structure of haptic rendering

strength, and found a linear correlation between the triaxial compressive strength and the drilling force. Abouzgia and James (1995) investigated the dependance of force on drill speed and measured the energy consumption during drilling. They found that drilling force increased slightly with increase in speed at low starting speeds and decreased with increase in speed at high starting speeds.

Allotta et al. (1996) developed an experimental model for the description of breakthrough during the penetration of a twist drill in a long bone as illustrated in Fig. 2.11. They presented an equation for the thrust force required to drill a hole and reported its good correlation with experimental data. The thrust force required to drill a bone is

$$T = K_s a \frac{D}{2} \sin \frac{\beta}{2} \qquad (2.1)$$

where T is thrust force, K_s is the total energy per unit volume, a is the feed rate expressed in unit length per revolution, D is diameter of the drill bit, and β is the convex angle between the main cutting lips (see Fig. 2.11). K_s represents the sum of shear energy required to produce gross plastic deformation. It is primarily the friction energy of the chip sliding past the tool plus other minor energies. K_s has been shown to vary between $4.8R_u$ and $6R_u$ (Allotta et al. 1996), where R_u is the unitary ultimate tensile load. $K_s = 5R_u$ is a practically acceptable value. During rotation and penetration across the bone, the drill bit is subject to the resistant torque (besides the thrust force) of

$$M_z = 5R_u a \frac{D^2}{8} \qquad (2.2)$$

Chi et al. (2005) presented another drilling force model by performing regression of measured drilling force versus process and material parameters. The obtained force model was validated by performing more experiments with different sets of parameter values. The thrust force model could be written as:

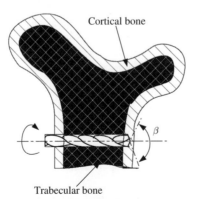

Fig. 2.11 Modeling force in drilling a long bone

$$T = 134.6 N^{-0.3327} v^{0.5189} \rho^{1.1841} \tag{2.3}$$

where T represents the thrust force, N is speed of drill bit in rotations per minute, v is feed-rate in mm/sec, and ρ is bone material density in g/cc.

Bone burring is also an important surgical procedure used in temporal bone surgery. Agus et al. (2002) presented a bone-burr interaction model. For a burr with a spherical bit of radius R rotating at angular velocity ω, they used Hertz's contact theory to derive the following elastic deformation force that exerts on the burr:

$$\vec{F}_e = C_1 R^2 \left(\frac{h}{R}\right)^{\frac{3}{2}} \hat{n} \tag{2.4}$$

where C_1 is a constant that depends on the elastic property of material, h is the tool embossing height, $\hat{n}$ is the normal direction of the contact surface. Also, the friction force can be obtained as

$$\vec{F}_\mu = \mu \int_\xi P(\vec{\xi}) \frac{\vec{r}(\vec{\xi}) \times \vec{\omega}}{\left|\vec{r}(\vec{\xi}) \times \vec{\omega}\right|} d\sigma \tag{2.5}$$

where μ is a friction coefficient, $\vec{\xi}$ represent a point on the contact surface, $P(\vec{\xi})$ is the pressure exerted by the burr on point $\vec{\xi}$, and $\vec{r}(\vec{\xi})$ is the displacement measured from the center of sphere burr bit to point $\vec{\xi}$, and $d\sigma$ represents a differential area on the contact surface.

The total force that should be provided by the haptic feedback device is, therefore,

$$\vec{F}_T = \vec{F}_e + \vec{F}_\mu \tag{2.6}$$

Other force models can also be applied in developing a virtual bone surgery system. For example, Eriksson et al. (2005) used an energy-based approach to determine how the force is as a function of material removal rate during the milling process. This model is same as the following simplified milling force model (Yang and Chen, 2003; Choi and Jerard, 1998):

$$F_t = K_t (MRR)/f \tag{2.7}$$

where F_t is the tangential cutting force, f is the feedrate, MRR is the material removal rate. The radial force is

$$F_r = K_r F_t \tag{2.8}$$

where K_t and K_r are constant and their values depend on workpiece material, cutting tool geometry and cutting conditions.

The spring-damping force model (Hua and Qin, 2002; McNeely et al. 1999; Avila and Sobierajski, 1996) can also be applied to virtual bone surgery. The force in this model can be expressed as

$$F = R(\vec{V}) + S(\vec{N}) \tag{2.9}$$

where $R(\vec{V})$ is the damping force, which is a motion retarding force proportional to velocity. $R(\vec{V})$ can be written as

$$R(\vec{V}) = -t_r(\rho, |\nabla\rho|)\vec{V} \tag{2.10}$$

where $\vec{V}$ is the velocity of the contacting point and t_r is a function of density ρ. $S(\vec{N})$ is the stiffness force normal to the object surface and is

$$S(\vec{N}) = \frac{\vec{N}}{\left\|\vec{N}\right\|} t_s(\rho, |\nabla\rho|) \tag{2.11}$$

where $\vec{N}$ is the surface normal at the contacting point and t_s is a function of density ρ. For volume rendering and isosurface rendering, t_r and t_s can be expressed differently (Hua and Qin, 2002; Avila and Sobierajski, 1996).

A haptic device can be used to give the user of the virtual bone surgery system a realistic force feedback by rendering the force and torque computed using cutting force models. At present, most virtual bone surgery systems use PHANToM device (SensAble Technology Corp.) and GHOST SDK for haptic rendering. Two examples of such a system are shown in Fig. 2.12. This PHANToM has three motors and six encoders to enable 6-DOF motion tracking and 3-DOF force feedback. The GHOST (General Haptics Open Software Toolkit) SDK is a C++ object-oriented software toolkit that enables application developers to interact with the haptic device and create a virtual environment at the object level. GHOST SDK provides a special

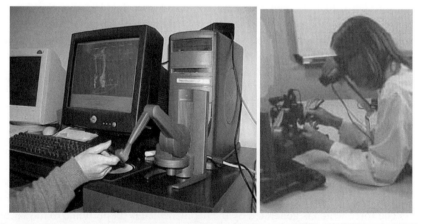

Fig. 2.12 Virtual bone surgery with haptic feedback: (a) Chi et al. (2005); (b) Agus et al. (2002)

class of functions called gstEffect, which allows adding "global" forces directly to the PHANToM. At each iteration of a servo loop the pointer of the Effect object is passed to a PHANToM node. By generating the Effect force when non-null intersection between the virtual tool and the virtual bone is detected, the system gives the user a realistic feel of force in real time.

In order to run the components of a virtual surgery system asynchronously, a multithreading virtual environment can be implemented. The multithreading computation environment allows maintaining suitable update rates for the various subsystems of the simulation system. The haptic loop must maintain an update rate of about 1000 Hz, while the graphics loop can get by with an update rate of about 30 Hz.

2.6.2 Collision Detection and Force Generation

In a bone surgery simulator, the haptic rendering consists two parts: collision detection and force generation. The goal of collision detection, also known as interference detection or contact determination, is to automatically report a geometric contact when it is about to occur or has just occurred (Lin and Gottschalk, 1998). Fast and accurate collision detection between geometric models is a fundamental problem in computer based surgery simulation. In developing a virtual bone surgery system, it is necessary to perform collision detection for purpose of simulating material removal and force feedback.

An early approach of haptic rendering used single-point representation of the tool for collision detection and penalty-based methods for force generation (Massie and Salisbury, 1994; Avila and Sobierajski, 1996). Collision detection was done by checking whether the point representing the tool is inside the object of consideration such as a bone. The surface information of an anatomic model can be obtained in terms of triangular facets using the marching cube algorithm previously described or by a method of surface reconstruction from dexel data (Peng et al. 2004).

Penalty-based methods generate a pre-computed force field based on the shortest distance from the interior point of an object to the object's surface. A problem of this approach is that there may be points in an object which have the same distance to the surface, e.g. Fig. 2.13(a). Another problem is that when pressing an object that has a sharp tip or fine feature such as that shown in Figure 2.13(b), the user will quickly feel the change of force direction from one side of the object to the other side and then feel no force at all. This may be a serious problem, especially when working with highly detailed models and small structures.

Constraint-based methods were introduced by Zilles and Salisbury (1995) and by (Ruspini et al. 1997; Ruspini and Khatib 1998). These methods use an intermediate object (representing the tool) which never penetrates a given workpiece such as a bone in the environment, as shown in Fig. 2.14. The intermediate object (called God-Object or Proxy) remains on the surface of the workpiece during the simulation process. The force which is generated by the haptic device is proportional to the vector difference between the physical position of the virtual tool and the proxy position

2 Virtual Bone Surgery

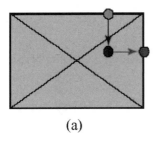 

(a) (b)

Fig. 2.13 Problems of penalty based haptic rendering

of this tool. The haptic rendering algorithm updates the proxy position in respect to the physical position by locally minimizing the distance from the proxy position to the physical position. Since these calculations have to be performed on-the-fly, contraint-based approaches are computationally more expensive than penalty-based approaches.

The single-point representation of an object for collision detection described above has many drawbacks including the following:

1. It is not suitable for inhomogeneous workpiece material, e.g., human bone;
2. It does not represent the 3D shape of a virtual tool;
3. It has discontinuities problem, e.g. sharp edges on the surface;

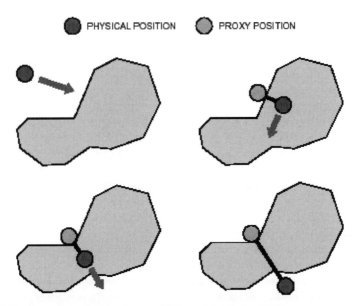

Fig. 2.14 Haptic rendering by virtual proxy (Zilles and Salisbury, 1995)

4. The virtual tool can reach points which may not be reachable by the real tool, e.g., entering a small hole with a large tool (Niu and Leu 2007).

Multi-point collision detection methods were developed more recently (McNeely et al. 1999; Petersik et al. 2002). These methods handle 3D shapes using multiple points on the surface of the tool. With these methods, more realistic simulations of tools and tool-object interaction can be achieved, and the drawbacks of the single-point approach can be overcome. However, multi-point collision detection is computationally more expensive. Moreover, this force feedback scheme may generate an unstable force in some cases (Nakao et al. 2003), especially when the number of points on the tool surface is not adequate.

2.7 Conclusion

Developing a bone surgery simulation system is a major undertaking and poses many technical challenges. The overarching objective of such a development is to build a high-fidelity simulation system which incorporates latest technologies in virtual reality including computer graphics and haptics rendering. This book chapter reviews the current bone surgery simulation systems, and the methods and techniques used to develop such systems.

The described virtual bone surgery system development consists of the following key tasks: image processing, geometric modeling, physical modeling, graphic rendering, and haptic rendering. A virtual bone surgery system usually takes pre-processed CT or MRI image data to construct a geometric model of the bone and soft tissue using volume or surface modeling methods, and update the geometric model continuously during the virtual surgery. Special data structures such as oc-tree or bounding volume plus quadtree are used to handle the large set of medical data. To perform graphic displays in real time, surface rendering with a marching cube algorithm is used in most virtual bone surgery systems. For force feedback, physics-based models are used to represent the interface forces between the surgical tools and the bone/soft tissue in deformation and material removal. Graphic rendering and haptic rendering are generated in real time using multithreading computations to provide realistic visualization and haptic feedback during the virtual surgery.

Research and development work on virtual bone surgery is far from mature. The R&D efforts thus far have rarely considered auditory rendering, which can play an important role in the generation of an immersive virtual environment. Sound cues can enhance haptic feedback when a user is interacting with an object in a virtual environment. In bone surgery, sound can provide information about the nature of the tool-bone contact region where the material removal operation occurs. For example, the change of sound from higher to lower pitches in bone drilling could signal reaching the boundary between the bone and the soft tissues. Thus it is very desirable to include auditory rending in the system development. An ideal virtual bone surgery system should be able to provide high-fidelity dynamic graphic displays with realistic force and sound feedback during the simulated surgery process.

References

Abouzgia MB, James DF (1995) Measurements of shaft speed while drilling through bone. J Oral Maxillofacial Surg 53:1308–1315

Agus M, Giachetti A, Gobbetti E, Zanetti G, Zorcolo A (2002) Real-time haptic and visual simulation of bone dissection. IEEE Virtual Reality Conference, pp 209–216

Allotta B, Belmonte F, Bosio L, Dario P (1996) Study on a mechatronic tool for drilling in the osteosynthesis of long bones: tool/bone interaction, modeling and experiments. Machatronics 6(4): 447–459

Astley O, Hayward V (2000) Design constraints for haptic surgery simulation. Proceedings of the 2000 IEEE international conference on Robotics & Automation, San Francisco, CA

Avila RS, Sobierajski LM (1996) A haptic interaction method for volume visualization. IEEE Visualization proceedings, San Francisco, MA, pp 197–204

Bærentzen A (1998) Octree-based volume sculpting. Proceedings of IEEE Visualization conference, Research triangle park, NC, pp 9–12

Bærentzen A (2001) Volume sculpting: intuitive, interactive 3D shape modelling. IMM, May 15

Barker VL (1999) Cathsim. In: Westwood JD, Hoffman HM, Robb RA, Stredney D (eds) Proceedings of MMVR IOS Press, San Francisco, USA, pp 36–37

Bentzen SM, Hvid I, Jorgensen J (1987) Mechanical strength of tibial trabecular bone evaluated by X-Ray Computer Tomography. J Biomech 20(8):743–752

Berkley J, Weghorst S, Gladstone H, Raugi G, Berg D, Ganter M (1999) Fast finite element modeling for surgical simulation. In: Westwood JD, Hoffman HM, Robb RA, Stredney D (eds) Proceedings of MMVR IOS Press, San Francisco, USA, pp 55–61

Boada I, Navazo I (2001) Multiresolution isosurface fitting on a surface octree. 6th International Fall Workshop Vision, Modeling and Visualization 2001, Stuttgart, Germany, pp 318–324

Bro-Nielsen M, Helfrick D, Glass B, Zeng X, Connacher H (1998) VR simulation of abdominal trauma surgery. Proceedings of Medicine Meets Virtual Reality 6 (MMVR-6), IOS Press, San Diego, California, pp 117–123

Bryan J, Stredney D, Wiet G, Sessanna D (2001) Virtual temporal bone dissection: A case study. IEEE Visualization 2001, October 21-October 26, San Diego

Cabral B, Cam N, Foran J (1994) Accelerated volume rendering and tomographic reconstruction using texture mapping hardware. Symposium on Volume Visualization, pp 91–98

Choi BK, Jerard RB (1998) Sculptured surface machining theory and applications. Kluwer Academic Publishers, Norwell, MA

Chi X, Leu MC, Ochoa J (2004) Modeling of haptic rendering for virtual bone surgery. Proceedings of ASME International Mechanical Engineering Congress and R&D Expo and Computers and Information in Engineering Conference, Anaheim, CA

Chi X, Niu Q, Thakkar V, Leu MC (2005) Development of a bone drilling simulation system with force feedback. Proceedings of ASME International Mechanical Engineering Congress and Exposition, Orlando, FL

Chi X, Leu MC (2006) Interactive soft tissue deformation simulation using physics-based modeling. Proceedings of ASME International Mechanical Engineering Congress and Exposition, Chicago, IL

Conditt M, Noble PC, Thompson MT, Ismaily SK, Moy G, Mathis KB (2003) Quantitative analysis of surgical technique in total knee replacement. Proceedings of the 49th Annual Meeting of the Orthopedic Research Society, pp 13–17

Delp SL, Loan P, Basdogan C, Rosen JM (1997) Surgical simulation: an emerging technology for training in emergency medicine. Presence 6(2):147–159

Edmond CV, Heskamp D, Sluis D, Stredney D, Wiet GJ, Yagel R, Weghorst S, Oppenheimer P, Miller J, Levin M, Rosenberg L (1997) Simulation for ENT endoscopic surgical training. Proceedings Medicine Meets Virtual Reality 5, San Diego, CA, pp 518–528

Eriksson M, Flemmer H, Wikander J(2005) Haptic simulation of the milling process in temporal bone operations. MMVR 13 Medicine Meets Virtual Reality Conference, Jan 2005

Foley JD, Dam AV, Feiner SK, Hughes JF (1996) Computer graphics: Principles and practice, 2nd edn. Addison Wesley, Boston

Galyean TA, Hughes JF (1991) Sculpting: an interactive volumetric modeling technique. Comput graph 4(25):267–274

Gibson S, Samosky J, Mor A, Fyock C, Grimson E, Kanade T, Kikinis R, Lauer H, McKenzie N, Nakajima S, Ohkami H, Osborne R, Sawada A (1997) Simulating arthroscopic knee surgery using volumetric object representations, real-time volume rendering and haptic feedback. Proceedings of Computer Vision and Virtual Reality in Medicine and Medical Robotics and Computer Assisted Surgery, pp 369–378

Gorman PJ, Meier AH, Krummel TM (2000) Computer-assisted training and learning in surgery. Comput Aided Surg 5:120–130

Haluck RS, Krummel TM (2000) Computers and virtual reality for surgical education in the 21st century. Arch Surg 135:786–792

Hobkirk J, Rusiniak K (1977) Investigation of variable factors in drilling bone. J Oral Surg, 35:968–973

Hua J, Qin H (2002) Haptic sculpting of volumetric implicit functions. Proceedings of the 2002 IEEE symposium on volume visualization and graphics, pp 55–64

Jackson A, John NW, Thacker NA, Gobbetti E, Zanetti G, Stone RJ, Linney AD, Alusi GH, Schwerdtner A (2002) Developing a virtual reality environment for petrous bone surgery: A "state-of-the-art" review. J Otology and Neurotol 23:111–121 March 2002

Jacob CH, Berry JT, Pope MH, Hoaglund FT (1976) A study of the bone machining process-drilling. J. Biomech 9:343–349

John NW, Thacker N, Pokric M, Jackson A, Zanetti G, Gobbetti E, Giachetti A, Stone RJ, Campos J, Emmen A, Schwerdtner A, Neri E, Franseschini SS, Rubio F (2001) An integrated simulator for surgery of the petrous bone. Proceedings Medicine mettes virtual reality, pp 218–224

Kaufman A, Cohen D, Yagel R (1993) Volume graphics. IEEE Computer 26(7):51–64

Karalis T, Galanos P (1982) Research on the mechanical impedance of human bone by a drilling test. Journal of Biomechanics 15(8): 561–581

Kovacevic D, Loncaric S, Sorantin E (1999) Deformable contour based method for medical image segmentation. First croatian symposium on computer assisted surgery, Zagreb, Croatia

Lacroute P, Levoy M (1994) Fast volume rendering using a shear-warp factorization of the viewing transformation. Proceedings of SIGGRAPH '94, pp 451–458

Levoy M (1988) Volume rendering – display of surfaces from a volume data. IEEE Computer Graphics & Applications, Los Alamitos, CA, 8(3):29–37

Levoy M (1990) Efficient ray-tracing of volume data. ACM Transactions on Graphics, New York, 9(3):245–261

Lin M, Gottschalk (1998) Collision detection between geometric models: A survey. In the Proceedings of IMA Conference on Mathematics of Surfaces, 1998

Lorensen WE, Cline HE (1987) Marching cubes: A high resolution 3D surface construction algorithm. Comput graph 21(4):163–169

Luis L, Schroeder W, Lydia N, Josh C (2003) ITK software guide: The insight segmentation and registration toolkit. Kitware Inc., Clifton Park, NY, USA

Maintz J, Viergever M (1998) A survey of medical image registration. Medical Image Analysis 2(1):1–36

Massie TM, Salisbury JK (1994) The phantom haptic interface: A device for probing virtual objects. ASME Haptic Interfaces for Virtual Environment and Teleoperator Systems 1: 295–301

Meißner M, Huang J, Bartz D, Mueller K, Crawfis R (2000) A practical evaluation of four popular volume rendering algorithms. In ACM Symposium on volume visualization

Mark WR, Randolph SC, Finch M, Verth JMV, Taylor II RM (1996) Adding force feedback to graphics systems: Issues and solution. Proceedings of the 23rd annual conference on computer graphics and interactive techniques, pp 447–452

McNeely WA, Puterbaugh KD, Troy JJ (1999) Six degree-of-freedom haptic rendering using voxel sampling. Proceedings of ACM SIGGRAPH, pp 401–408

Morris D, Sewell C, Blevins N, Barbagli F (2004) A collaborative virtual environment for the simulation of temporal bone surgery. Proceedings of Medical Image Computing and Computer-Assisted Intervention Conference, Saint-Malo, FRANCE, pp 319–327

Nakao M, Kuroda T, Oyama H (2003) A haptic navigation system for supporting master-slave robotic surgery. Proceedings of ICAT 2003, Tokyo, Japan

Niu Q, Chi X, Leu MC (2005) Large medical data manipulation for bone surgery simulation. Proceedings of ASME International Mechanical Engineering Congress and Exposition, Orlando, FL

Niu Q, Leu MC (2007) Modeling and rendering for a virtual bone surgery system. Proceedings of Medical Meets Virtual Reality Conference, Long Beach, CA

Okamura AM (1998) Literature survey of haptic rendering, collision detection, and object modeling

Peng X, Chi X, Ochoa J, Leu MC (2003) Bone surgery simulation with virtual reality. Proceedings of ASME Design Engineering Computers and Information in Engineering Conferences, Chicago, IL

Peng X, Zhang W, Asam S, Leu MC (2004) Surface Reconstruction from Dexel Data for Virtual Sculpting. Proceedings of ASME International Mechanical Engineering Conference, Anaheim, CA

Petersik A, Pflesser B, Tiede U, Hoehne KH, Leuwer R (2002) Haptic volume interaction with anatomic models at sub-voxel resolution. Haptics, Orlando, Florida, pp 66–72

Pflesser B, Petersik A, Tiede U, Hohne HK, Leuwer R (2002) Volume cutting for virtual petrous bone surgery. Comput Aided Surg 7:74–83

Plaskos C, Hodgeson A, Cinquin P (2003) Modeling and optimization of bone-cutting forces in orthopedic surgery. Medical Image Computing and Computer-Assisted Intervention – MICCAI 2878:254–261

Ritter L, Burgielski Z, Hanssen N, Jansen T, Lievin M, Sader R, Zeilhofer HF, KeeveE (2004) 3D interactive segmentation of bone for computer-aided surgical planning. Proceedings 4th Annual Conference of the International Society for Computer Assisted Orthopedic Surgery, CAOS'04, Chicage, IL

Røtnes JS, Kaasa J, Westgaard G, Grimnes M, Ekeberg T (2002) A tutorial platform suitable for surgical simulator training (SimMentor^{TM}). Medicine Meets Virtual Reality 2002

Ruspini DC, Kolarov K, Khatib O (1997) The haptic display of complex graphical environments Proceedings of ACM SIGGRAPH, pp 345–352

Ruspini D, Khatib O (1998) Dynamic models for haptic rendering systems. Advances in Robot Kinematics: ARK98, Strobl/Salzburg, Austria, pp 523–532

Schroeder W, Martin K, Lorensen B (2002) The visualization toolkit—an object-oriented approach to 3D graphics, 3rd edn. Prentice Hall, New Jersey

Shekhar R, Fayyad E, Yagel R, Cornhill J (1996) Octree based decimation of marching cubes surfaces. Visualization 96:335–342

Sutton P, Hansen DC (1999) Isosurface extraction in time-varying fields using a temporal branch-on-need tree (T-BON). IEEE Visualization '99, pp 147–153

Udilijak T, Ciglar D, Mihoci K (2003) Influencing parameters in bone drilling. In Proceedings of 9th International Scientific Conference on Production Engineering CIM, pp 133–142

Velasco F, Torres JC (2001) Cells octree: a new data structure for volume modeling and visualization. Proceedings of the VI Fall Workshop on Vision, Modeling and Visualization, Stuttgart, Germany, pp 151–158

VRMedLab, 2003, http://www.bvis.uic.edu/vrml/

Wang SW, Kaufman AE (1995) Volume sculpting. Proceedings of Symposium on Interactive 3D

Graphics, Monterey, CA, pp 151–156

Westermann R, Kobbelt L, Ertl T (1999) Real-time exploration of regular volume data by adaptive reconstruction of isosurfaces. The Visual Computer, 15(2):100–111

Westover L (1990) Footprint evaluation for volume rendering. Proceedings of ACM SIGGRAPH, pp 367–376

Wiet GJ, Bryan J, Dodson E, Sessanna D, Stredney D, Schmalbrock P, Welling B (2000) Virtual temporal bone dissection. In: Westwood et al (eds) Proceedings of MMVR8. IOS Press Amsterdam, pp 378–384

Wiggins KL, Malkin S (1976) Drilling of bone. J Biomech 9:553–559

Wilhelms J, van Gelder A (1990) Octrees for faster isosurface generation. Proceedings of the 1990 workshop on volume visualization, pp 57–62

Yang Z, Chen Y (2003) Haptic rendering of milling. In Proceedings of Eurohaptics Conference

Zhu, Q, Chen Y, Kaufman AE (1998) "Real-time biomechanically-based muscle volumedeformation using FEM," Proceedings of EUROGRAPHICS '98

Zilles C, Salisbury JK (1995) A constraint-based god object method for haptics display. Proceedings of IEEE/RSJ International Conference on Intelligent Robots and Systems (IROS)

Chapter 3
Medical Imaging Challenges Photogrammetry

Petros Patias

3.1 Introduction

Medical imaging is the most important source of anatomical and functional information, which is indispensable for today's clinical research, diagnosis and treatment, and is an integral part of modern health care. Current imaging modalities provide huge floods of data, which can only be transformed to useful information if automated. That is, the inherent (highly resolved both spatially and radiometrically) information cannot be recovered, understood and exploited, unless highly sophisticated algorithms can be devised. Therefore, both medical imaging and the automated interpretation of imagery are of central importance today.

Most of the clinically evolving abnormal situations are actually evolving in space, i.e. they are 3D processes. There are numerous parameters inherent to these 3D processes, that have in general been understudied and have the potential to better delineate the actual pathologic process and probably contribute significant prognostic information. Therefore, accurate 3D-information extraction is fundamental in most of the situations.

Such conclusions inevitably underline the importance of the technology transfer across disciplines. Indeed, it is this technology fusion among different disciplines that has changed enormously medical diagnosis and treatment in recent years.

In this quest for more accurate 3D reconstructions and more robustness of the automated processes, photogrammetry has an important role to play. Although, up to now, photogrammetric contributions were aiming mainly at the reconstructions of the outer body using optical sensors, the mapping of the internal body presents a much greater research area; both in terms of scientific challenge and in terms of the wider range of applications.

Photogrammetry's comparative advantage is the ability to produce, process and exploit in 3D big amounts of high-resolution data in a geometrically consistent, robust and accurate way. During its course, photogrammetry has developed and fine-tuned its techniques, driven by the necessity to identify, measure and reconstruct natural and man-made objects from satellite, aerial and close-range digital images. This gave a considerable growth to photogrammetry and established it as the method of choice for a wide range of applications.

In many respects, current challenges in medical imaging show remarkable similarities to usual photogrammetric problems. Towards their solution, photogrammetry has the potential to highly contribute and bring new level of understanding, once a cross-disciplinary understanding is established.

Usual barriers to such inter-disciplinary cooperation have been long ago identified as differences in education, background knowledge, used terminology and disciplinary-focused line of thinking. The major aim and intended contribution of this paper is to smooth such differences and attempt to bridge the gap between photogrammetrists and experts involved in medical imaging.

In doing that, the paper is structured in two main parts. In the first part, a general account of the principal ideas underlying the photogrammetric practice is presented. The purpose is to provide the non-photogrammetrists with a basic understanding of both the main course as well as the terminology of photogrammetry. Additionally, it aims at establishing a basic and realistic background of what can be expected from these techniques.

The second part is devoted to the presentation of major challenges in medical imaging. The major apparent aim is to familiarize photogrammetrists, this time, with the basic problems, terminology, literature, and research trends in medical imaging. On purpose, these problems are presented in their basic configuration and in generic terms, in order to provoke photogrammetrically understood research questions. By no means this presentation is exhaustive. It is only meant to highlight some major problems and pinpoint the possible contributions from the part of photogrammetry.

3.2 Meeting Photogrammetry

3.2.1 Generic Photogrammetric Problems

In generic terms, photogrammetry tries to reconstruct 3D objects from their 2D images. Supposing that an image consists of the "projections" of all object points through "lines-of-sight", the image plane is merely a section of this bundle of rays in space. Along every line a corresponding object point can be determined, but since all the space points along this line will project at the same point on the image plane, this determination is not unique and the 3D coordinates of the object point cannot be recovered. However, if we introduce another image (Fig. 3.1) the object point will be uniquely defined as the intersection of the two corresponding rays. Additional images will add additional rays, but the point will still be the unique intersection of all corresponding rays.

The important steps now are the following three: First, find the correspondences of rays (or alternatively of the "homologous" points on the two images). Second, measure their image coordinates in a consistent way. Third, associate or transform these measurements to the object space. At the end of the third step we will have recovered the 3D coordinates of all object points, therefore the object will be reconstructed in 3D.

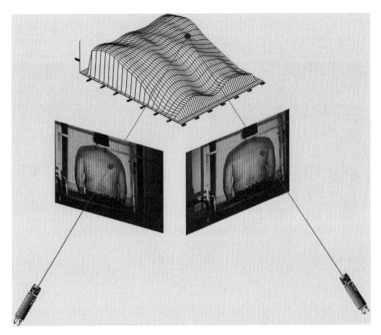

Fig. 3.1 The object point is uniquely determined as the intersection of corresponding lines

3.2.2 Image Matching

The first step clearly involves the correct "matching" of all image points of the left image to their counterparts in the right image. In doing that, photogrammetry uses a number of sophisticated techniques, which involve both radiometric (e.g. measures of similarity of grey values) and geometric constraints (e.g. relative orientation of one image with respect to the other, see Fig. 3.2) in order to find all matches, minimizing the "false alarms" at the same time. Additionally, it uses a least squares estimation procedure, which also accounts for measurement errors and provides statistically valid indications of the "correctness" of the matches.

3.2.3 Internal Sensor Geometry

The second step requires first the definition of a consistent coordinate system on each image, to which each image point measurements will refer. Such a coordinate system also relates the image points to the centre of the sensor system. This way all image points will have plane coordinates (x, y) as per measured and a "virtual" third dimension (c) equal to the "focal length" of the sensor system. Additionally in this step possible image distortions, due to imperfect sensors, will be corrected ac-

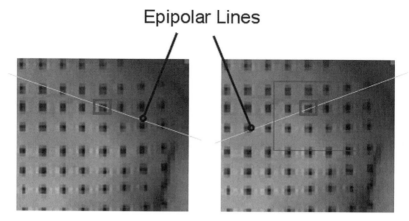

Fig. 3.2 Geometric constraints are also used in searching for correspondences

cording to sensor calibration information. In case such information is not provided, these errors will be modelled later on (see 2.4).

3.2.4 Sensor Model and Object Reconstruction

Thus, for each object point we have a set of image coordinates (x_1, y_1, c), (x_2, y_2, c) corresponding to (at least) two homologous points on the related images. The third step now involves the transformation of these coordinates to a unique triplet (X, Y, Z) for each object point. In doing that we need to know how the image has been formed, by what mechanism the bundle of rays have been generated; in other words, how does the sensor "map" the real world. We call it the "sensor model" and it relates the "image space" to the "object space".

$$x = x_0 - c \frac{R_{11} \Delta X + R_{12} \Delta Y + R_{13} \Delta Z}{R_{31} \Delta X + R_{32} \Delta Y + R_{33} \Delta Z}$$
$$y = y_0 - c \frac{R_{21} \Delta X + R_{22} \Delta Y + R_{23} \Delta Z}{R_{31} \Delta X + R_{32} \Delta Y + R_{33} \Delta Z} \quad (3.1)$$

$$x = A_1 X + B_1 Y + C_1 Z$$
$$y = A_2 X + B_2 Y + C_2 Z \quad (3.2)$$

$$x = \frac{L_1 X + L_2 Y + L_3 Z + L_4}{L_9 X + L_{10} Y + L_{11} Z + 1}$$
$$y = \frac{L_5 X + L_6 Y + L_7 Z + L_8}{L_9 X + L_{10} Y + L_{11} Z + 1} \quad (3.3)$$

$$x = A_1X + B_1Y + C_1Z + D_1XY + E_1XZ + F_1YZ + G_1X^2 + H_1Y^2 + I_1Z^2 + \ldots$$
$$y = A_2X + B_2Y + C_2Z + D_2XY + E_2XZ + F_2YZ + G_2X^2 + H_2Y^2 + I_2Z^2 + \ldots$$

(3.4)

Figure 3.3 shows typical such models: Fig. 3.3a shows a parallel projection (as for example in ultrasound images), Fig. 3.3b shows the most used general perspective projection, including sensor tilts (i.e. all images acquired through a lens system), Fig. 3.3c shows a variation of 3b, where the image is formed beyond the object (i.e. as in X-raying) and finally Fig. 3.3d refers to images formed by cross-sections of the images (i.e. as in CT or MRI. Note that many biomedical 3D reconstructions originate from 2D slices in various orientations (axial, sagittal, coronal, etc.), which are reconstructed by back-projection (e.g. CT, PET or SPECT) or direct 3D reconstructions (e.g. MRI).

In all different cases the same object (in our example: a perfect square) is imaged distorted but in a different manner each time. To these distortions, one should also add distortions caused by imperfections of the sensor system itself (see 2.3), which, to a first approximation and for the economy of our discussion, can be considered linear (i.e. affine). Therefore, the real image each time will have the form shown at the bottom line of Fig. 3.3. And we are called to "deduce" the 3D coordinates of a perfect square from "correct" (plus measurement errors) measurements on "imperfect" images.

This is a central problem in photogrammetry, which is being solved by describing the geometry of the sensor model through analytic formulation. For example Eq. (3.1) describes the perspective projection of Fig. 3.3b, 3.3c; Eq. (3.2) describes the geometry of Fig. 3.3a accounting also for affine sensor distortions; similarly, Eq. (3.3) describes the geometry of Fig. 3.3b, 3.3c accounting also for affine sensor distortions. If we pretend that we know nothing about the way the sensor forms the image, or if the actual sensor geometry is too complex to be analytically described, then the only option left is to use a generic and rather "blind" formulation, like the

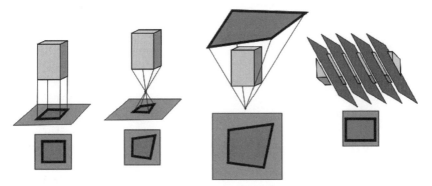

Fig. 3.3 Ways a 3D object (solid parallelepiped with square basis is assumed) is imaged by different sensor geometries. Last row shows the acquired image (3a-paralell projection, 3b-general perspective, 3c-general perspective, image formation beyond the object, 3d-cross sectioning)

one in Eq. (3.4). In this we admit that the image is formed in an unknown to us way and thus any kind of distortion is possible (linear plus non-linear). To the degree that this is not actually the case, that assumption may lead to irregular solutions and awkward figures. Thus, it is always much safer to explore and correctly model the actual geometry of the involved sensor.

The interested reader can refer to either standard photogrammetric textbooks (like Mikhail et al., 2001), or to more specific ones, dealing with close-range photogrammetry (like Atkinson, 1996 or Karara, 1989). A useful reference is also the Theme Issue "Photogrammetry and Remote Sensing in Medicine, Biostereometry and Medical Imaging" of the ISPRS Journal of Photogrammetry and Remote Sensing, Vol. 45, No. 4, 1990.

3.3 Challenges in Medical Imaging

Modern healthcare practices are increasingly depended on data and information captured and processed by a wide range of medical imaging modalities, like Digital Radiography Imaging (X-Rays), Ultrasonic Imaging, Digital Subtraction Angiography (DSA), Computed Tomography (CT), Optical Coherence Tomography (OCT), Magnetic Resonance Imaging (MRI), Functional Magnetic Resonance Imaging (fMRI), Magnetic Resonance Angiography (MRA), Nuclear Magnetic Resonance (NMR), Positron Emission Tomography (PET), and Single Photon Emission Computed Tomography (SPECT), see e.g. (Klingenbeck and Rienfelder, 1990). To this list, other sensors not directly producing images but used along with imaging, like Magnetic Resonance Spectroscopy (MRS), Electro-Encephalo-Graphy (EEG), Magneto-Encephalo-Graphy (MEG), etc., should be added as well. These modalities are classified to anatomical ones (e.g. X-rays, MRI, CT) or to functional ones (e.g. PET, SPECT, MRS, fMRI).

There are great differences among these techniques, in terms of sensor geometries, type of measurements, physical properties of measured quantity, resolution, acquisition time etc. The choice of the best-suited imaging technique to solve any particular clinical problem is mainly based on geometric accuracy, discrimination ability, versatility, speed and degree of invasion. However, the imaging procedure itself, although important, is only the first step in this process. Converting data to useful and meaningful information is the next equally important step towards diagnosis and treatment.

In the next sections we will try to group and analyse, in generic terms, the major problems and challenges Medical Imaging is facing, while pursuing this second step. We recognize that this classification is neither easy nor exhaustive. However, it highlights, in generic terms, the otherwise vast and detailed medical research efforts and the related literature.

The major aim of this analysis is to pin-point the grand challenges in medical imaging practice and describe them in photogrammetric terms in a productive way—meaning a way that attracts photogrammetric interest and also been subjected to solutions along the photogrammetric line of thought.

3.3.1 Multi-sensor Image Fusion

3.3.1.1 Problem Statement

Fusion and synthesis of images from different sensors is a major challenge in medical practice. The problem mainly stems from the fact that the selection of the "most appropriate" imaging modality is, quite often, not feasible if existing at all. Therefore, optimal quality of information can only be obtained by combining the sensitivity and specificity of different sensors.

A classical example in terms of image contents is CT/MRI fusion: CT images provide excellent contrast of bones and other dense structures (high intensity values) to the surrounding tissue, whereas MRI images soft tissues very well and only poorly the bones. However, the visualization of the spatial relationships of both is very useful in many clinical areas (e.g. spinal pathologies—vertebral bodies and disks, nerves, etc.). The benefits are even more profound in combining anatomical imaging modalities with functional ones, e.g. PET/CT in lung cancer, MRI/PET in brain tumors, SPECT/CT in abdominal studies, Ultrasonic Images/MRI for vascular blood-flow, etc.

However, the maximum clinical value of such image fusion can be realized only if an accurate multi-image registration is available. Clearly, this issue involves a wide range of technical problems to be solved, such as:

1. Equalization of images' radiometric appearance (which are caused by different contrast agents, acquisition parameters, sensor specificities, etc)
2. Correction of geometric distortions (through the sensors' calibration models).
3. Cooperativeness of different resolutions (e.g. 3mm in MRI, 1mm in CT, 0.5mm in Ultrasound).
4. Development of surface/volume matching strategies.
5. Development of qualitative and quantitative "similarity" measures (geometry, mutual information content, etc).
6. Geometric transformation between different sensors' imaging models (which should account for differences in imaging geometries) - 2D (pixels) or 3D (voxels) registration (depending on the image content and the issue at hand).
7. Automatic realization of the procedure (in order to minimize user induced errors and maximize timely operations, especially when real-time requirements are imposed).

3.3.1.2 State-of-the Art

Image fusion or co-registration can be based either on external targets (either implanted or co-imaged) or on internal fiducials (anatomically recognizable landmarks common in all images). The first case normally provides much better results and external quality measures, but requires implantation or co-imaging of control targets, which is not always possible. Classical examples of the first (implanted) are Börlin (2000) and [URL7, in Appendix A] (Fig. 3.4) for use of dental and orthopaedic implants in registration of X-ray images, or the use of titanium screws (Tate and

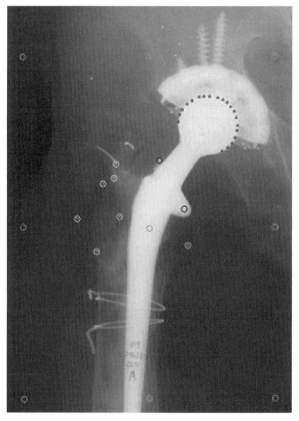

Fig. 3.4 The RSA system uses implanted markers and/or calibration cages to register X-Ray images (from URL7)

Chapman, 2000) (Fig. 3.5) for fusion of MRI and CT spinal images. Examples of the second (co-imaged) are the use of ear rods in X-ray cephalograms (Aoki et al., 2000) (Fig. 3.6), (Thomas et al., 1996) for fusion of X-ray and optical images in orthognathic surgery, the use of optical targets (Schewe and Ifert, 2000) for fusion of optical face images with plaster-cast optical image measurements for orthodontic applications, the use of optical targets (Fig. 3.7), (Zawieska, 2000) for optical image registration for scoliosis treatment or for face reconstruction (Jansa et al., 2000), and the use of optical targets for registration of blood vessels (D'Apuzzo, 2001).

The second case uses highly recognizable standard anatomical landmarks, (e.g. carotid, visual motion area V5 of the brain or the hand section of the rolandic motor area (Turner and Ordidge, 2000)) or solely information generated by the patient's anatomy, which is recognizable in all images. For examples, the reader can refer to (Kiskinis et al., 1998) (Fig. 3.8) for fusion of pre-operative DUS (Duplex Ultrasonic Scanning) images with post-operative optical images of specimens of carotid artherosclerotic plaque and post-operative optical microscopic images of dyed plaque slices, to [URL2] for fusion of MRI, PET and SPECT images, to [URL1] for fusion of optical video image with MRI scans for image-guided

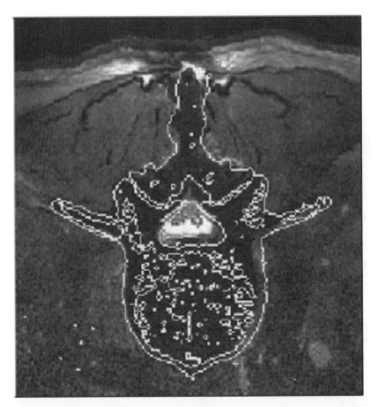

Fig. 3.5 MR image overlaid with boundary extracted from CT (from Tate and Chapman, 2000)

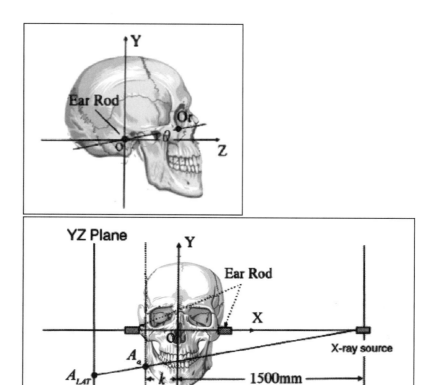

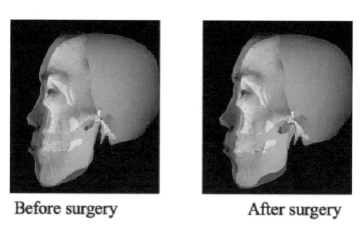

Fig. 3.6 Use of ear rods for fusion of Digital Radiographs with optical images for orthognathic applications (from Aoki et al., 2000)

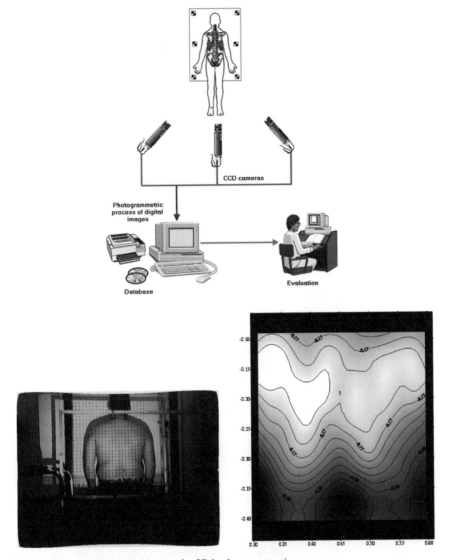

Fig. 3.7 Use of optical digital images for 3D back reconstruction

neurosurgery, and fusion of pre-operative MRI images and intra-operative ultrasound images (Fig. 3.9).

In both cases, most of the current research efforts are directed more on maximizing qualitative similarities rather than establishing geometric correspondences. The reader may refer to Van den Elsen et al. (1993) for a review of image/volume registration techniques, to Woods et al. (1993), Hill et al. (1994) and Collignon et al. (1995) for description of "similarity" measures and to Wells et al. (1996), Studholme

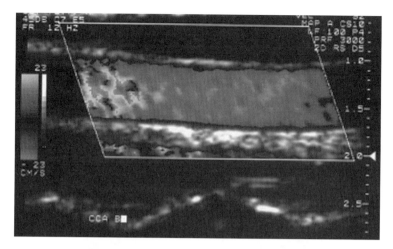

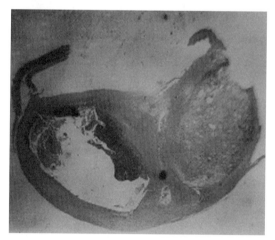

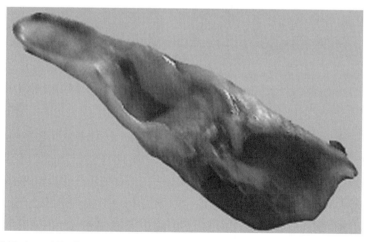

Fig. 3.8 Fusion of DUS images, microscopic images and optical images of specimen of the artherosclerotic plaque (from Kiskinis et al., 1998)

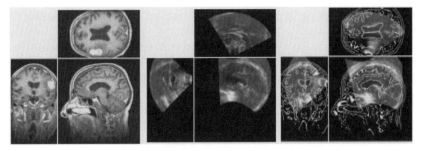

Fig. 3.9 Co-registration of MRI images (left) and ultrasound images (middle) (from Roche et al., 2000)

et al. (1996) and Meyer et al. (1997) for registration algorithms maximizing qualitative similarities.

Generally the methods used are based on grey-level information and use either correlation techniques or mutual information minimization. Another suite of techniques is based on surface (or volume) registrations and in most of the cases a previous segmentation is assumed, which, as we will see next, is a serious problem itself. Such techniques, however, very often fail. For example, MRI images suffer from a number of geometric distortions and artefacts up to 4mm (e.g. resonance offsets caused by chemical shifts, "potato-chip" effects, varying slice thickness, "bow-tie" effect, etc., see e.g. (Tate and Chapman, 2000) or [URL3]), which cannot be corrected by the above techniques. Ultrasonic images, from the other hand, are fundamentally "noisy" and "speckled", which presents a major problem in above techniques. Additionally, generally such techniques cannot provide quantitative indications of the registration performance. On the analytical side, the above techniques use merely generic 2D-affine transformations.

3.3.1.3 Possible Contribution of Photogrammetry

The major shortcoming of the non-photogrammetric techniques, developed in image registration, is their inability to take into account the actual imaging geometry; they replace it by a simplistic affine model. Thus, they are left to cope only with the images' similarities in appearance, that is they only impose radiometric constraints, when searching for correspondences. These constraints, however, if strictly imposed, cannot cope with eventual differences in appearances when using different imaging modalities; if loosely imposed they are of little help.

In photogrammetry also, it has been realized, long ago, that mere radiometric constraints are not adequate for achieving high accuracies and high reliability (meaning reduced number of false matches) of results. This is the reason why the quest for geometrical constraints has led to modelling of sensor geometry. In the usual photogrammetric practice, both radiometric and geometric constraints are adjusted together. Additionally, sensor orientation parameters (i.e. imaging angles), which can be rather easily recorded during imaging, can impose additional strict geometric constraints (e.g. epipolar lines), which will highly ease and robustify the matching process, even under poor radiometric similarities. Such sensor models can be easily extended to accommodate for additional sensor imperfections (through self-calibration or test-field calibration procedures) according to usual photogrammetric procedures. It should be also noted that photogrammetry has devised techniques, which can be of local nature and are thus able to produce locally valid matches, despite of possible global dissimilarities.

Photogrammetric research has reported accuracies of 1mm for 3D body reconstruction (e.g. Jansa et al., 2000) and 0.1mm for 3D facial reconstruction (e.g. Schewe and Ifert, 2000) or 3D reconstruction of orthodontics wires (Suthau et al., 2000). Finally, experience from other photogrammetric applications shows expected accuracies of 1/10-1/3 of a pixel, which translates to sub-millimetre accuracies in medical images of the internal body.

Therefore, this knowledge should be transferred to medical imaging for a better quality of registration. Actually, the situation in many cases is rather easier here, since the necessary sensor models are simpler than the ones routinely used in other photogrammetric applications (see Fig. 3.3).

3.3.2 Matching of Sequences of Images

3.3.2.1 Problem Statement

Very often, cross-registration of data, acquired from the same patient using different acquisition methods under variable conditions, is required. Such matching of sequences of images is used for assessing regional physiological changes over time or for post-therapeutic assessment. In its 3D version, the problem can be stated as the presentation of previous studies re-computed to match the current study's slice locations.

3.3.2.2 State-of-the Art

In principle (and in practice) the technical problems involved here are very similar to those mentioned for image fusion. In order to co-register a number of images, again, the alternatives are two: either use external targets, or recognize anatomical landmarks. Then one should proceed with the matching process.

The major difference between the two problems is that matching a sequence of images is generally easier, since they are images taken with the same sensor, and characterized thus by the same internal geometry. This simplifies matching considerably, since more information (and thus external geometric constraints) can be entered into the matching process, making the solution more robust against to mismatches and noise.

3.3.2.3 Possible Contribution of Photogrammetry

On the technical side, this problem is not much different than the "Multi-sensor image fusion" (see Section 3.1), since the latter does not distinguish the imaging time. Therefore the notes in Section 3.1 are valid here as well, with the additional argument that, assuming the image sequences more often involve a single imaging modality, the matching process is rather easier.

One should also note that "image stacks" are very often processed in photogrammetric applications to assess changes over time. These include aerial and satellite imagery, as well as archived photography for e.g. land-use, forest, water resource management, or use of old images for architectural reconstructions. Experiences gained from such applications, involving also the use of "un-calibrated" sensors or matching of structures (lines, area patches etc.) instead of single points can be very useful in medical imaging.

3.3.3 Image Segmentation

3.3.3.1 Problem Statement

Segmentation of images in homogeneous areas is usually practised in order to distinguish the organs of interest against their surroundings and it aims at efficient identification of clinically important morphological information about the patient's anatomy and pathology. In the majority of instances today, a radiologist edits manually an image by enhancing organs of interest or by removing structures that obscure them. This manual editing, besides the time- and workload it requires, is highly depending on the user experience and thus it is non-reproducible; not to mention that without automation, large sequences of images (e.g. in order to get statistically significant results in population studies, or when a large number of organs have to be identified) cannot be processed. Therefore, until now, image segmentation is a major bottleneck in the use of medical image data.

3.3.3.2 State-of-the Art

Medical image segmentation, in principle, is not different from any other image segmentation problem. Image segmentation is one of the primary problems in image processing, and as such it attracted research interest long ago. There is quite an extensive literature on image segmentation, which roughly can be split into two basic approaches: region-based and gradient-based. The first are looking for similarities and use the homogeneity of localized features and other pixel statistics (e.g. simple thresholding, split-and-merge techniques (e.g. Burt et al., 1981; Kittler and Illingworth, 1985), energy minimizing techniques (e.g. Mumford and Shah, 1985), texture classification etc.) The second focus on differences, e.g. edge detection and following (Canny, 1986; Marr and Hildreth, 1980).

More specifically in medical imaging, segmentation is usually based either on the physical/chemical characteristics of the organ of interest or purely on the differences in their appearance in the image. Recent robustification of the above techniques include integration of procedures (e.g. Chakraborty, 1996) and use of all available information like gradients, curvatures, homogeneity of intensity, texture, shapes, etc., model-based operators as in deformable surfaces (e.g. Montagnat et al., 2000), or neural-network techniques (e.g. Kondo et al., 2000). More recently, model-based techniques in the form of morphological "filters" utilizing prior knowledge about the structure to segment, appear as possible alternatives (see e.g. Fig. 3.10 and Chakraborty, 1996).

3.3.3.3 Possible Contribution of Photogrammetry

Image segmentation, as the generic problem of information extraction from images, was and still is, in one form or the other, a central problem in photogrammetry and remote sensing. Numerous applications call for extraction, aggregation and generalization of the information, which is rather easily and intuitively extracted by the human vision system, while, at the same time, is very difficult to be algorithmically devised and thus automatically processed by machine vision systems.

Due to this necessity, numerous techniques have been developed and tested in photogrammetry, and are successful to various degrees. Clustering and classification techniques of satellite imagery, using principal components or co-factor analysis, and region-growing techniques, for automatic recognition of natural and man-made objects in aerial imagery, belong to this suite.

Again non-photogrammetric image segmentation techniques, in their initial form, involve similarity indices based on radiometric similarities of clusters, ignoring the mechanism through which these clusters have been imaged. Photogrammetry's contribution should be the additional introduction of valid geometric constraints, based on sensor models. This fact reduces the search space and eliminates misgroupings, leading thus to higher quality results. Additionally, it devised a number of parametric statistical tests and indicators of quality, based on sound statistical principles.

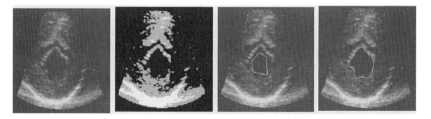

Fig. 3.10 Heart ultrasound image segmented using various segmentation techniques (from Chakraborty, 1996)

3.3.4 Motion Tracking and Analysis

3.3.4.1 Problem Statement

Motion monitoring is very important in many medical instances. In some instances, motion is the "signal", i.e. the characteristic that has to be monitored and analyzed. In others, motion is the "noise", i.e. the entity that masks or even corrupts other important characteristics and introduces artifacts in their analysis. Examples of the first kind are: gait analysis, study/simulation of normal and abnormal motion of joints and skeletal relationships, real-time, in-vivo applications of ultrasounds in ophthalmology, DUS colour blood-flow imaging in arteries, MRI and fMRI monitoring of brain's blood and neuronal activity, monitoring of cardiac or respiratory motion cycles using different imaging modalities, etc. Examples of the second kind are as many and important too: tiny motions of patient during MRI imaging (e.g. head motion of 0.3mm in brain imaging, see e.g. [URL8]) can result in severe diffuse artefacts distributed over the entire image, or completely mask the effect of the used contrasts, tissue movements around large arteries (associated to heart beats) (e.g. 1–2 mm) or thoracic motions (associated to pressure changes of respiration) cause intensity variations and ghost artefacts in MRI (Turner and Ordidge, 2000), changes in angle between the directions of blood flow and the beam cause important changes to the observed magnitude and direction of blood flow velocity in ultrasonic colour imaging (Wells, 2000), etc.

As it is understood, detailed and accurate motion tracking and analysis is very important in all above situations. This involves, depending on the problem at hand, the solution of diverse technical problems, like:

1. Subject's pose estimation and recovery in an either geometrically controlled (through external fiducials or markers, in cases of outer body measurements) or uncontrolled (in cases of inner body measurements) environments.
2. Recovery of multi-view sensors' attitude and position.
3. Estimation of 3D movement trajectories and recognition of movement patterns.
4. Automatic realization of the procedure (in order to minimize user induced errors and maximize timely operations, especially when real-time requirements are imposed).

3.3.4.2 State-of-the Art

Generally, automatic motion estimation and tracking follows two main approaches: Statistical and model-based. In the first, typically, the subject is reduced to binary silhouettes and statistical measures are derived through techniques like Principal Component Analysis, Linear Discriminant Analysis, etc. Statistical approaches are more often based on 2D than on 3D analysis and are best suited for recognition and discrimination of motion, rather than for its exact estimation. Model-based approaches, on the other hand, are more often based on 3D analysis and are more demanding. They are depending on accurate measurements of either single characteristic points or linear structures and, in their advanced form, couple the geometric information (e.g. sensor orientation, epipolar constraints, etc.) with radiometric data (e.g. evidence gathering procedures).

To many respects, many of the technical problems involved here are similar to those in image registration or segmentation. Motion tracking in a rigid (e.g. Roche et al., 2000) or non-rigid body fashion (e.g. Chui, 2001 and [URL12] or Zhang,

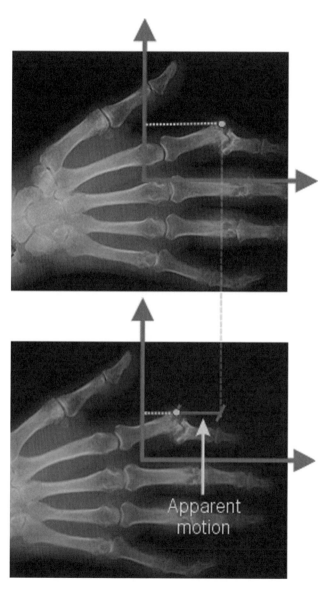

Fig. 3.11 Motion can be deduced from parallax measurements

1992) are among the most popular approaches, very similar to the ones used in matching of free-form curves.

Non-contact motion measurement systems with high-resolution CCD cameras have been developed, which track (infra-red, or red light) either passive retro-reflective or active markers (e.g. Vicon system [URL11], Optotrak and Polaris system [URL9], or the Qualisys system [URL10]). In image-guided surgery, which is one of the most demanding real-time applications, video camera trackable LEDs on the patient (e.g. Leventon, 1997; Youmei, 1994, [URL4]) have been used to relate real-world object motion to image recognised motion. Additionally, 2D Ultrasonic scanning probes, fitted with positional locators (either electromagnetic or ultrasonic), have been developed for spatially registered 2D scans (e.g. [URL5], [URL6]).

3.3.4.3 Possible Contribution of Photogrammetry

Photogrammetry is principally a 3D positioning technique and it can operate in three different modes: (a) steady sensor—moving object, (b) moving sensor—steady object, and (c) moving sensor—moving object. In all cases, the motion of either one (or the relative motion of both) causes apparent movement (along the x-axis) of any object point in the stereoscopic pair of images (see Fig. 3.11). From this, so-called, x-parallax both motion and the third dimension can be inferred.

Using this line of thinking, photogrammetry has up to now shown numerous applications with very high quality of results. Besides, the systems mentioned above, and which are based on photogrammetric principles, numerous publications exist (e.g. Anai and Chikatsu, 1999; Boulic et al., 1998; Gravrila and Davis, 1996; D'Apuzzo et al., 2000; Sharman et al., 2000; Carter and Nixon, 2000 to name a few recent ones).

The majority of these applications is based on optical methods and refers to measurements of the exterior body, however. Baring in mind the range of the motion problems, we have already mentioned, we should admit that photogrammetry has not, up to now, used all its resources towards their solution. It is true that the potentiality of the photogrammetric procedures is much higher than what has been realized to date.

3.4 Conclusion

Multi-sensor image registration and fusion, matching of image sequences, image segmentation, or motion tracking and analysis are some of the problems, which currently challenge medical imaging. By analysing these problems and reviewing the up-to-date proposed solutions, possible contribution areas from the part of photogrammetry have been identified.

In conclusion, the major contribution, expected from the part of photogrammetry, is to offer its rigor in geometric modelling and enhance, this way, the existing techniques. It is anticipated that this will contribute to a much higher accuracy and much higher quality of the results, meaning higher reliability and higher robustness to "false alarms" and noise levels. This is increasingly important now that new sensors are being developed, which provide higher resolutions and finer information.

It is true that the potential of the photogrammetric procedures is much higher than what has been realized to date. It is also true that differences in education, background knowledge, used terminology and disciplinary-focused line of thinking build up barriers to inter-disciplinary cooperation. And this needs to be changed.

Looking into the future, it is hoped that cross-disciplinary knowledge fusion and technology transfer will be very fruitful, beneficial and rewarding to all parties involved.

References

Anai, T., Chikatsu, H., 1999. Development of human motion analysis and visualisation system. Int'l Archives of Photogrammetry and Remote Sensing 32 (5-3W12), 141–144.

Aoki, Y., Hashimoto, S., Terajima, M., Nakasima, A., 2000. Simulation of postoperative 3D facial morphology using physics-based head model. Int'l Archives of Photogrammetry and Remote Sensing 33 (Supplement B5), 12–19.

Atkinson, K. B. (Ed.), 1996. Close range Photogrammetry and Machine Vision. Whittles Publishing, Latheronwheel.

Börlin, N., 2000. Model-based Measurements in Digital Radiographs, PhD Thesis, Department of Computing Science, Umeå University, Sweden.

Boulic, R., Fua, P., Herda, L., Silaghi, M., Monzani, J.-S., Nedel, L., Thalmann, D., 1998. An anatomic human body for motion capture. Proc. EMMSEC'98, September, Bordeaux, France. Available at http://ligwww.epfl.ch/flthalmann/papers.dir/EMMSEC98.pdf (accessed 22 April 2002).

Burt, P., Hong, T., Rosenfeld, A., 1981. Segmentation and estimation of region properties through co-operative hierarchical computation. IEEE Transactions on System, Man and Cybernetics, 11(12), 802–809.

Canny, J., 1986. A computational approach to edge-detection, IEEE Transactions on Pattern Analysis and Machine Intelligence 8(6), 679–698.

Carter, J., Nixon, M., 2000. On measuring trajectory-invariant gait signatures, Int'l Archives of Photogrammetry and Remote Sensing 33(B5/1), 114–121.

Chakraborty, A., 1996. Feature and Module Integration for Image Segmentation, PhD dissertation, Image Processing and Analysis Group, Departments of Diagnostic Radiology and Electrical Engineering, Yale University

Chui, H., 2001. Non-Rigid Point Matching: Algorithms, Extensions and Applications, PhD Dissertation, Image Processing and Analysis Group, Departments of Diagnostic Radiology and Electrical Engineering, Yale University.

Collignon, A., Vandermeulen, D., Suetens, P., Marschal, G., 1995. 3D multimodality medical image registration using feature space clustering. In: N. Ayache (Ed.), Proc. CVRMed'95, Lecture Notes in Computer Science, Vol. 905, pp. 195–204, Springer-Verlag, Berlin.

D'Apuzzo, N., 2001. Photogrammetric measurement and visualization of blood vessel branching casting: A tool for quantitative accuracy tests of MR-, CT- and DS-Angiography. Proc. SPIE, Vol. 4309, pp. 34–39.

D'Apuzzo, N., R. Plaenkers, P. Fua, 2000. Least squares matching tracking algorithm for human body modeling, Int'l Archives of Photogrammetry and Remote Sensing 33(B5/1), 164–171.

Gravrila, D. M., Davis, L., 1996. 3D model-based tracking of humans in action: a multi-view approach. Proc. Conference on Computer Vision and Pattern Recognition, San Francisco, CA, USA, pp. 73–80.

Hill, D., Studholme, C., Hawkes, D., 1994. Voxel similarity measures for automated image registration. In: R. Robb (Ed.), Visualization in Biomedical Computing, Proc. SPIE, Vol. 2359, pp. 205–216.

Jansa, J., Melykuti, B., Öhreneder, Ch., 2000. Videometric system for surface determination in Medical applications, Int'l Archives of Photogrammetry and Remote Sensing 33(B5/1), 79–85.

Karara, H. M. (Ed.), 1989. Non-Topographic Photogrammetry, 2nd edition, American Society for Photogrammetry and Remote Sensing, Science and Engineering Series, Behtesda, Maryland, USA.

Kiskinis, D., Patias, P., Megalopoulos, A., Kostopoulou, E., Gymnopoulos, K., Tsioukas, V., Gemenetzis, D., Kousoulakou, A., Styliadis, A., 1998. 3-D Carotid artery reconstruction from DUS images, Int'l Archives of Photogrammetry and Remote Sensing 32(5), 453–457.

Kittler, J., Illingworth, J., 1985. On threshold selection using clustering criteria. IEEE Transactions on System, Man and Cybernetics 15(5), 652–655.

Klingenbeck, K., Rienfelder, H., 1990. Medical Imaging Techniques. ISPRS J Photogrammetry and Remote Sensing 45(4), 203–226.

Kondo, H., Zhang, L., Koda, T., 2000. Computer aided diagnosis for pneumoconiosis radiographs using neural network. Int'l Archives of Photogrammetry and Remote Sensing 33(B5/1), 453–458.

Leventon, M., 1997. A Registration, Tracking, and Visualization System for Image-Guided Surgery. MSc Thesis, Department of Electrical Engineering and Computer Science, MIT.

Marr, D., Hildreth, E., 1980. Theory of edge detection. Proc. Royal Society London, Series B 207(1167), 187–217.

Meyer, C., Boes, J., Boklye, K., Bland, P., Zasadny, K., Kison, P., Koral, K., Frey, K., Wahl, R., 1997. Demonstration of accuracy and clinical versatility of mutual information for automatic multimodality image fusion using affine and thin-plate spline warped geometric deformations, Medical Image Analysis 1(3), 195–206.

Mikhail, E.M., Bethel, J.S., McGlone, J.C., 2001. Introduction to Modern Photogrammetry. John Wiley & Sons Inc., New York.

Montagnat, J., Delingette, H., Scapel, N., Ayache, N., 2000. Representation, shape, topology and evolution of deformable surfaces. Application to 3D medical image segmentation. INRIA Report N° 3954.

Mumford, D., Shah, J., 1985. Boundary detection by minimizing functionals. Proc. IEEE Conf. Computer Vision and Pattern Recognition, pp. 22–26.

Roche, A., Pennec, X., Malandain, G., Ayache, N., Ourselin, S., 2000. Generalized Correlation Ratio for Rigid Registration of 3D Ultrasound with MR Images. INRIA Report N° 3980.

Schewe, H., Ifert, F., 2000. Soft tissue analysis and cast measurement in orthodontics using Digital Photogrammetry, Int'l Archives of Photogrammetry and Remote Sensing 33(B5/2), 699–706.

Sharman, K., Nixon, M., Carter, J., 2000. Towards a marker-less human gait analysis system. Int'l Archives of Photogrammetry and Remote Sensing 33(B5/2), 713–719.

Studholme, C., Hill, D., Hawkes, D., 1996. Automated 3D registration of MR and CT images of the head. Med Imag Anal 1(2), 163–175.

Suthau, T., Hemmleb, M., Zuran, D., Jost-Brinkmann, P.-G., 2000. Photogrammetric measurement of linear objects with CCD cameras - Super-elastic wires in orthodontics as an example. Int'l Archives of Photogrammetry and Remote Sensing 33(B5/2), 780–787.

Tate, P., Chapman, M., 2000. The assessment of magnetic imagery for computer assisted spinal surgery. Int'l Archives of Photogrammetry and Remote Sensing 33(B5/2), 809–816.

Thomas, P., Newton, I., Fanibuda, K., 1996. Evaluation of a low-cost digital photogrammetric system for medical applications. Int'l Archives of Photogrammetry and Remote Sensing 31(B5), 405–410.

Turner, R., Ordidge, R., 2000. Technical challenges of functional Magnetic Resonance Imaging. IEEE Eng Med Biol 19(5), 42–54.

Van den Elsen, P., Pol, M., Viergever, M., 1993. Medical Image matching – A review with classification. IEEE Eng Med Biol 12(1), 26–39.

Wells, P., 2000. Current status and future technical advances of ultrasonic imaging. IEEE Eng Med Biol 19(5), 14–20.
Wells, W., Viola, P., Atsumi, H., Hakajima, S., Kikinis, R., 1996. Multimodal volume registration by maximization of mutual information. Med Imag Anal 1(1), 35–51.
Woods, R., Mazziotta, J., Cherry, S., 1993. MRI-PET registration with automated algorithm. J Comp Assist Topogr 17(4), 536–546.
Youmei, G., 1994. Non-contact 3D biological shape measurement from multiple views. M.Sc. Thesis, Dept. of Computer Science, University of Western Australia.
Zawieska, D., 2000. Topography of surface and spinal deformity. Int'l Archives of Photogrammetry and Remote Sensing 33(B5/2), 937–942.
Zhang, Z., 1992. Iterative point matching for registration of free-form curves. INRIA Report N° 1658.

Appendix A. Useful WEB addresses

Image-guided Surgery laboratory, http://www.bic.mni.mcgill.ca/research/groups/igns/igns_proj.html#intro, Image-guided neurosurgery project, http://www.irus.rri.on.ca/igns/ (accessed 29 April 2002).
Bidaut, L., Laboratory of Functional and Multidimensional Imaging, Department of Radiology and Surgery, University Hospitals of Geneva, Switzerland, http://www.expasy.ch/LFMI/ (accessed 22 April 2002).
Hornak, J.P., The Basics of MRI, http://www.cis.rit.edu/htbooks/mri/inside.htm (accessed 22 April 2002).
P. Munger, 1994. Accuracy considerations in MR image guided neurosurgery, MSc Thesis, McGill University, http://www.bic.mni.mcgill.ca/users/patrice/msthesis/msthesis.html (accessed 22 April 2002).
Freehand 3D Ultrasound, Dept. of Engineering, University of Cambridge, http://svr-www.eng.cam.ac.uk/research/biomed/cam_3dus.html (accessed 22 April 2002).
Dept. of Engineering, University of Cambridge, Speech, Vision and Robotics Group Publications, http://svr-www.eng.cam.ac.uk/reports/index-biomed.html (accessed 22 April 2002).
The Digital RSA System, RSA BioMedical Innovations AB, Umeå, Sweden, http://www.rsabiomedical.se (accessed 22 April 2002).
Computational Imaging Science Group in the Division of Radiological Sciences of Guy's Hospital, London, http://www-ipg.umds.ac.uk/cisg/index.htm (accessed 22 April 2002).
Northern Digital Inc, http://www.ndigital.com (accessed 22 April 2002).
Qualisys, http://www.qualisys.com (accessed 22 April 2002).
Vicon, http://www.vicon.com (accessed 22 April 2002).
Yale University Ph.D. archive, http://noodle.med.yale.edu/thesis/ (accessed 22 April 2002).

Chapter 4
Computer Aided Tissue Engineering Scaffold Fabrication

M. W. Naing, C. K. Chua and K. F. Leong

Abstract A novel CAD system of structures based on convex polyhedral units has been created for use with Rapid Prototyping (RP) technology in tissue engineering applications. The prototype system is named the *Computer Aided System for Tissue Scaffolds or CASTS.*

CASTS consists of a basic library of units that can assemble uniform matrices of various shapes. Each open-cellular unit is a unique configuration of linked struts. Together with an algorithm which allows the designer to specify the unit cell and the required dimensions, the system is able to automatically generate a structure that is suitable for the intended tissue engineering application. Altering the parameters can easily change the desired shape and spatial arrangement of the structures.

The main advantage of CASTS is the elimination of reliance on user skills, much unlike conventional techniques of scaffold fabrication. From a small range of basic units, many different scaffolds of controllable architecture and desirable properties can be designed. The system interface of CASTS in Pro/ENGINEER is user friendly and allows complete transfer of knowledge between users without the need for complex user manuals.

A femur implant was successfully fabricated using Selective Laser Sintering (SLS) and a standard commercial material Duraform^{TM} Polyamide. Following the successful fabrication of the scaffold, further investigations were planned to verify the viability of the system. A disc shape scaffold was designed for this purpose. In this disc-shaped design, four different strut lengths were used, resulting in four different pore sizes and porosity. Scaffolds of the one-unit layer cells built using Duraform^{TM} Polyamide were then examined under a light microscope to check the consistency and reproducibility of the microstructures.

Three types of biomaterials were tested on CASTS: PEEK, PEEK-HA biocomposite, and PCL. The scaffolds built showed very good definition of the pre-designed microarchitecture and were readily reproducible. While delamination occurred at larger unit cell sizes, this had no effect on the overall shape or the structural integrity of the scaffolds.

The potential of this system lies in its ability to design and fabricate scaffolds with varying properties through the use of different unit cells and biomaterials to suit different tissue engineering applications.

4.1 Introduction

For the past decade, the focus of Tissue Engineering (TE) has been on the aspect of culturing organs in the hope of replacing damaged or diseased organs in the body. In TE, expertise from the fields of biological and material sciences, and engineering are combined to develop viable biological substitutes of a tissue which helps to restore, maintain and/or improve the functions of that tissue. For matrix-producing connective tissues, the cells are anchorage-dependent and the presence of 3-D scaffolds with interconnected pore networks are crucial to aid in the proliferation and reorganisation of the cells [1, 2]. The scaffolds, which can be fabricated from natural or synthetic biomaterials [3–5], facilitate the creation of functional and structurally appropriate biological replicas of healthy versions of the required tissues. The ultimate aim is for these fabricated structures to be the basis of tissue regeneration such that patients can readily obtain implants which will not put them at risk for infection nor require them to be on life-long medication for organ rejection.

At present, scaffold fabrication is mostly done using a variety of conventional methods such as salt and particulate leaching [1, 4–6], which have been successfully applied to establish the viability of various TE applications; engineered skin has been used clinically [7] and organs successfully tested in preclinical studies include blood vessels [8] and the bladder [9]. However, these techniques rely heavily on users' skills and the applied procedure. Due to the difference in skills and procedures between individual users, the fabricated scaffolds are subject to variations and are not easily reproducible. Also, such techniques can only produce scaffolds with a range of pore sizes. As a result, these scaffolds render the researcher incapable of making consistent analysis.

Since its inception, Rapid Prototyping (RP) technology has created a significant impact on the medical community. By combining medical imaging, Computer Aided Design (CAD) and RP, it is possible to create accurate patient-specific anatomical models [10–13] and customised prototypes of devices for a wide variety of medical applications [14–19]. This has prompted researchers to experiment with RP techniques to fabricate scaffolds which can give controlled microarchitecture and higher consistency than those fabricated using conventional techniques. However, the internal microarchitecture of scaffolds built using the original building styles and patterns supplied by RP system manufacturer are limited [20]. As such, further improvements are needed to augment the range of possible pore sizes, their accuracy and the consistency in their distribution and density. Since RP processes begin with the creation of a 3-D CAD model of the scaffold structure, one possible solution to achieve such improvements is to design the scaffold's internal microarchitecture during the CAD modelling stages before committing the scaffold designs to RP fabrication [20–27].

To alleviate the difficulty encountered in creating scaffolds with designed internal architectures, Hollister et al. [21] introduced an image-based approach for designing and manufacturing TE scaffolds. In their work, a general program for building up a scaffold's internal architecture by repeating a unit structure of cylinders, spheres, and other entities was developed. Hollister et al. also developed CAD techniques [22–26] for creating sacrificial moulds with designed internal channels or cavities which resembled the negative image of the final required scaffold. The moulds were used to cast hydroxyapatite scaffolds for bone TE applications from a highly loaded "reactive ceramic suspension" following the lost mould shape forming process.

This project takes a different approach and aims to make significant improvement by employing CAD data manipulation techniques to develop a novel algorithm. This algorithm can be used to design and assemble a wide range of internal scaffold architectures from a selection of open celled polyhedron shapes. It is used in conjunction with an integrated manufacturing approach [10] that combines medical imaging and RP technologies to achieve rapid automated production of pre-designed 3-D tissue scaffolds that are not only consistent and reproducible, but also patient-specific. The system is named CASTS or Computer Aided System for Tissue Scaffolds.

In contrast to the fabrication approach adopted by Hollister et al., which is based on the lost mould casting process, CASTS is aimed at direct fabrication of scaffolds on a RP system such as Selective Laser Sintering (SLS). By adopting a direct fabrication approach, many disadvantages associated with casting can be eliminated. Examples of such disadvantages include additional production stages (lead time) incurred in developing a suitable suspension of the required material and a mould, increased costs, increased material wastage, and risk of material contamination.

Scaffolds are necessary for growing bone tissue as they act as temporary substrates for the anchorage-dependent osteoblasts. Bone scaffolds should ideally assume the shape of the defect, provide mechanical support to the defect while healing occurs, and allow cell proliferation and tissue ingrowth into the scaffold. Seeded cells adhere to the scaffold in all three dimensions, proliferate and produce their own ECM which takes over the function of the biomaterial scaffold [28, 29].

4.2 Requirements of Tissue Engineering Scaffolds

Functional properties of the scaffold depend on the characteristics of the scaffold material, the processing techniques, and the scaffold design, which in turn decides how the cells interact with the scaffold. The scaffold must be designed to cater to the conflicting needs of tissues in terms of mechanical strength, porosity, uniformity in pore size, and complexity in three dimensions [30]. Listed below are the desired characteristics of a scaffold [6, 26, 31, 32]:

1. 3-dimensional, highly porous with an interconnected pore network for cell growth and flow transport of nutrients and metabolic waste,
2. Biocompatibility and biodegradability,
3. Suitable surface chemistry and topography for cell attachment, proliferation, differentiation, and also to encourage formation of ECM,

4. Pre-defined microarchitecture,
5. Mechanical properties to match those of the tissues at the site of implantation; for load-bearing tissue, the scaffold must provide additional mechanical support during regeneration of tissue,
6. Sterilisablity.

4.3 Application of RP in TE Scaffold Fabrication

Conventional techniques of scaffold fabrication rely heavily on user skills and experience such that there is poor repeatability in between users. Processing parameters are often inconsistent and inflexible thus resulting in highly inconsistent micro and macro structural properties in the scaffold. Use of organic solvents may have harmful effects on the cells and cause the cells to die or mutate. Porogen particles employed to induce pores may not be completely removed, making the process inefficient in terms of porosity. Most conventional processes are also limited to producing scaffolds with simple geometry which may not satisfy the geometric requirements of the defect [33].

Although RP techniques can potentially address most of the macro and micro structural requirements of TE scaffolds, only a few of the RP systems have been used for scaffold fabrication to date [33]. These systems include: 3-dimesional Printing (3DP), Fused Deposition Modeling (FDM), Selective Laser Sintering (SLS), and inkjet printing [34].

3DP was used to fabricate poly(lactic-co-glycolic acid) or PLGA scaffolds through the use of moulds and particulate leaching using sucrose as porogen. While this method was proven to produce viable scaffolds, the lengthy process makes it tedious and ineffectual [35]. FDM fabricated polycaprolactone (PCL) scaffolds had pore sizes of 160–700 μm and 48–77% of porosity. Fibroblast cells seeded onto these scaffolds showed complete ingrowth after a few weeks of culture. Degradation tests proved that the process did not affect the material properties [36]. Sintering of Poly-Ether-Ether-Ketone (PEEK), poly-L-lactic acid (PLLA) and PCL powder stocks had different degrees of success. Microporosity of the specimens was varied by changing laser power (or energy density), scan speed and part bed temperature. High porosity and interconnectivity were obtained for all specimens [37]. Model Maker II was used to produce moulds for scaffold manufacturing. The lost mould method was used to cast out the scaffolds which are made of naturally occurring materials such as tricalcium phosphate [38] and collagen [39].

4.4 Methodology

Even with the incorporation of RP systems, the work carried out on TE scaffolds has been restricted due to many limitation factors, one of which is the lack of variety in patterns. What is required is a comprehensive system which can:

1. Provide the users with a database of designs to choose from,
2. Generate scaffolds of different parameters, and
3. Customise scaffolds according to patients' specifications.

The aim of this project, therefore, is to develop such a system which can satisfy these requirements. The prototype system is named the ***COMPUTER AIDED SYSTEM FOR TISSUE SCAFFOLDS*** or ***CASTS***. Figure 4.1 shows the process flow.

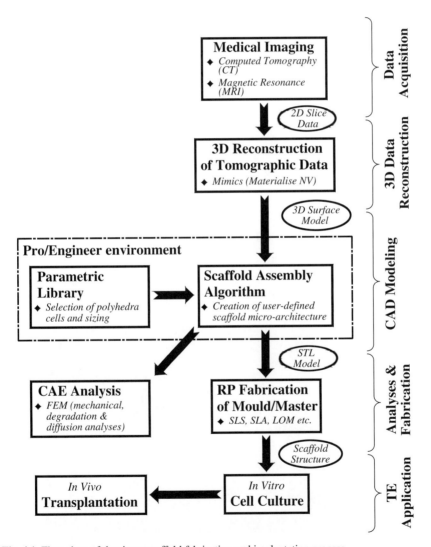

Fig. 4.1 Flow chart of the tissue scaffold fabrication and implantation process

4.5 Concept verification

To verify the system, a scaffold in the shape of a femur was generated and fabricated. The surface profile of the femur was first extracted using MIMICS™ (Materialise, Belgium NV) from CT scan data (as shown in Fig. 4.2a). This surface profile was imported into Pro/ENGINEER software and further manipulated to obtain a closed volume (Fig. 4.2b). The internal architecture was generated using the algorithm and merged into the volume using Boolean operations (Fig. 4.2c).

The Selective Laser Sintering system (Sinterstation 2500) was used to fabricate the generated scaffold. The material used was Duraform™ Polyamide. The laser power was set to 4W. The scan speed and part bed temperature were set to default values for Duraform™, which were 200 in/s (5080 mm/s) and 165 °C respectively. The powder layer thickness of each layer was set to 0.006 in (0.152 mm). Figure 4.3 shows the fabricated femur scaffold together with a RP fabricated model of the femoral head.

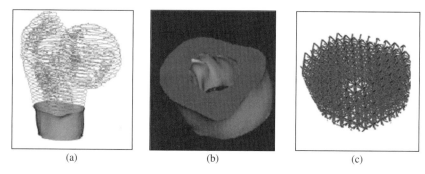

Fig. 4.2 CAD generation of a parent-specific scaffold

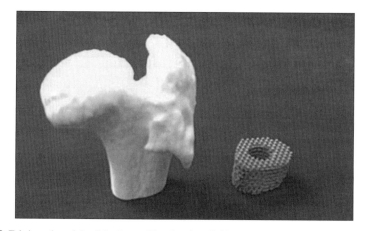

Fig. 4.3 Fabricated models of the femoral head and scaffold structure for the femoral bone segment

The scaffold implant fabricated had good interconnectivity and the features of the microarchitecture could be clearly seen. It is also noted that the powder inside the scaffold was easily removed using manual means. Polishing the scaffold however, was not possible because of the intricate nature of the structure.

4.6 Validation of CASTS

In order to check the repeatability of the system, disc-shaped scaffolds were generated using the algorithm and fabricated using the SinterStation 2500. The configuration tested was octahedron-tetrahedron configuration from the scaffold library. Each specimen sample fabricated was 16 mm in diameter. This was intended to facilitate the cytotoxicity testing that will follow once the specimens are fabricated using biomaterials.

Using the user interface of the algorithm in Pro/ENGINEER, the configuration desired was first selected. After the selection, the layout was regenerated to activate the prompts for the secondary input. The other main inputs were the length of struts and the diameter of strut which together control the pore size and porosity. In the case of the octahedron-tetrahedron configuration, the strut length was the same throughout the unit cell and only one value needed to be specified.

For consistency of results, the diameter of the strut was fixed at 0.25 mm (approximately two times the resolution of SLS) and only the strut length was varied to produce variation in the structures in terms of pore sizes and porosity. Table 4.1 is a summary of properties of the scaffolds generated.

Each scaffold was generated to a size larger than the desired final scaffold size (see Fig. 4.4a). Concurrently, a set of 16 mm discs of surfaces was created in Pro/ENGINEER. The discs were varied in height from 1.0 to 2.5 mm, to match one layer of unit cell for differing strut lengths (see Fig. 4.4b). Once the scaffold structure had been generated, the appropriate surface model was appended to the disc and a Boolean subtraction was performed which gave the scaffold the shape of the disc (see Fig. 4.4c & 4.4d).

The scaffold files (see Fig. 4.5) were exported into .STL format and checked for errors in edges and contours after which, they were sent to the SLS system to be fabricated.

Table 4.1 Properties of scaffolds generated

Strut length (mm)	Pore size (mm)	Porosity
1.0	0.327	0.583
1.5	0.616	0.815
2.0	0.905	0.896
2.5	1.193	0.933

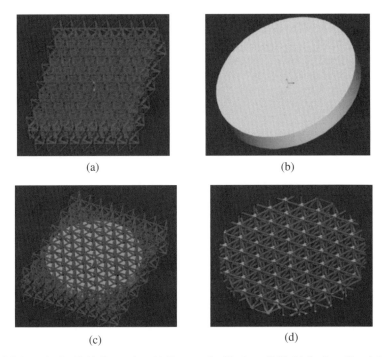

Fig. 4.4 Steps in Scaffold Generation (a) Rectangular block scaffold (b) Surface file of the disc (c) Scaffold with surface model embedded and (d) Final scaffold in the shape of the disc

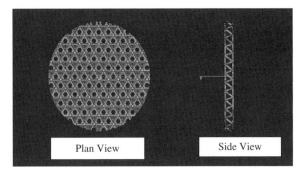

Fig. 4.5 An example of the generated scaffolds

4.7 Duraform™ Polyamide Scaffolds

Nylon 12, with the commercial name Duraform^{TM} Polyamide, a standard material for SLS, was used for this investigation. As this material has been widely studied [18, 40–43] with regards to its application in SLS, this will reduce the number of variables that need to be dealt with. The advantages of Durafrom^{TM} are uniform powder size and consistent melting point (see material specifications in Table 4.2).

4 Computer Aided Tissue Engineering Scaffold Fabrication

Table 4.2 Material specifications – Duraform™ Polyamide

Particle shape	Irregular
Particle size range, μm	25–92
Average particle size, μm	60
Powder density, kg/m³	590
Solid density, kg/m³	970
Melting Point, °C	186

(a) Strut length = 2.5mm

(b) Strut length = 2.0mm

(c) Strut length = 1.5mm

(d) Strut length = 1.0mm

Fig. 4.6 Micrographs of fabricated scaffold samples

Figure 4.6 shows the micrographs of a set of fabricated scaffolds. The features of the scaffolds were examined under microscope using Nikon SMZ-U Stereoscopic microscope with light unit Photomic PL 3000. Scaffolds with strut length 1.5 mm and 2.0 mm have reasonable pore sizes (0.616–0.905 μm) with little trapped powder. This shows that the scaffold library and the algorithm, together with the SLS system, can produce viable scaffolds with consistent and reproducible microarchitecture.

4.8 Biomaterial Scaffolds

Two types of biomaterials were tested with regards to the feasibility of fabricating pre-designed scaffolds using CASTS.

4.8.1 Poly-ether-ether-ketone (PEEK) and Hydroxyapatite (HA)

The same PEEK and Hydroxyapatite (HA) powders as well as the PEEK-HA composite blend used by Tan et al. [37] was used for scaffold fabrication. The micrographs of as-received PEEK and HA powders were taken using Scanning Electron Microscope (SEM) to study the shape of the powders (Fig. 4.7). PEEK particles are irregular in shape whereas HA particles are relatively more spherical. This difference in morphology is important as it makes it possible to easily distinguish between the two types of materials by visual inspection.

4.8.2 Polycaprolactone (PCL)

PCL used in this project was purchased from Solvay Interox Pte Ltd, UK under the brand name CAPA® 6501. The powder has a molecular weight of 50,000. Scanning Electron Micrography shows the particles have irregular shape (Fig. 4.8).

Table 4.3 shows the specifications of the three materials used for the validation of CASTS. HA powders used in the research meet the requirements of the ASTM F 1185–88 and have a particle size distribution of at least 90 wt% of particles below the size of 60 µm as determined by Coultier Counter analysis.

4.9 Results and Discussion

4.9.1 Pure PEEK Scaffolds

Previous PEEK and PEEK-HA composite samples built in the SLS are only in the form of solid discs. Therefore, it is necessary to ascertain the feasibility of incorporating these materials into CASTS. As a start, pure PEEK scaffolds were fabricated before experimenting with the composite material.

(a)   (b)

Fig. 4.7 SEM micrographs of as-received (a) PEEK and (b) HA powders [31]

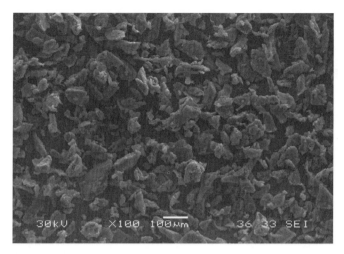

Fig. 4.8 SEM Micrographs of as-received PCL [44]

The scaffolds built were based on the same STL files used for Duraform™ samples. As seen with Duraform™ samples, scaffolds with strut length 2.0 mm had very large pores (>1000 μm) which are not suitable for cell seeding. The large pores may cause the cells to slip through and settle at the bottom instead of attaching to the scaffold surfaces. Also, in the event that some cells do adhere to the surfaces and proliferate, this will result in a weak cell mass as cells only tend to grow along surfaces. Therefore, for this stage, only three varying strut lengths were considered: 1.0, 1.5, and 2.0 mm.

For this investigation, a laser power of 17W was used at part bed temperature 140 °C and scan speed 5080 mm/s (200 in/s) [37]. The reason for using one of the lower laser powers is to ensure that the energy density (i.e. temperature) incident on the part during sintering does not become too high such that the properties of the material are altered.

The first batch of specimens was built at a layer thickness of 0.152 mm (0.006 in). The parts built were found to be fragile and hard to handle. This could be due to

Table 4.3 Material Specifications

	PEEK	HA	PCL
Brand Name	Victrex	Camceram	Solvay
Particle shape	Irregular	Spherical	Irregular
Average particle size, μm	25	5-60	<100
Powder density, kg/m^3	1320*	3050	1.15
Glass transition temperature/ °C	143	N.A	−60
Melting temperature/ °C	343	1500	60

*The density of the powder is assumed to be of the same value as the solid from which it is crushed.

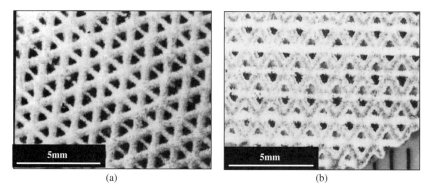

Fig. 4.9 Scaffolds with strut length 1.5 mm (a) top view (b) bottom view

the inability of the laser power to sinter each layer to the next completely at such thickness. Therefore, a second batch was fabricated with the layer thickness value reduced to 0.10 mm (0.004 in).

In general, the fabricated scaffolds exhibit well-defined microarchitecture, indicating the possibility of incorporating biomaterials into CASTS, even though the structural integrity of the PEEK scaffolds was observed to be not as good as those fabricated with Duraform^{TM}. This is expected since Duraform^{TM} is a commercial material with optimised settings in SLS while PEEK is a new material. The advantage of PEEK scaffolds over Duraform^{TM} scaffolds is in the reduction of powder trapped within the pores and also the ease of powder removal. This may be attributed to the small particle size of PEEK powders, which at 25 μm, is less than half the average size of the Duraform^{TM} powders.

Upon close inspection, the scaffolds show favourable results with intact struts and well-defined pores (Fig. 4.9). The intended architecture was clearly visible under the microscope. This is further shown in Fig. 4.10 where the top few layers have been removed to expose the interconnected struts within the specimen.

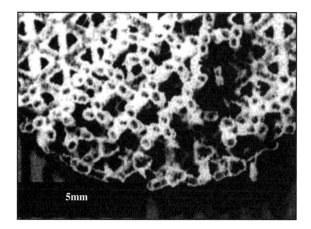

Fig. 4.10 Sample with top layers exposed to show the inner structure

Fig. 4.11 Delaminated first layer

Delamination was observed for the first layer (0.10 mm) in some samples with larger pore sizes (>900 μm). This indicated that the bond between the first layer and the next is weak. However, the delamination of the first layer did not affect the overall structure of the scaffold as seen in photos taken under the microscope. Figure 4.11 shows the delaminated first layer of a scaffold.

4.9.2 PEEK-HA Composite Scaffolds

After successfully sintering the PEEK scaffolds, PEEK-HA composite scaffolds were fabricated. The composite blend is made up of 90% weight PEEK with 10% weight HA. The physical blend was produced by using PEEK as a base material and adding in HA gradually in the roller mixer [37]. Figure 4.12a shows the distribution of HA particles (spherical structures) in the composite after blending.

(a)

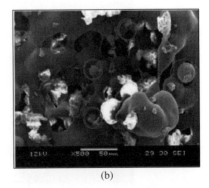
(b)

Fig. 4.12 SEM micrographs of PEEK-10% wt HA (a) before and (b) after sintering [37]

Fig. 4.13 Top and bottom views of the PEEK-HA composite scaffold

After processing in SLS, it was seen that the HA particles were trapped in the interconnected PEEK matrix (see Fig. 4.12b). This is only possible because PEEK has a much lower melting point compared to HA and the temperatures that the composite is exposed to in the SLS are not high enough to affect the HA particles.

Experiments showed that scaffolds with strut length 1.5 mm were most promising in terms of structural shape as well as theoretical pore size (~600 μm) and porosity (>80%). Therefore, a set of scaffolds with strut length 1.5 mm was fabricated using the composite. The same parameters used for sintering PEEK were used to process the PEEK-10% wt HA composite i.e. laser power of 17W at part bed temperature 140 °C, scan speed 200 in/s (5080 mm/s), and layer thickness of 0.004 in (~0.10 mm).

The scaffolds showed similar structure integrity and pore sizes as the pure PEEK scaffolds. The pores were also clearly visible under microscope (see Fig. 4.13).

One observation made was the "balling effect" present in the fabricated specimens. This was not observed in the Duraform ™ scaffolds or the pure PEEK samples. Nelson [45] attributed this to surface tension present in the powder bed.

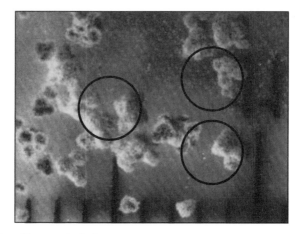

Fig. 4.14 Balling effect seen in the inner struts

As PEEK particles act as the binder and the HA particles remain in the solid phase, droplets of binder form on the scanned surface. Figure 4.14 shows the inner struts which have been removed from the scaffold. The balling effect was visible in some of the clusters of struts (highlighted in circles).

4.9.3 Polycaprolactone (PCL) scaffolds

Polycaprolactone, being a biodegradable material, has a natural advantage over PEEK. Manually-fabricated PCL scaffolds also been used to grow bone [6]. The scaffold model was fabricated using the following set of parameters: laser power = 3W, fill scan speed = 5080 mm/s (200 in/s) (default), powder layer thickness = 0.102 mm (0.004 in), warm up height = 6.35 mm (0.250in), and cool-down height = 2.54 mm (0.100 in). PCL disc scaffolds were fabricated using the same set of parameters. Similar to the biocomposite scaffolds, the strut length was 1.5 mm with strut diameter 0.25 mm.

A sample of the fabricated PCL scaffold is shown in Fig. 4.15. In general, it was found that PCL scaffolds exhibited better structural integrity and higher strengths than PEEK or biocomposite scaffolds. There was also a lower percentage of delamination. When observed under the microscope, there was evidence of crystallinity in the scaffolds. However, due to the much larger particle size of PCL, the struts were found to be much thicker than those of PEEK scaffolds.

One challenge with scaffolds fabricated by SLS is the removal of powder. With PCL scaffolds, powder removal was relatively easy compared to materials such as

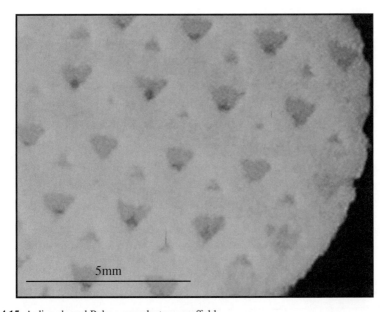

Fig. 4.15 A disc-shaped Poly-ε-caprolactone scaffold

Fig. 4.16 A PCL Scaffold sample

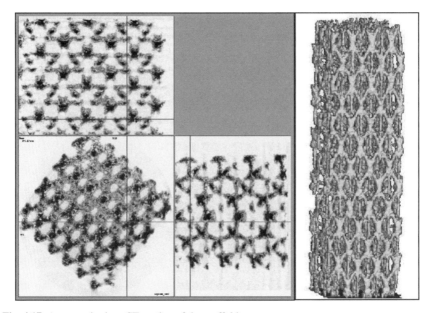

Fig. 4.17 A scanned micro-CT section of the scaffold

Duraform™. As the PCL parts fabricated were elastic, powder trapped within the scaffolds were easily removed by using a sieve shaker.

As the final step m the verification process, 3-dimensional scaffolds of 12.5 in × 12.5 in × 12.5 in were fabricated, and the scaffolds were analysed through imaging techniques. (Fig. 4.16). To check for broken struts and residual trapped powder within the scaffolds, microcomputed tomography (micro CT) was carried out using SkyScan-1074 Portable x-ray micro-CT scanner. Samples were first scanned in their entirety (slice thickness of 16 µm) and then zoomed in to the view the volume at the centre (slice thickness of 8 µm). Results showed that there was a through network of pores within the scaffolds. Figure 4.17 shows a scanned slice of the whole cross section of the scaffold built in XZ-plane and on the right, a portion of the reconstructed three dimensional scaffold.

4.10 Conclusion

The main advantage of this system is the elimination of the reliance on user skills that are necessary in conventional techniques of scaffold fabrication. From a small range of basic units, many different scaffolds of different architecture and properties can be designed and built. The system interface of CASTS in Pro/ENGINEER is user friendly and allows complete transfer of knowledge between users without the need for complex user manuals.

Biomaterial scaffolds fabricated using the system showed much potential. Not only was there consistency in structure, there was little problem with powder trapped within the scaffolds fabricated using selective laser sintering. Hence, CASTS is a viable system for generating and producing scaffolds for tissue engineering applications.

References

1. Thomson, R.C., et al. (2000). Polymer scaffold processing, in *Principles of Tissue Engineering*, R.P. Lanza, R. Langer, and J. Vacanti (eds), Academic Press: San Diego, pp. 251–262.
2. Ingber, D.E. (2000). Mechanical and chemical determinants of tissue development, in *Principles of Tissue Engineering*, R.P. Lanzer, R. Langer, and J. Vacanti (eds), Academic Press: San Diego, pp. 101–110.
3. Bell, E. (2000). Tissue engineering in perspective, in *Principles of Tissue Engineering*, R.P. Lanzer, R. Langer, and J. Vacanti (eds), Academic Press: San Diego.
4. Widmer, M.S. and A.G. Mikos (1998). Fabrication of biodegradable polymer scaffolds for tissue engineering, in *Frontiers in Tissue Engineering*, C.W. Patrick, A.G. Mikos, and L.V. Mcintyre, (eds) Elsevier Sciences: New York, pp. 107–120.
5. Yang, S.F., et al. (2001). The design of scaffolds for use in tissue engineering. Part 1. Traditional factors. *Tissue Engineering*, **7**(6), 679–689.
6. Hutmacher, D.W. (2000). Scaffolds in tissue engineering bone and cartilage. *Biomaterials*, **21**(24), p. 2529–2543.

7. Niklason, L.E. and R. Langer (2001). Prospects for organ and tissue replacement. *Jama-Journal of the American Medical Association*, **285**(5), 573–576.
8. Niklason, L.E., et al. (1999). Functional arteries grown in vitro. *Science*, **284**(5413), 489–493.
9. Oberpenning, F., et al. (1999). De novo reconstitution of a functional mammalian urinary bladder by tissue engineering. *Nature Biotechnology*, **17**(2), 149–155.
10. Chua, C.K., et al. (1998). An integrated experimental approach to link a laser digitiser, a CAD/CAM system and a rapid prototyping system for biomedical applications. *International Journal of Advanced Manufacturing Technology*, **14**(2), 110–115.
11. Chua, C.K., et al. (1998). Biomedical applications of rapid prototyping systems. *Automedica*, **17**(1), 29–40.
12. Chua, C.K., et al. (1998). Rapid prototyping assisted surgery planning. *International Journal of Advanced Manufacturing Technology*, **14**(9), 624–630.
13. Low, F.H., et al. (1999). Prototyping of patellar components using morphometrical data, principal component analysis and CAD/CAE techniques. *Automedica*. **18**(1): p. 27–45.
14. Chua, C.K., et al. (2000). Fabricating facial prosthesis models using rapid prototyping tools. *Integrated Manufacturing Systems - The International Journal of Manufacturing Technology Management*, **11**(1), 42–53.
15. Levy, R.A., et al. (1997). CT-generated porous hydroxyapatite orbital floor prosthesis as a prototype bioimplant. *American Journal of Neuroradiology*, **18**(8), 1522–1525.
16. Ono, I., et al. (1999). Treatment of large complex cranial bone defects by using hydroxyapatite ceramic implants. *Plastic and Reconstructive Surgery*, **104**(2), 339–349.
17. Curodeau, A., E. Sachs, and S. Caldarise (2000). Design and fabrication of cast orthopedic implants with freeform surface textures from 3-D printed ceramic shell. *Journal of Biomedical Materials Research*, **53**(5), 525–535.
18. Cheah, C.M., et al. (2002). Characterization of microfeatures in selective laser sintered drug delivery devices. *Proceedings of the Institution of Mechanical Engineers, Part H: Journal of Engineering in Medicine*, **216**(6), 369–383.
19. Porter, N.L., R.M. Pilliar, and M.D. Grynpas (2001). Fabrication of porous calcium polyphosphate implants by solid freeform fabrication: A study of processing parameters and in vitro degradation characteristics. *Journal of Biomedical Materials Research*, **56**(4), 504–515.
20. Leong, K.F., C.M. Cheah, and C.K. Chua (2002). Building scaffolds with designed internal architectures for tissue engineering using rapid prototyping. *Tissue Engineering*, **8**(6), 1113.
21. Hollister, S.J., et al. (2000). An image-based approach for designing and manufacturing craniofacial scaffolds. *International Journal of Oral and Maxillofacial Surgery*, **29**(1), 67–71.
22. Chu, T.M.G., et al. (2001). Hydroxyapatite implants with designed internal architecture. *Journal of Materials Science-Materials in Medicine*, **12**(6), 471–478.
23. Hollister, S.J., et al. (2001). Design and manufacture of bone replacement scaffolds, in *Bone Mechanics Handbook*, S. Corwin, (ed.), CRC Press: Boca Raton, pp. 1–14.
24. Feinberg, S.E., et al. (2001). Image-based biomimetic approach to reconstruction of the temporomandibular joint. *Cells Tissues Organs*, **169**(3), 309–321.
25. Chu, T.M.G., et al. (2002). Mechanical and in vivo performance of hydroxyapatite implants with controlled architectures. *Biomaterials*, **23**(5), 1283–1293.
26. Hollister, S.J., R.D. Maddox, and J.M. Taboas (2002). Optimal design and fabrication of scaffolds to mimic tissue properties and satisfy biological constraints. *Biomaterials*, **23**(20), 4095–4103.
27. Chua, C.K., et al. (2003). Development of a tissue engineering scaffold structure library for rapid prototyping. Part 1: Investigation and classification. *International Journal of Advanced Manufacturing Technology*, **21**(4), 291–301.
28. Temenoff, J.S., L. Lu, and A.G. Mikos (1999). Bone tissue engineering using synthetic biodegradable polymer scaffold, in *Bone Engineering*, J.E. Davies (ed.), Em Incoporated: Toronto.

29. Chen, G.P., T. Ushida, and T. Tateishi (2001). Development of biodegradable porous scaffolds for tissue engineering. *Materials Science & Engineering C-Biomimetic and Supramolecular Systems*, **17**(1–2), 63–69.
30. Hollister, S.J. (2005). Porous scaffold design for tissue engineering. *Nature Materials*, **4**(7), 518–524.
31. Agrawal, C.M. and R.B. Ray (2001). Biodegradable polymeric scaffolds for musculoskeletal tissue engineering. *Journal of Biomedical Materials Research*, **55**(2), 141–150.
32. Mooney, D.J. and R.S. Langer (1995). Engineering biomaterials for tissue engineering, in *The Biomedical Engineering Handbook*, J.D. Bronzino (ed.), CRC Press: Boca Raton, pp. 109/1–109/8.
33. Leong, K.F., C.M. Cheah, and C.K. Chua (2003). Solid freeform fabrication of three-dimensional scaffolds for engineering replacement tissues and organs. *Biomaterials*, **24**(13), 2363–2378.
34. Chua, C.K. and K.F. leong (2003). *Rapid Prototyping: Principles and applications in manufacturing* (2nd ed.), World Scientific: Singapore.
35. Lee, M., J.C.Y. Dunn, and B.M. Wu, (2005). Scaffold fabrication by indirect three-dimensional printing. *Biomaterials*, **26**(20), 4281–4289.
36. Lam, C.X.F. (2002). In vitro degradation studies of customised PCL scaffolds fabricated via FDM, in *ICBME*. Singapore.
37. Tan, K.H., et al., (2003). Scaffold development using selective laser sintering of polyetheretherketone-hydroxyapatite biocomposite blends. *Biomaterials*, **24**(18), 3115–3123.
38. Limpanuphap, S. and B. Derby (2002). Manufacture of biomaterials by a novel printing process. *Journal of Materials Science-Materials in Medicine*, **13**(12), 1163–1166.
39. Sachlos, E., et al., (2003). Novel collagen scaffolds with predefined internal morphology made by solid freeform fabrication. *Biomaterials*, **24**(8), 1487–1497.
40. Phung, W.J. (2002). Investigation of building orthopaedic implantation models using Rapid Prototyping, in *School of Mechanical & Production Engineering*. Nanyang Technological University: Singapore.
41. Tontowi, A.E. and T.H.C. Childs, (2001). Density prediction of crystalline polymer sintered parts at various powder bed temperatures. *Rapid Prototyping Journal*, **7**(3), 180–184.
42. Childs, T.H.C. and A.E. Tontowi, (2001). Selective laser sintering of a crystalline and a glass-filled crystalline polymer: Experiments and simulations. *Proceedings of the Institution of Mechanical Engineers Part B-Journal of Engineering Manufacture*, **215**(11), 1481–1495.
43. Ho, H.C.H., I. Gibson, and W.L. Cheung, (1999). Effects of energy density on morphology and properties of selective laser sintered polycarbonate. *Journal of Materials Processing Technology*, **90**, 204–210.
44. Liu, E., (2005). Development of customised biomaterial powder composite for Selective Laser Sintering, in *Mechanical and Aerospace Engineering*. Nanyang Technological University: Singapore.
45. Nelson, J.C., (1993). *Selective Laser Sintering: A Definition of the Process and an Empirical Sintering Model*. UMI: Ann Arbor, Mich, pp. xvi, 231.

Chapter 5
CAD Assembly Process for Bone Replacement Scaffolds in Computer-Aided Tissue Engineering

M. A. Wettergreen, B. S. Bucklen, M. A. K. Liebschner and W. Sun

5.1 Background and Significance

Guided tissue regeneration is gaining importance in the field of orthopaedic tissue engineering as need and technology permits the development of site-specific engineering approaches. Computer Aided Design (CAD) and Finite Element Analysis (FEA) hybridized with manufacturing techniques such as Solid Freeform Fabrication (SFF) is hypothesized to allow for virtual design, characterization, and production of scaffolds optimized for tissue replacement. A design scope this broad is not often realized due to limitations in preparing scaffolds both for biological functionality and mechanical longevity. To aid designers in production of a successful scaffold, characterization and documentation of a library of micro-architectures is proposed, capable of being seamlessly merged according to the flow perfusion characteristics (porosity permeability), and mechanical properties (stiffness, strength), determined by the designer based on application and anatomic location.

Successful scaffolds must fulfill three basic requirements: (1) provide architecture conducive to cell attachment, (2) support adequate fluid perfusion, and (3) provide mechanical stability during healing. The first two of these concerns have been addressed successfully with standard scaffold fabrication techniques. The use of solvent casting or gas leaching [1] and melt molding [2] has produced scaffolds that are globally porous with void architectures large enough to support cellular adhesion, migration, and differentiation. At high enough porosities (>60%), these scaffolds provide adequate fluid flow to deliver nutrients and remove degradation products [3–6]. In instances where load-bearing implants are required, such as the spine, application of these normal design criteria is not always feasible. The scaffold may support tissue invasion and fluid perfusion but with insufficient mechanical stability; are likely to collapse upon implantation as a result of the contradictory nature of the design factors involved. It is believed that the random architecture imbued by the fabrication process contributes to low mechanical integrity. Also, a compensatory increase of porosity for fluid perfusion considerations innately results in reduced mechanical properties. Using solvent leaching with NaCl particulates, porosity must be at least 60% by volume to support adequate fluid perfusion, a level which seriously compromises the mechanical stability of all but the strongest materials [4], none of which are resorbable [7].

The idea of designing synthetic scaffolds for guided tissue regeneration [8, 9] is attractive for several reasons. First, by supplying a base material, the locations upon which tissue will integrate with the surrounding environment are known a priori. As a consequence, several schemes may be developed based on the degradation of the scaffold or changes in morphology over time [10–13] which can influence how subsequent tissue may form. Secondly, by controlling the shape, or at least the surface contour, one can manipulate the mechanical environment in which tissue exists [14–16]. This was shown to be influential on cellular metabolism [17–19]. Additionally, the morphology indicates, based on a superposition of isostress/isostrain measurements, which areas of the scaffold would be overloaded once implanted at an osteoporotic site and could thus lead to tissue necrosis [20–22]. Finally, with control over the scaffold domain comes control over the scaffold interfaces, or transitions from one type of architecture to another, which if joined incongruously can lead to disjoint stress profiles as well as cellular sparsity.

To date, many of the biomaterials used as scaffolding have mechanical behavior inferior to the constituency of bone [23–25]. Therefore their design cannot always be biomimetic (or at least derived biomimetically). Given a case where the architecture of the scaffold needs to deviate from natural tissue, as in fulfilling the demand of exceptional stiffness, an assembly of derived analytical shapes is recommended [26, 27]. In the absence of this necessity, a biomimetic approach using bone-derived architectures is beneficial for the many of the non-mechanical tangibles (e.g. volume fraction) and qualitative intangibles (e.g. fluid perfusion, metabolic waste removal, and those listed previously).

Addressing mechanical stability of a resorbable implant requires more specific control over the scaffold design. With design and manufacturing advancements, such as rapid prototyping (28–30) and other fabrication methods, research has shifted towards the optimization of scaffolds with both global mechanical properties matching native tissue [31], and micro-structural dimensions tailored to a site-specific defect [32]. Recent work by Kelsey et al. [33], proposed a selection process for composite implants based on patient specific parameters coupled with a finite element code. The authors purport that a criterion based selection system will improve the matching of patient defect to implant size and reduce the failure rate of the implant. While this study demonstrates the need to design patient-specific implants, it only highlights the ability to select the most adequate implant for entire prostheses. Hollister et al. [34] generated an optimization scheme for regular, repeating architectures, which have the advantage of being described by constitutive equations relating microstructure to global structure based on homogenization theory. For example, a cube was generated with intersecting, hollow, orthogonal, embedded cylinders. Indices describing the architecture were related to modulus and stiffness and optimized in a way to maximize porosity whilst maintaining stiffness [34]. The advantage of these systems is that the architecture is designed to promote strength independently of material, thus maximizing material arrangement.

While these studies have demonstrated improved research into the role of material organization of bioengineered scaffolds, certain deficiencies still exist which make direct application of these improvements impossible [27]. The previous research has demonstrated the ability to create architectures of repetitious microstructures and characterize them. However, the ideal implant is one that would readily be assembled in series or parallel, each location corresponding to a specific mechanical and fluid property. This implant would aid in load transfer as well as match the already existent geometry of the defect. Furthermore, current databases or libraries of architectural building blocks are not helpful if they are used to describe an environment in which load transfer is incomplete, as calculated strengths and stiffness would not be representative of actual directional load transfer. This lack of a common interface can result in stress lines and border fractures, but can be overcome with well planned Computer Aided Tissue Engineering (CATE) [35].

In this chapter, the authors present a library of primitives (unit building blocks and interfaces) to be implemented in Computer Aided Tissue Engineering (CATE) [36, 37]. Here, two systems are presented: [1] an analytically derived library of regular polyhedra and [2] a tissue-derived, trabecular bone library. These unit blocks may be merged according to various qualities, some of which are illustrated in this manuscript. Density or volume fraction is a suitable metric determined from image intensity, though the authors expect that more significant metrics related to tissue regeneration will need to be identified. Nevertheless, the correlation between bone mechanical properties (stiffness and strength) to apparent density is well established [25, 38]. Using principles of an assembled library, scaffolds may be created in a site-specific manner with conformations that resemble bone. This process may be preferred over current scaffold generation techniques that make use of random processes of solvent casting or gas leaching to create the necessary void spaces [39, 40]. Our study incorporated a linear, isotropic, finite element analysis on a series of various micro-structures to determine their material properties over a wide range of porosities [41]. Furthermore, an analysis of the stress profile throughout the unit blocks was conducted to investigate the effect of the spatial distribution of the building material.

5.2 Concept of Computer Aided Tissue Engineering

5.2.1 Unit Cubes and Interfaces

There are two types of design approaches to consider when creating scaffold building blocks. The first is to create a variety of scaffold shapes which are characterized throughout the spectrum of the independent design variable (Unit cube method). This is particularly useful if a specific shape or repeated pattern is recognized from the outset. For example if one preferred a design favoring the volume fraction of a given shape, experience indicates that scaffolds need at least 60% porosity to account for nutrient delivery and below approximately 90%, the mechanical integrity will be suspect [3, 5, 26]. On the other hand, if the shape is the unknown, but the interfaces and contact angles of adjacent unit blocks are known (interface method), then the inverse problem, usually a shape/topology problem, may be solved as long as a design rule or objective is provided. Figure 5.1 illustrates the two conceptual differences.

Though the focus of this chapter is on shape recognition of tissue primitives, i.e. a known shape, the authors use other examples to illustrate the interface method. The solution of the inverse problem is a topology optimization problem. These types of problems are difficult to solve. One has to find a representation of the shape that can capture the fine scales of tissue and that allows the application of mathematical optimization. Among the available approaches are density functions [42, 43], level-set methods [44], evolutionary structural optimizations (ESO) [45–47], and other implicit function based methods [48].

For demonstrative purposes and ease of implementation, a heuristic method was chosen, known as the modified method of intelligent cavity creation (ICC) which is a subclass of ESO. This method has the advantage of dictating the number of cavities (voids) created during the process, and is relatively easy to implement. The major disadvantages are that the optimization goal is not well formulated and the computational order is of the scaffold dimension cubed. Briefly, a completely solid material is used as the starting geometry and reduced to a final shape through a combination of surface erosion (nibbling ESO) and cavity creation. The reduction serves to minimize the addition of peak stresses caused by material reduction, by removing only "unneeded" material. A metric is used to determine "unneeded" material and when the internal stress state may tolerate the addition of a cavity (for a complete description see [45]). The fixed number of cavities regularizes the problem and the numerical instabilities often associated with the inverse description.

The interfaces and contact angles of the bone cube in Fig. 5.1 were estimated to create boundary conditions (Fig. 5.2A). Assumed displacement boundary conditions of equal magnitude were applied along the trabecular axes. The elastic modulus of the solid material was 1×10^6 times stronger than void material and had a Poisson's ratio of 0.3. The created routine used an interactive routine to retrieve the von Mises stress (ABAQUS, Inc., Pawtucket, RI) (Fig. 5.2B) with optimization rejection

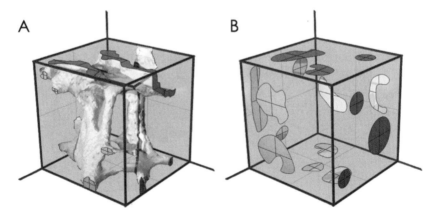

Fig. 5.1 Unit cube and interface methods of primitive design. The unit cube architecture is known (A), or the interface (B) and loading conditions are known

ratios of 0.01 and a maximum of two internal cavity initiations. The final internal architecture had a porosity of 64% (Fig. 5.2C).

5.2.2 Biomimetic Design Theory and Implementation

Bone and other architectures contain complex geometry with mechanical properties that vary spatially and with anatomic site. In a region of interest (ROI) at least two continuous phases (bone matrix and interstitial fluid) are responsible for the global mechanical properties. Subdivisions of this ROI will contain smaller regions of discrete architecture and thus mechanical properties which when summed together result in the global properties. If for example, the defect site contained both cortical and trabecular bone, then homogenization theory would correctly assume the global properties but would fail in the determination of the properties of the subregions of the ROI. To complete and replicate load transfer upon implantation of an implant, especially in a region containing varied architecture and properties, an engineered scaffold must mimic the variants with respect to direction. In this regard, the interior properties of the bone under study can be obtained by a quantitative computed tomography based approach (QCT). The CT slices of the ROI can be queried to obtain the CT# for a discrete number of voxels within the ROI of volume V_m, given by:

$$\varphi_k(x, y, z, CT\#)^* N_k^* t = V_m, \quad k = 1, 2, Ln \qquad (5.1)$$

where x, y, z represent the position of the voxel within the coordinate space; N_k represents the number of voxels within the slice; t represent the thickness of each slice; n represents the total number of slice planes. Mechanical property characterization can be achieved by correlating the CT# to density by a linear interpolation using relations available in published literature. This density can in turn be then related to E, the Young's modulus of the tissue structure, allowing the heterogeneous elasticity of the bone to be defined [49, 50]. The ROI is then subdivided into discrete units, with each unit associated with its own characteristic Young's Modulus, E.

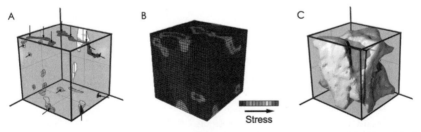

Fig. 5.2 Interface method. The boundary conditions (A) determine the stress state (B) which dictates the final shape (C) of any optimization method (in this case ESO)

The intended 3D scaffold that replaces V_m, will be composed of discrete sub unit architectures V_i:

$$V_m = V_1(P_1, S_1) \cup V_2(P_2, S_2) \cup V_3(P_3, S_3) \cup L \cup V_i(P_i, S_j) \tag{5.2}$$

where sub-volume V_i denotes the unit cell, P_i denotes the spatial position of the unit cell and S_j denotes the characteristic unit cell assembled in V_m. Each unit cell has specific mechanical properties and will be matched based on the initial mechanical property characterization using data set available in (5.1). Varying mechanical properties can be achieved by either altering the porosity of the unit cell or changing the internal architecture of the unit cell while keeping the porosity the same. Thus, subdivisions of the engineered scaffold must mimic the select architectural properties of these regions. It is therefore proposed that the micro-architecture be designed to replicate the site specific mechanical properties at a resolution that match the achievable feature size of current solid freeform fabrication methods.

5.3 Identification of Unit Cubes

5.3.1 Tissue Primitives as Unit Cubes

The goal of this study was to identify and implement patterns of trabecular bone into a descriptive stiffness library composed of unit cubes and interfaces. Each library unit is referred to as a primitive, and is the combination of one unit cube and six interfaces. The feasibility of joining two primitives is handled by a library subset of seven dissimilar interfaces, which share a common surface area (Fig. 5.3B). The stiffness of unit cubes was calculated and was used, along with its morphology (Fig. 5.3A) to present an example of the application of the library to a partial scaffold assembly.

In an effort to isolate repeating patterns within trabecular architectures, 20 sections from 10 T-9 human vertebral bodies were scanned (μCT Scanco80, Basserdorf, Switzerland) at 30 μm isotropic resolution. The image sets were reconstructed into segmented, binary, trabecular bone datasets of voxel slice dimensions 2048 × 2048. Architectures were loaded into an image processing suite (Analyze Direct, Inc., Lenexa, KS) and viewed in the three physiological planes in >2 mm thick sections. In most cases, trabecular subsections were translated into stereolithography files for 3-D viewing (Fig. 5.4A). Because each individual bone contains many complex repeating patterns, often joined with other repeating patterns, the number of each unit shape could not be documented. Nevertheless, there was enough evidence to construct computer models of derived tissue interfaces (Fig. 5.3A) and unit cubes (Fig. 5.3B).

The architecture of bone was examined on three length scales. The first and simplest length scale representing a junction of two single rods or a junction of rod with a plate was not considered significant. The second length scale was unit blocks witnessed on at least two trabecular lengths. The final scale, lengths of more than four trabecular lengths were scarcely found with any consistency, possibly due to difficulties in manual identification, but most likely because trabecular bone is seemingly a random conglomeration of simpler repeating units. Therefore, this length was not included at any point in the library. Repeated patterns, such as the highlighted regions in Fig. 5.4A, were translated into tissue primitives (Fig. 5.4B) for the library. Emergent patterns could be found in numerous locations illustrated by the abridged version of the Sunglasses shape portrayed in Fig. 5.4C.

The primitive library consists of five Archimedean, five tissue-derived, two Johnson, and two Catalan solids. The solids are respectively: (Truncated Icosahedron, Truncated Dodecahedron, Truncated Octahedron, Truncated Tetrahedron, Cuboctahedron), (Armchairs, Archway, Folding Chairs, Ball and Stick, Sunglasses), (Bi-Augmented Triangular Prism, Augmented Hexagonal Prism), and (Deltoidal Icositetrahedron, Triakistetrahedron). Incidentally, the Catalan solids are duals of the Small Rhombicuboctahedron, and Truncated Tetrahedron, respectively. Mor-

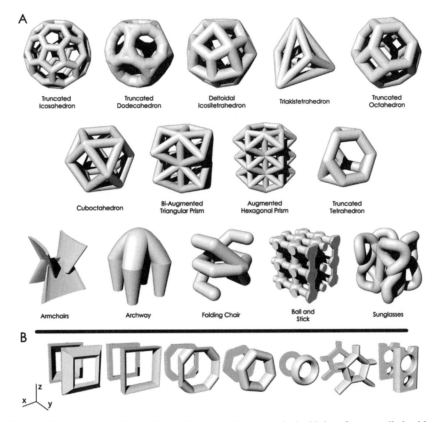

Fig. 5.3 Tissue primitive library. The unit blocks (A) are attached with interfaces on all six sides (B) resulting in a tissue primitive. Unit blocks are limned at their "native" volume fraction (see Fig. 5.6)

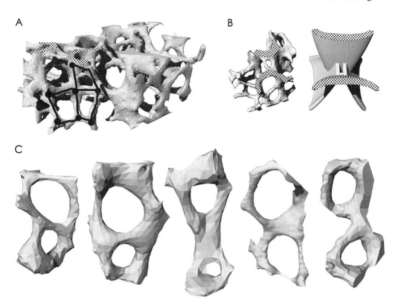

Fig. 5.4 Identification of tissue primitives. Tissue primitives were identified on several length scales in several physiological directions from μCT scans. They were identified (A), derived into CAD representations (B), from emergent, repetitious patterns (C) within the samples

phological features of the non-wireframe versions of these polyhedra have been well documented [51]. In all cases, architectural features were noted repeatedly within bone samples. In most cases, closed 3-D versions did not exist, but were included in closed form for space-filling and regularity purposes. For example, in Fig. 5.4A a partial, wire frame version of a combined face of the Deltoidal Icositetrahedron (highlighted, bottom-left) can be clearly seen, with the appropriate angles (~98° face edge, ~21° protrusion), but does not exist in closed form as pictured in Fig. 5.3.

Tissue-derived shapes which bore no resemblance to previously recorded polyhedra were also documented. The Armchairs unit cube consists of four curvilinear plates joined at a central junction. The Archway represents a distorted "X" interface extruded in a curvilinear fashion along the height. The Folding Chairs is a spring-like shape which when projected in two of three directions forms a commonly occurring hexagonal pattern. The Ball and Stick model is a basic orthogonal rod model, where the created rhombi exist at 60° angles, also corresponding to a regularly observed trend. The final Sunglasses shape was a 3-D version of a curvilinear figure-eight pattern (Fig. 5.4C). The locations of the primitives within the bone was location dependent, but not exclusive. For example, the hexagonal projections tended to exist in axially cut sections while in coronal views, the orientation of the struts was in-line with loading, resulting in more rectangular interfaces and rectangular unit cubes. Often the transition from pentagonal shape to a rectangular shape could be observed by a diminutive or unused strut, within the coronal view.

5.3.2 Regular Polyhedra as Unit Cubes

In situations when assembling tissue-derived unit cubes is cumbersome, incorporation of more regular architectural configurations can prove beneficial. Unit cell polyhedral models were generated using Rhinoceros 3D (McNeel Associates, Seattle, WA). All polyhedra were generated within the same bounding box of 3 mm × 3 mm × 3 mm which placed constraints upon the possible size of each structural element. All generated architectures exhibit symmetry along the three axes and were thus considered orthotropic with respect to geometry, while isotropic with respect to material behavior. Two types of models were generated, the first based on solid geometry with the inclusion of void spaces to create porosity and the second based on geometric regular polyhedra.

The first type of architecture was a space filling solid structure, such as a sphere or cube. A void structure was superimposed onto a solid structure using standard Boolean processes [52], resulting in hollow or shelled structure exhibiting the desired geometry. Void structures were applied in such a manner as to result in an architecture which was symmetric in three directions. Porosity of these solid architectures was determined from the ratio of solid volume to the global bounding box volume. The porosity of the solid architectures was adjusted to fixed values by modifying the volume of the void elements until the porosities matched.

The second type of architecture generated were wire frame approximations of the basic set of geometric polyhedra, the Archimedean and Platonic solids [53]. The advantage of using these polyhedra as models is that they are regular, that is both equiangular and equilateral, and thus exhibit the desired symmetry. These models were initially employed as volumetric representations, as illustrated in Fig. 5.5A for the case of the rhombicubeoctahedron. Each edge was converted to a beam of the corresponding length. As shown in Fig. 5.5B, the resulting shape contains the same number of beams as the original shape had edges. Each beam is equilateral and has the same diameter. The porosity of the architectures was determined from the ratio of the summation of the beam volumes to the volume of the bounding box. Porosity adjustment was accomplished by globally modifying the beam diameters until the desired value was obtained.

The final porosity of each architecture was resized to 80% volumetric porosity, a value corresponding to the porosity of trabecular bone. The 12 architectures generated including their common interface are displayed in Fig. 5.6. Briefly, the square holes architecture is a solid cube with square void placed in orthogonal directions. The plumber's nightmare was taken from a common structural organization of lipid bilayers and is composed of hollow orthogonal pillars with a shelled sphere central to the architecture. The hollow sphere is the superposition of a solid sphere with a void sphere applied to the center of the shape, while the hollow corners architecture has four void spheres applied to the corners. The fire hydrant is composed of three orthogonally arranged beams. The eight spheres architecture is composed of eight shelled spheres arranged adjacent to each other. The cross beams scaffold consists of four beams equi-spaced around the center of the volume copied

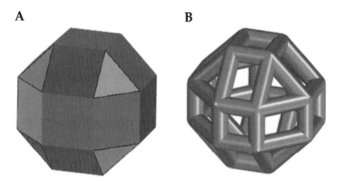

Fig. 5.5 (A). Rhombicubeoctahedron, space filling solid model. (B). Wireframe approximation of rhombicubeoctahedron. Each beam has the same length and diameter

orthogonally to each other. The truncated octahedron and the rhombitcubeoctahedron are exact approximations of platonic solids. The curved connectors is composed of three, quarter tori at each corner connecting the common interface. The previously listed three shapes were also modified to include cross beams in the center of the architecture. The cross beam architecture was generated at additional porosities of 50, 60 and 70% porosity to illustrate the effect of volume on material properties.

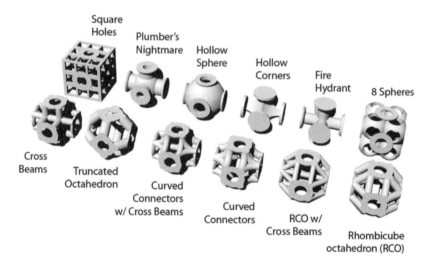

Fig. 5.6 Polyhedra generated as unit cubes. All polyhedra contain the same material volume (20%) and are bounded by the same dimensions. Only the architectural arrangement of material differs between architectures. The torus are the same dimensions across all shapes

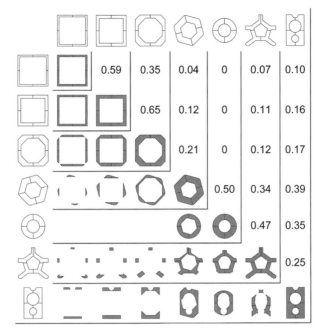

Fig. 5.7 Interface matching. Seven interfaces can be matched with each other according the relative fraction of intersecting areas. Each interface forms a mathematical basis which has a spectrum of values

5.4 Interfaces Between Unit Cubes

5.4.1 Tissue Primitives

Tissue primitives connect to other tissue primitives through their designed interfaces (Fig. 53B) Fig. 5.7 portrays interface matching and the common surface area between adjoining interfaces. The diagonal entries represent a perfect joining with a relative surface area of one. The off-diagonal elements reflect the common area and symmetrically the relative percent of cross-section filling, calculated as the fraction of the common portion over the superposition of both areas (intersection/union). Each interface has a basis of values which describe how well a given interface relates to another. For example, the circular interface matches only with itself or the hexagonal interface with any mechanical integrity, reflecting the limitations of the library, and the necessity of multiple interfaces.

Ideally, each primitive's interface would provide a perfect matching. However, because unit cube architectures were derived, and not designed, a universal interface is not possible. Moreover, simple solid plate attachments or similar constructs would hinder the permeability and porosity of the scaffold. For that reason, native bone interfaces were derived as well. Architectures with drastically dissimilar or similar stiffness values are readily attachable in order to avoid further stress concentrations.

5.4.2 Regular Polyhedra

In this unique subset of the former case, a single type of interface may be used to connect all unit cubes. When matching two continuums of differing micro-architectures, stress discontinuities will occur. For regular polyhedra, a common interface in the form of a torus can be applied. This shape was added to each side of the polyhedra to add regularity to each architecture and aid with load transfer when connected in series and in parallel. The torus was sliced axially so that each side had one half torus, thus when two shapes were combined, a whole torus was produced. An illustration of the torus concept of matching sides is depicted in Fig. 5.8. The dimensions of each torus were kept constant for complete load transfer during joining. Following the addition of the torus, all architectures were again resized to the appropriate porosity and subsequently exported as .igs files in preparation for finite element analysis (FEA).

5.5 Finite Element Analysis

5.5.1 Tissue Primitives

Finite element analysis was conducted on each unit cube in order to characterize the spectrum of modulus values. The unit block modulus could be later used as a parameter to select shapes for the assembly of a site-specific scaffold. The analysis was completed in ABAQUS. A seeding density of 0.75 was used in most cases providing on average 37,000, linear tetrahedral elements. Each architecture was assigned isotropic material properties of $E = 2\text{GPa}$ and $\nu = 0.3$. The stiffness and elastic modulus in the loading direction were calculated in the standard manner through the summation of the reaction forces. The normalized values (with respect to the isotropic modulus) and the orientation in which they were tested are reported in Fig. 5.9.

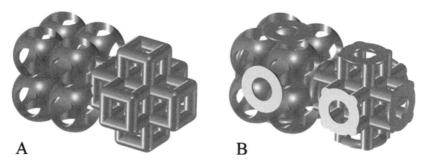

Fig. 5.8 Illustration of improved matching between dissimilar architectures (cross beams and eight spheres) with the addition of an axially sliced torus to each of the six sides of an architecture

Sorted modulus values fell within the same neighborhood which introduced a modicum of error in the method of ranking the shapes in absolute terms. Modulus values corresponded with the volume fraction only weakly. The Arm Chair is a notable example which has the smallest material volume range (and material volume mean value), yet the third largest modulus value. At the other extreme, the Folding Chair and Cuboctahedron have large volume fractions, but behave poorly mechanically. Intuitively, the Ball and Stick fares well mechanically and has 65% material volume.

5.5.2 Regular Polyhedra

Finite element analysis was completed on the architectures to determine their material properties due to specific material arrangement. Each polyhedra was imported into ABAQUS from the prepared .igs file and subjected to a linear, prescribed displacement test. An illustration of the finite element procedure is provided in Fig. 5.10. Isotropic material properties for each polyhedra were assigned with an elastic modulus of $E = 2$ GPa and a Poisson's ratio of $v = 0.3$. Each polyhedra was displaced the equivalent of 1% and the reaction force (RF) was calculated from the

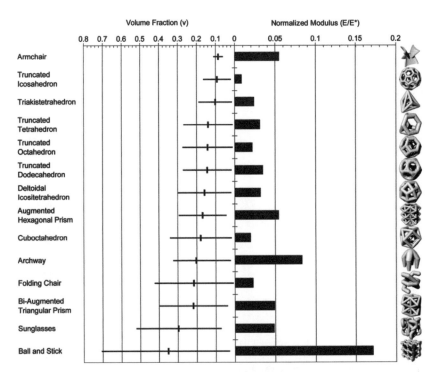

Fig. 5.9 Volume fraction and scaled modulus. The range in volume fraction of each unit cube (left) and normalized elastic modulus, with respect to the average volume fraction (right)

top nodes of that polyhedra. Young's Modulus (E) was calculated by dividing RF over the bounding box area. First, a convergence study was completed with the cross beam architecture to determine the required mesh density for the tests. The architecture was meshed at varying seed densities with tetrahedral elements and subjected to the linear displacement test. Consequently, a seeding density of 0.075 (ABAQUS CAE option) for all polyhedra at 80% porosity was chosen which resulted in meshes with roughly 75,000 elements.

Linear displacement of the polyhedra was completed for two cases, confined and unconfined displacement. Confined displacement included boundary conditions on the vertical faces which restricted a bulging effect during displacement. The confined case illustrates the properties of the polyhedra as it may act when placed within a continuum in a global scaffold. The unconfined case represents the deformation characteristics of the polyhedra without the influence of adjacent cells. The load subjected to a unit block in a large scaffold will most likely be in between the investigated cases.

Figure 5.11 illustrates the calculated elastic moduli for each architecture evaluated. These values represent the average of the confined and unconfined case for each unit block. Despite the fact all shapes, for a given porosity, contain the same amount of material and material properties, there was a large range of modulus values obtained, with the weakest architecture (curved connectors) having a modulus of 0.96 MPa, while the strongest architecture (square holes) had an elastic modulus of 174 MPa.

Elemental principle stress was evaluated for each polyhedra as a method to characterize the loading on each architecture as a result of the material arrangement. For each prescribed displacement, the maximum principle stress was calculated and outputted as a histogram. Selected results of this analysis are displayed in Fig. 5.12. As can be seen from Fig. 5.12A, the material volume plays a direct role in the resulting stress profile for each architecture. The majority of the elements in the 80% porous cross beams architecture experienced little to no stress at all. This is the first trend that was illustrated from the analysis of the stress profiles. As the material

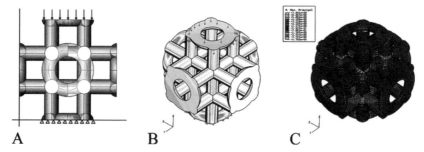

Fig. 5.10 Illustration of finite element procedure. (A), unit cell, seen in plane with bottom face fixed in translation in the y direction and the top face displaced in the y-direction. All polyhedra were subjected to 1% strain. (B), boundary conditions applied to the unit cell in the y-direction. (C), contour plot of maximum principal stress within the architecture. The inner central regions of the architecture (light regions) are the regions of high stress while elsewhere (dark color) stress is near zero

5 CAD Assembly Process for Bone Replacement Scaffolds in CATE

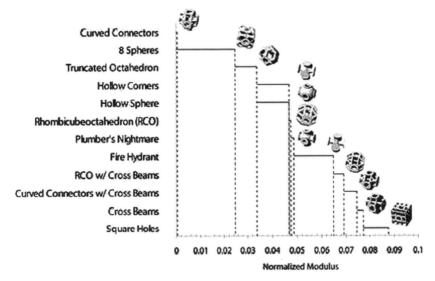

Fig. 5.11 Results of prescribed displacement tests for unit architectures at 80% porosity. All polyhedra were displaced to 1% strain to obtain reaction force and modulus. Results from combined and unconfined compression were averaged to obtain displayed values

volume increases, the peak stress that is experienced on the cross beams architecture increases and also the stress profile spreads out to encompass more values as a result of increased elements contained in the architecture. All of the shapes exhibit a distinct final peak that occurs at approximately −25.0 MPa. This outer peak is a trend that was observed in other architectures as seen in Fig. 5.12C such as the fire hydrant architecture. Also to be observed from this figure the similarities in stress profile at and around zero stress for the two dissimilar architectures of the fire hydrant and the hollow corners. In Fig. 5.12B, it can be observed that the cross beams 80 architecture has twice as many elements that are near zero stress as the other four architectures. The cross beams architecture also exhibits two distinct peaks while the remaining shapes may be more widely dispersed or have more peaks.

5.6 Morphological Analysis

5.6.1 Tissue Primitives

Morphological analysis was conducted on each unit cube in order to assist in characterizing the properties of the library. A "native" volume fraction for each unit cube was determined by averaging the range of possible volumes (Fig. 5.9). The lower bound was calculated in each unit cube with similar strut diameters, signifying that it is a function of the number of struts and spatial arrangement. The upper bound was determined by expanding each diameter until a unique feature of the shape was obscured (such as a hole closing in on itself). The only exception was the Arm Chair

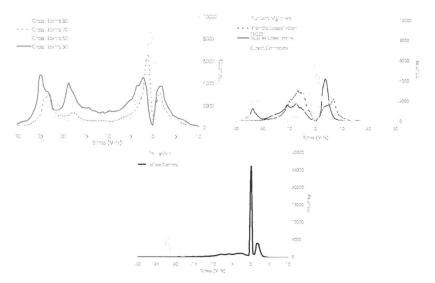

Fig. 5.12 Histogram of finite element results. Top left, stress distribution of the cross beam architecture at four porosities. Top right, stress distribution of four additional architectures evaluated against the cross beams 80% porosity architecture. The majority of the elements in these four architectures are loaded with few elements if any experiencing null levels of stress. Bottom, stress distribution of the fire hydrant architecture displayed against the hollow corners architecture. The stresses of both architectures near zero are identical. Several of the architectures exhibited a second peak near 1 MPa

which was given an average plate thickness commensurate to the strut diameter. The Armchair and Truncated Icosahedron were the most porous while the Ball and Stick was the least. Similarly the range in material volumes tended to increase with respect to the mean value. In other words the least porous material also existed over the widest range (Ball and Stick). Table 5.1 portrays a more robust analysis of each architecture. Volume fractions range from 0.09–0.35, though most of the shapes are within 80–90% porosity, which is suitable for scaffold engineering. In general, a high surface to volume ratio (S/V) is advantageous as the available area for fluid flow over attached cells will be higher. Of the identified shapes, the Armchair has the largest surface to volume ratio, and Archway the smallest. The average Connectivity Index, or number of struts connecting to a vertex, and length to diameter ratio is also reported (Table 5.1).

5.6.2 Regular Polyhedra

The structural properties of regular polyhedra will vary with porosity and were examined. Figure 5.13 displays the results of the displacement test for the Cross Beam polyhedra at the varying porosities. At the highest material volume, the elastic modulus of the architecture was 652 and 600 MPa for confined and unconfined, respectively. As material volume decreased, so did the modulus in both cases

5 CAD Assembly Process for Bone Replacement Scaffolds in CATE

Table 5.1 Morphology of tissue primitive unit blocks

	Volume Fraction (v)	Surface / Volume
Armchair	0.09	1.87
Truncated Icosahedron	0.09	1.65
Triakistetrahedron	0.10	1.22
Truncated Tetrahedron	0.14	0.90
Truncated Octahedron	0.14	1.05
Truncated Dodecahedron	0.14	0.99
Deltoidal Icositetrahedron	0.16	1.08
Augmented Hexagonal Prism	0.17	1.55
Cuboctahedron	0.18	1.04
Archway	0.20	0.61
Folding Chair	0.22	0.75
Bi-Augmented Triangular Prism	0.22	1.27
Sunglasses	0.30	1.10
Ball and Stick	0.35	0.95

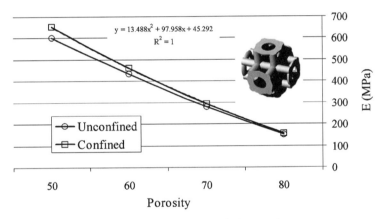

Fig. 5.13 Effect of unit cell material volume on mechanical properties

according to a squared polynomial equation. At 80% volumetric porosity, the elastic modulus between confined and unconfined compression differed by less than 3%.

5.7 Scaffold Assembly

5.7.1 Generation of a Density Map

A density map was constructed which represents the local volume fraction of material within 3 mm sub-volumes, chosen to be the bounding dimension of the each primitive unit. Each section of this three-dimensional matrix maps an averaged volume fraction to that region. Through this procedure, apparent properties may be used to identify the placement of primitives which were designed at a length scale of several trabeculae based on manufacturability and available materials.

The modulus (stiffness) of each primitive was assumed to correlate well with apparent density. Only the relative distribution of values were important for this proof of principle, though any number of correlation relationships, linear or power law, have been used in the literature. The density map was created by averaging a series of binary slices of one vertebral body, progressively, in order to reduce memory requirements needed for processing an entire vertebral body. The map provided a means in which patient/site specific properties of the bone may be assigned to the scaffolding.

5.7.2 Tissue Primitives

The modulus values were ranked according to their magnitude, and the density map subdivided into 14 ranges, representing each of the 14 shapes. The available matching of an interface onto a unit cube is depicted through Table 5.2. The fraction reported is the contact volume of the internal side of the interface and unit cube, divided by the volume of the interface. In practice, Boolean operations in Rhinoceros provided this functionality. Each row of the table corresponds to one of six faces. Values below a certain threshold provide decidedly poor contact volume for load transfer through the interface and into the unit cube, and therefore, should

Table 5.2 Interface matching to tissue primitives. The seven interfaces can be matched to the primitive unit blocks based on the percentage of common contact volume

Primitive	Interfaces 1	2	3	4	5	6	7	Face	Primitive	Interfaces 1	2	3	4	5	6	7
Armchair	0.06	0.02	0.08	0.15	0.18	0.15	0.09	+z	Augmented Hexagonal Prism	0.15	0.02	0.03	0.32	0.43	0.31	0.29
	0.06	0.02	0.08	0.15	0.18	0.15	0.10	-z		0.15	0.02	0.03	0.32	0.43	0.31	0.29
	0.03	0.05	0.04	0.02	0.01	0.02	0.04	+y		0.05	0.00	0.00	0.08	0.11	0.10	0.07
	0.03	0.05	0.05	0.02	0.01	0.02	0.04	-y		0.05	0.00	0.00	0.08	0.11	0.10	0.07
	0.03	0.05	0.05	0.02	0.01	0.04	0.03	+x		0.11	0.01	0.02	0.21	0.24	0.20	0.32
	0.03	0.05	0.05	0.02	0.01	0.04	0.02	-x		0.11	0.01	0.02	0.21	0.24	0.20	0.32
Truncated Icosahedron	0.06	0.00	0.00	0.05	0.10	0.09	0.11		Cuboctahedron	0.23	0.20	0.14	0.53	0.63	0.43	0.33
	0.05	0.00	0.00	0.05	0.08	0.07	0.10			0.23	0.20	0.14	0.53	0.63	0.43	0.33
	0.06	0.00	0.00	0.08	0.19	0.15	0.11			0.23	0.20	0.14	0.53	0.63	0.43	0.33
	0.07	0.00	0.00	0.08	0.20	0.16	0.12			0.23	0.20	0.14	0.53	0.63	0.43	0.33
	0.06	0.00	0.00	0.06	0.13	0.11	0.08			0.23	0.20	0.14	0.53	0.63	0.43	0.33
	0.06	0.00	0.00	0.05	0.12	0.11	0.09			0.23	0.20	0.14	0.53	0.63	0.43	0.33
Triakis - tetrahedron	0.12	0.05	0.04	0.17	0.27	0.24	0.15		Archway	0.09	0.00	0.00	0.09	0.28	0.25	0.27
	0.12	0.05	0.04	0.17	0.27	0.27	0.15			0.23	0.01	0.00	0.27	0.21	0.07	0.15
	0.06	0.03	0.02	0.07	0.09	0.06	0.09			0.30	0.09	0.16	0.37	0.43	0.42	0.28
	0.06	0.03	0.02	0.07	0.09	0.06	0.09			0.30	0.09	0.17	0.36	0.42	0.41	0.27
	0.06	0.03	0.02	0.07	0.09	0.12	0.05			0.30	0.09	0.18	0.38	0.42	0.41	0.27
	0.06	0.03	0.02	0.07	0.09	0.12	0.05			0.30	0.09	0.16	0.37	0.43	0.41	0.28
Truncated Tetrahedron	0.02	0.00	0.00	0.05	0.16	0.15	0.22		Folding Chair	0.29	0.14	0.17	0.32	0.19	0.14	0.37
	0.42	0.13	0.16	0.64	0.55	0.38	0.40			0.04	0.00	0.00	0.02	0.08	0.06	0.12
	0.08	0.01	0.02	0.08	0.10	0.09	0.18			0.04	0.00	0.00	0.02	0.06	0.06	0.12
	0.25	0.08	0.11	0.30	0.21	0.17	0.23			0.30	0.14	0.17	0.32	0.24	0.20	0.48
	0.18	0.04	0.05	0.23	0.27	0.20	0.23			0.30	0.14	0.17	0.32	0.24	0.20	0.48
	0.15	0.03	0.05	0.20	0.28	0.20	0.24			0.24	0.13	0.16	0.32	0.24	0.20	0.48
Truncated Octahedron	0.00	0.00	0.00	0.03	0.12	0.12	0.17		Bi-Augmented Triangular Prism	0.28	0.13	0.16	0.34	0.46	0.39	0.39
	0.00	0.00	0.00	0.03	0.12	0.12	0.17			0.24	0.13	0.16	0.34	0.46	0.39	0.39
	0.03	0.00	0.00	0.05	0.11	0.10	0.14			0.24	0.06	0.14	0.29	0.17	0.13	0.24
	0.04	0.00	0.00	0.05	0.11	0.10	0.14			0.24	0.06	0.14	0.29	0.17	0.13	0.24
	0.03	0.00	0.00	0.05	0.11	0.10	0.14			0.28	0.14	0.25	0.42	0.30	0.23	0.37
	0.04	0.00	0.00	0.05	0.11	0.10	0.14			0.12	0.04	0.10	0.23	0.39	0.32	0.30
Truncated Dodecahedron	0.06	0.00	0.00	0.07	0.27	0.24	0.18		Sunglasses	0.12	0.04	0.10	0.23	0.39	0.32	0.30
	0.06	0.00	0.00	0.07	0.27	0.24	0.18			0.14	0.04	0.09	0.18	0.22	0.20	0.24
	0.05	0.00	0.00	0.04	0.12	0.12	0.12			0.14	0.04	0.09	0.18	0.22	0.20	0.24
	0.06	0.00	0.00	0.04	0.12	0.12	0.12			0.14	0.04	0.09	0.17	0.22	0.20	0.23
	0.05	0.00	0.00	0.05	0.13	0.11	0.12			0.14	0.04	0.09	0.17	0.22	0.20	0.23
	0.06	0.00	0.00	0.05	0.13	0.11	0.12									
Deltoidal Icosa - tetrahedron	0.06	0.00	0.00	0.07	0.19	0.18	0.20		Ball and Stick	0.45	0.45	0.53	0.56	0.50	0.51	0.51
	0.04	0.00	0.00	0.07	0.19	0.18	0.19			0.45	0.45	0.54	0.56	0.51	0.51	0.51
	0.06	0.00	0.00	0.07	0.21	0.19	0.17			0.47	0.38	0.38	0.39	0.26	0.32	0.29
	0.06	0.00	0.00	0.06	0.15	0.19	0.17			0.47	0.32	0.38	0.39	0.26	0.32	0.29
	0.06	0.00	0.00	0.08	0.21	0.19	0.17			0.47	0.43	0.46	0.38	0.26	0.32	0.42
	0.06	0.00	0.00	0.07	0.21	0.20	0.17			0.47	0.47	0.46	0.38	0.27	0.32	0.42

5 CAD Assembly Process for Bone Replacement Scaffolds in CATE   105

be avoided. A threshold of 0.1 was chosen; however the exact number may be modified by the user depending upon the input criteria. The scaffold assembly, then, was produced in the following fashion:

- Unit cubes were placed according to their ranking (modulus) in Cartesian space.
- Interfaces were selected between each face by applying the threshold value for Table 5.2, below which an interface is not considered suitable to transfer the load from its designated unit cube. A perfect match is always chosen between adjacent unit cells when available, under the assumption no perfect match has preference over another (i.e. each has an adequate area for load transfer).
- When no perfect match is available, select between the available interfaces (Table 5.2), those which provide the largest relative match (Fig. 5.7).

An example of a portion of an assembled scaffold for human vertebral body is presented in Fig. 5.14.

5.7.3 Regular Polyhedra

Assembly of regular polyhedra is done through a simple modulus ranking. Complexities due to varying interfaces are not introduced in this simplified system. Figure 5.15 depicts the manufactured result for a portion of human vertebral body.

5.8 Discussion

The authors have presented two libraries based on [1] the native architecture of trabecular bone and [2] the assembly of regularly oriented unit cubes, in an attempt to improve the techniques of scaffold design. Though many of the factors and appropriate cues of what makes a good scaffold have yet to be elucidated, the authors

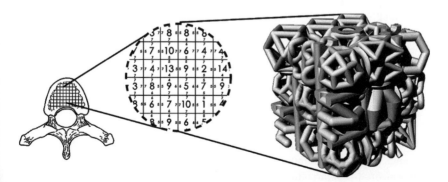

Fig. 5.14 Assembly of scaffold using density map. Primitive unit blocks are assigned based on the density map and Fig. 5.9. Interfaces are assigned based on Table 5.2 and Fig. 5.7

Fig. 5.15 Approximation of a human lumbar vertebral body generated using computer aided tissue engineering principles and the PatternMaster. Included in the model are 46 truncated octahedrons, 209 rhombitruncated cuboctahedrons, 188 hexahedrons, and 61 architectures which are a combination of a rhombitruncated cuboctahedron and a hexahedron

believe that [1] can be effective even in the absence of this knowledge because its derivation utilizes the architecture of bone. Because of manufacturing limitations and ease of implementation, [2] may be preferred. In this chapter, the authors have introduced a library in which the apparent properties, not tissue properties, may be matched in a patient/site specific manner, yet the architecture maintains much of the same tissue level shape (porosity, permeability) that are essential for its biological functionality. The shortcomings of this technique are that a single scalar relationship (density, modulus) is used as the scaffold assembly mapping, yet it is known that architecture as well as the density is necessary for a strength description. Moreover, the modulus value was obtained only in one loading direction, and a more thorough analysis would involve finding all the elasticity constants, and, therefore, defining a very specific map. Nevertheless, even in this case, there would be the possibility of multiple architectural configurations that have similar elastic properties [14].

Of the designed unit cubes, the majority had porosities native to bone and within the window suitable for scaffold design. The S/V of each shape existed over a relatively small range (0.61–1.87). The Archway had the second largest normalized modulus value, but with a poor S/V, denoting that it is most useful as a support element, while the Arm Chair had the largest S/V, suitable for fluid perfusion, yet its l/d value of 11.50 could signify buckling, except for the strongest materials. Most l/d values were low, and safely under the buckling threshold. The Truncated Tetrahedron exhibited two distinct ratios within the same unit cube, signaling that bending or partial buckling is a possible failure mode.

The need for regularizing the domain into cubic volumes is necessary from an assembly and property matching perspective. A classical analog is splitting up a mechanics problem into finite elements that can be solved individually. Obviously, this engineered regularization does not occur naturally in bone. For example, a common question when identifying primitives was: "Where does one border end and the next begin?" Additionally, many of the unit cubes did not exist in the three orthogonal planes, yet they are assembled as such. These types of problems are not easily avoided with an architecture as complex as bone.

There are gross differences in the global properties of several of the regular polyhedra which are directly related to the architecture. The curved connectors shape contains material only in the corners of the shape thus offering no structural support to any mechanical deformation. Even with a 90% increase in modulus from the unconfined to the confined case, the shape is still the weakest by two orders of magnitude. As seen in Fig. 5.13, there is little difference in the elastic moduli between a confined and unconfined case for the cross beams. This may be due to the uniform material arrangement which positions the majority of the load to be transferred directly through the architecture. As a result, as seen from the histogram in Fig. 5.12, the majority of the elements for the 80% porosity were completely unloaded. This supports the previous claim that the arrangement of the material supports the stress in all three directions.

The stress profile serves to demonstrate the contribution of architecture to the global loading. Counter-intuitively, the cross beams architecture at 80% porosity has more elements that are not loaded than the other three architectures. This means that the higher the porosity of an architecture, the greater dependence is placed upon elements that truly add to the strength of the architecture whereas in the lower porosity architectures, less dependence is placed upon the material organization and thus more elements are able to carry higher stress values overall. The contribution to specific stress profiles can be seen in Fig. 5.12B in comparison of the RCO, RCO w/ cross beams and the cross beams architecture. The RCO exhibits a two peak stress profile but the RCO w/ xbeams includes a third peak which is reminiscent of the cross beams architecture. It is speculative that this peak is due to the cross beams which are present in the center region of these architectures but which are missing from the original RCO architecture. Additionally in Fig. 5.12C, it can be seen that hollow corners exhibits a related, yet optimized stress profile to the fire hydrant, despite their very distinguishable material organization.

Selecting the best interface for each face of a given unit cube is a problem that has multiple solutions. Load transfer from one primitive to the next requires that enough volume is present between the contact area of the unit cube and interior portion of the interface and that there is enough common surface area between interfaces, so that localized stress concentrations do not occur. Our currently methodology of selecting a tolerance to apply to Table 5.2, in order to decide which interfaces may be placed on a unit cube, is the factor responsible for allowing multiple solutions. Auxiliary methodologies, such as a mathematical optimization which maximizes the weighted average of the (unit cube—interface) and (interface—interface) contact values (Table 5.2, Fig. 5.7) is plausible, and could induce a unique solution for any modulus map; though negative consequences to any optimization objective function do exist, i.e. a maximum average connection area does not imply anything about individual unit cubes; Moreover, one stress concentration may be sufficient to trigger the scaffold's failure.

Lastly, other alternatives such as interface methods were discussed and an example provided (Fig. 5.2). The main shortcoming is the computational order of such optimization strategies. Because an entire continuum is used, the order is of the dimension cubed, which is not yet conducive to large bone samples. Moreover, it is generally not clear what internal loading vectors are present on the face of a unit cube at this time.

Studies have shown that mineral deposition and cell dependent growth is correlated to the mechanical environment of cells in question [54]. It is hypothesized that stress profiles with uniform distributions may do better at promoting cell adhesion. A comparison of a cutaway profile for the fire hydrant and plumber's nightmare can be seen in Fig. 5.16. Stress distributions in the dominant loading direction for the plumber's nightmare reveal an axial transfer of stress to the non loading direction as evidenced through elements in tension on the periphery due to buckling. Additionally, there are zero stress-free elements in the plumber's nightmare which possibly provides a biologically favored architecture, as evidenced through the Von Mises stress distribution. The solid framed fire hydrant exhibits a more predictable stress profile with minimal off axis stress distribution.

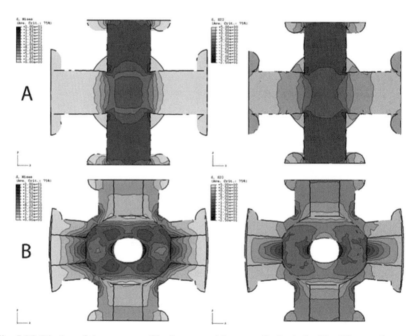

Fig. 5.16 Display of the stress profiles in two architectures for both the Von Mises and stress in the loading direction. (A), the fire hydrant architecture experiences most of its loading along the dominant axis and elsewhere stress is near zero. (B), the plumber's nightmare distributes more stress in non-loading directions as a result of the spherical, hollow, center chamber

5.9 Conclusion

In this chapter, the authors have illustrated the creation of a unit library of architectures that could be used to assemble a complex scaffold of individual, well characterized microstructures. This strategy allows the designer to tailor the mechanical properties of localized regions while maintaining the integration between adjacent microstructures. The designer is offered the choice of keeping to the native architecture of the biological system with the compromise of an advanced system of interface attachment or selecting regular polyhedra with similar apparent properties but no interface difficulties. The mechanical properties of microstructures of a different porosity or even shape from those outlines may be intelligently interpolated through supplied figures. Additionally, design characteristics such as strength instead of modulus can be also be deduced, as linear finite element analyses were conducted. As was illustrated with the cross beam polyhedra, the mechanical properties vary with material volume. Due to demands imposed by the need fortissue ingrowth, solid materials will not allow adequate tissue ingrowth or fluid perfusion. For these reasons, a porous material is required which may support such demands.

References

1. Behravesh E, Timmer MD, Lemoine JJ, Liebschner MA, Mikos AG (2002) Evaluation of the in vitro degradation of macroporous hydrogels using gravimetry, confined compression testing, and microcomputed tomography. Biomacromolecules 3(6):1263–1270.
2. Hutmacher DW (2001) Scaffold design and fabrication technologies for engineering tissues—state of the art and future perspectives. J Biomater Sci Polym Ed 12(1):107–124
3. Agrawal CM, McKinney JS, Lanctot D, Athanasiou KA (2000) Effects of fluid flow on the in vitro degradation kinetics of biodegradable scaffolds for tissue engineering. Biomaterials 21(23):2443–2452
4. Fisher JP, Holland TA, Dean D, Engel PS, Mikos AG (2001) Synthesis and properties of photocross-linked poly(propylene fumarate) scaffolds. J Biomater Sci Polym Ed 12(6): 673–687
5. Zeltinger J, Sherwood JK, Graham DA, Mueller R, Griffith LG (2001) Effect of pore size and void fraction on cellular adhesion, proliferation, and matrix deposition. Tissue Eng 7(5): 557–572
6. Nauman EA, Fong KE, Keaveny TM (1999) Dependence of intertrabecular permeability on flow direction and anatomic site. Ann Biomed Eng 27(4):517–524
7. Buckwalter, JA, Hunziker EB, Orthopaedics. (1996) Healing of bones, cartilages, tendons, and ligaments: a new era. Lancet 348(Suppl 2):sII18
8. Widmer MS, Gupta PK, Lu LC, Meszlenyi RK, Evans GRD, Brandt K et al (1998) Manufacture of porous biodegradable polymer conduits by an extrusion process for guided tissue regeneration. Biomaterials 19(21):1945–1955
9. Hutmacher DW (2000) Scaffolds in tissue engineering bone and cartilage. Biomaterials 21(24):2529–2543
10. Hasirci V, Lewandrowski K, Gresser JD, Wise DL, Trantolo DJ (2001) Versatility of biodegradable biopolymers: degradability and an in vivo application. J Biotechnol 86(2): 135–150
11. Holland TA, Tessmar JKV, Tabata Y, Mikos AG (2004) Transforming growth factor-beta 1 release from oligo(poly(ethylene glycol) fumarate) hydrogels in conditions that model the cartilage wound healing environment. J Control Release 94(1):101–114

12. Alsberg E, Kong HJ, Hirano Y, Smith MK, Albeiruti A, Mooney DJ (2003) Regulating bone formation via controlled scaffold degradation. J Dent Res 82(11):903–908
13. Cheah CM, Chua CK, Leong KF, Cheong CH, Naing MW (2004) Automatic algorithm for generating complex polyhedral scaffold structures for tissue engineering. Tissue Eng 10(3–4):595–610
14. Lin CY, Kikuchi N, Hollister SJ (2004) A novel method for biomaterial scaffold internal architecture design to match bone elastic properties with desired porosity. J Biomech 37(5): 623–636
15. Mattheck C (1994) Design in nature. Interdisciplin Sci Rev 19(4):298–314
16. Mattheck C, Bethge K, Tesari I, Scherrer M, Kraft O (2004) Is there a universal optimum notch shape? Materialwissenschaft Und Werkstofftechnik 35(9):582–586
17. Mullender M, van Rietbergen B, Ruegsegger P, Huiskes R (1998) Effect of mechanical set point of bone cells on mechanical control of trabecular bone architecture. Bone 22(2): 125–31
18. Mullender M, El Haj AJ, Yang Y, van Duin MA, Burger EH, Klein-Nulend J (2004) Mechanotransduction of bone cells in vitro: mechanobiology of bone tissue. Med Biol Eng Comput 42(1):14–21
19. Mullender MG and Huiskes R (1995) Proposal for the regulatory mechanism of Wolff's law. J Orthop Res 13(4):503–512
20. Frost HM (1999) Why do bone strength and "mass" in aging adults become unresponsive to vigorous exercise? Insights of the Utah paradigm. J Bone Miner Metab 17(2):90–97
21. Marks SC, Cielinski MJ, Sundquist KT (1996) Bone surface morphology reflects local skeletal metabolism. Microsc Res Tech 33(2):121–127
22. Bucklen B, Wettergreen M, Liebschner MA (2005) Mechanical aspects of tissue engineering. Seminars in Plastic Surgery 19(3):217–228
23. Niebur GL, Yuen JC, Burghardt AJ, Keaveny TM (2001) Sensitivity of damage predictions to tissue level yield properties and apparent loading conditions. J Biomech 34(5):699–706
24. Keaveny TM, Morgan EF, Niebur GL, Yeh OC (2001) Biomechanics of trabecular bone. Annu Rev Biomed Eng 3:307–333
25. Nicholson, PHF, Cheng XG, Lowet G, Boonen S, Davie, MWJ, Dequeker J et al (1997) Structural and material mechanical properties of human vertebral cancellous bone. Med Eng Phys 19(8):729–737
26. Wettergreen M, Bucklen B, Starly B, Yuksel E, Sun W, Liebschner MA (2005) Creation of a unit block library of architectures for use in assembled scaffold engineering. Computer-Aided Des 37(11):1141–1149
27. Wettergreen M, Bucklen B, Sun W, Liebschner MA (2005) Computer-aided tissue engineering of a human vertebral body. Ann Biomed Eng 33(10):1333–1343
28. Taboas JM, Maddox RD, Krebsbach PH, Hollister SJ (2003) Indirect solid free form fabrication of local and global porous, biomimetic and composite 3D polymer-ceramic scaffolds. Biomaterials 24(1):181–194
29. Yang S, Leong KF, Du Z, Chua CK (2002) The design of scaffolds for use in tissue engineering. Part II. Rapid prototyping techniques. Tissue Eng 8(1):1–11
30. Sachlos E, Reis N, Ainsley C, Derby B, Czernuszka JT (2003) Novel collagen scaffolds with predefined internal morphology made by solid freeform fabrication. Biomaterials 24(8): 1487–1497
31. Griffith LG (2002) Emerging design principles in biomaterials and scaffolds for tissue engineering. Ann NY Acad Sci 961:83–95
32. Simon JL, Roy TD, Parsons JR, Rekow ED, Thompson VP, Kemnitzer J et al (2003) Engineered cellular response to scaffold architecture in a rabbit trephine defect. J Biomed Mater Res 66A(2):275–82
33. Kelsey D, Goodman SB (1997) Design of the femoral component for cementless hip replacement: the surgeon's perspective. Am J Orthop 26(6):407–412
34. Hollister SJ, Maddox RD, Taboas JM (2002) Optimal design and fabrication of scaffolds to mimic tissue properties and satisfy biological constraints. Biomaterials 23(20): 4095–4103

35. Sun W, Starly B, Darling A., Gomez C. (2004) Computer-aided tissue engineering: application to biomimetic modelling and design of tissue scaffolds. Biotechnol Appl Biochem 39: 49–58
36. Sun W, Darling A, Starly B, Nam J (2004) Computer-aided tissue engineering: overview, scope and challenges. Biotechnol Appl Biochem 39(Pt 1):29–47
37. Sun W, Starly B, Darling A, Gomez C (2004) Computer-aided tissue engineering: application to biomimetic modelling and design of tissue scaffolds. Biotechnol Appl Biochem 39(Pt 1):49–58
38. Van Rietbergen B, Odgaard A, Kabel J, Huiskes R (1998) Relationships between bone morphology and bone elastic properties can be accurately quantified using high-resolution computer reconstructions. J Orthop Res 16(1):23–28
39. Hutmacher DW, Goh JCH, Teoh SH (2001) An introduction to biodegradable materials for tissue engineering applications. Ann Acad Med Singapore 30(2):183–191
40. Hutmacher DW, Sittinger M, Risbud MV (2004) Scaffold-based tissue engineering: rationale for computer-aided design and solid free-form fabrication systems. Trends Biotechnol 22(7):354–362
41. Keaveny TM, Guo XE, Wachtel EF, McMahon TA, Hayes WC (1994) Trabecular bone exhibits fully linear elastic behavior and yields at low strains. J Biomech 27(9): 1127–1136
42. Bendsoe MP, Kikuchi N (1988) Generating optimal topologies in structural design using a homogenization method. Comp Meth Appl Mecha Eng 71(2):197–224
43. Bendsoe MP, Sigmund O (2003) Topology optimization. Theory, methods and Applications. Springer Verlag, Berlin, Heidelberg, New York
44. Sethian JA, Wiegmann A (2000) Structural boundary design via level set and immersed interface methods. J Comput Phys 163(2):489–528
45. Kim H, Querin OA, Steven GP, Xie YM (2002) Determination of an optimal topology with a predefined number of cavities. AIAA J 40(4):739–744
46. Kim H, Querin OM, Steven GP, Xie YM (2000) A method for varying the number of cavities in an optimized topology using evolutionary structural optimization. Structural and Multidisciplinary Optimization 19(2):140–147
47. Xie YM, Steven GP (1997) Evolutionary structural optimization. Springer, London
48. Belytschko T, Xiao SP, Parimi C (2003) Topology optimization with implicit functions and regularization. Int J Numer Methods Eng 57(8):1177–1196
49. Rho JY, Hobatho MC, Ashman RB (1995) Relations of mechanical properties to density and CT numbers in human bone. Med Eng Phys 17(5):347–355
50. Rice JC, Cowin SC, Bowman JA (1988) On the dependence of the elasticity and strength of cancellous bone on apparent density. J Biomech 21(2):155–168
51. http://mathworld.wolfram.com/ArchimedeanSolid.html
52. Sun W, Hu X (2002) Reasoning boolean operation based CAD modeling for heterogeneous objects. Computer Aided Design 34:481–488
53. Cromwell PR (1997) Polyhedra. Cambridge University Press, Cambridge 451
54. Cowin SC (1983) The mechanical and stress adaptive properties of bone. Ann Biomed Eng 11(3–4):263–295

Chapter 6
Computational Design and Simulation of Tissue Engineering Scaffolds

Scott J. Hollister, Chia-Ying Lin, Heesuk Kang and Taiji Adachi

6.1 Introduction

The primary goals of tissue engineering scaffolds are to:

- Provide guidance for anatomical tissue shape/volume
- Provide temporary function within the tissue defect
- Enhance tissue regeneration through mass transport and biologic delivery

To achieve these goals, we must first define quantitative measures that characterize these goals. The first goal is general to all tissues and is readily defined by the shape of the tissue defect that needs to be filled. This defect is defined by CT or MR patient scanning. The next two goals are tissue specific and not as readily defined. Function could be mechanical, electrical, and/or chemical. Quantitative measures can be defined through constitutive models, including elastic or other mechanical coefficients, conductivity coefficients, or diffusion/permeability coefficients. However, there is little definition as to what coefficients are most relevant for which tissue, let alone the target values of these coefficients. In addition, since most scaffolds degrade over time, we must be able to characterize the time dependent nature of these constitutive coefficients.

Still more nebulous is the definition of enhanced tissue regeneration. Many factors impact tissue regeneration, including mass transport, cell-surface interaction and biologics delivered from scaffolds, defined here as cells, proteins, and/or genes. Permeability and diffusion coefficients govern mass transport of proteins and other nutrients like oxygen to cells. Permeability and diffusion coefficients are also likely related to cell migration into the scaffold. Cell-surface interaction will be determined by ions and proteins on the cell surface as well as surface roughness. Biologic delivery will also depend upon material binding and surface characteristics. In summary, we can propose the following list of quantitative measures to characterize scaffold design and performance that is by no means complete:

- *Mechanical Function Measures*: Linear Elasticity, Nonlinear Elasticity, Poroelasticity
- *Electrical Function Measures*: Electrical Conductivity

- *Mass Transport/Enhanced Regeneration Measures*: Permeability, Diffusivity, Hydrophilicity, Surface Roughness

Defining the best scaffold for a particular reconstruction application depends on two important components. First, we must be able engineer scaffolds by combining hierarchical design with biomaterial fabrication to achieved desired properties. Second, we must experimentally test the engineered scaffolds in models ranging from in vitro cell models to large functional animal models to determine relevant and critical scaffold characteristics. This chapter will deal primarily with the first component, and will specifically focus on computational scaffold design and simulation of scaffold performance.

To begin designing scaffolds, however, we must first recognize the important design and simulation issues:

- Computational representation of scaffold topology
- Design for Initial Scaffold Characteristics
- Simulation/Design of Time Dependent Scaffold Performance

6.2 Computational Representation of Scaffold Topology

How to actually represent the scaffold topology is the basis for any scaffold design or simulation. This representation must account for the hierarchical nature of scaffold features, the resulting numerical methods that will be used for design/simulation, and the need to integrate scaffold designs with patient data.

Currently, there are two methods for representing scaffold topology: traditional Computer Aided Design (CAD) techniques utilizing computational geometry (Sun et al., 2004a; 2004b), and Image-Based Design (IBD) (Hollister et al., 2000; 2002; Hollister, 2005). CAD techniques use solid models bounded by Non-Uniform Rational B-Spline (NURBS). IBD uses the same structured voxel dataset as imaging modalities (hence the name "Image-Based") where topology is represented by density distribution within the voxel dataset. The techniques are not mutually exclusive, as data is often converted from one representation to the other. However, the final design and subsequent numerical analyses are strongly influenced by the choice of topology representation.

We primarily use IBD, and will focus on its use for representing scaffold topology. Detailed descriptions of CAD methods for topology representation are given in Sun et al. (2004a) and Wettergreen et al. (2005). An important component of either technique is the ability to represent scaffold features from the sub-micron scale to the multiple centimeter scale. In nature, or course, these scales are one continuum. However, in the computational world this scale must be represented as a segmented hierarchy due to the massive data storage requirements.

Using image-based techniques, each distinct hierarchical scale is represented by a separate set of image datasets. Thus, a CT or MRI scan, with a resolution of 0.5–1 mm of patient anatomy would be the coarsest or most global dataset. Another

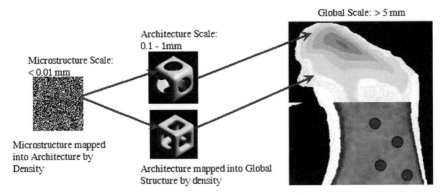

Fig. 6.1 Schematic of hierarchical image-based design (IBD) process using multiple datasets. One dataset is used to represent different features on each different scale. Each feature is mapped to the more global level using voxel density to specify 3D location

set of datasets would represent the 0.05–0.1 mm scale, another the 0.005–0.01 mm scale and so on (Fig. 6.1).

Within each dataset, density represents the feature topology. By segmenting each image into different density levels, we may identify different materials within a given image, and further use density as a flag to indicate where features from a finer resolution dataset map into a coarser dataset. By this method, we can generate designed features on as fine a scale as we desire, and then map these features into the correct tissue/anatomic location.

By splitting features into such a spatial hierarchy, we can ease the computational burden of multi-scale design, and utilize specially develop mathematical theories to compute multi-scale characteristics. Specifically, we can determine how architectural design features from one level determine physical characteristics of a more global level. Once all the features of scaffold design are set for each structural hierarchy, we then interpolate all image datasets to the same resolution and then use Boolean operations to combine the features into one image design database. Within this image dataset, the final scaffold design is determined by thresholding the density into scaffold and void. Multiple materials can easily be incorporated by setting each material to have a different density above the void threshold density. However, this is typically beyond the precision of most current scaffold fabrication technologies. Once the final designed is achieved within a thresholded image dataset, it can be readily converted into a triangular surface representation in a .STL format for fabrication.

6.3 Design of Initial Scaffold Characteristics

Scaffolds are naturally porous structures with one or more materials making up the solid portion. It is intuitive that the base scaffold materials coupled with the arrangement of these base materials and pores in three dimensional (3D) space will determine the effective scaffold mechanical, electrical and mass transport characteristics. These effective scaffold characteristics can then be related to tissue regeneration.

To effectively design scaffold material and pore arrangement, however, we need a mathematical theory that allows us to compute effective physical characteristics like elasticity, electrical conductivity, permeability, and diffusion based on 3D architecture. Homogenization is one such rigorously defined theory specifically developed to analyze multiscale materials (Sanchez-Palencia, 1980; Guedes and Kikuchi, 1990). It has also been utilized to analyze multiscale tissue mechanics by our own group (Hollister et al., 1991; 1994) and others (Aoubiza et al., 1996; Yin and Elliot 2005). Since 2000, our research group (Hollister et al. 2000; 2002; Hollister 2005; Lin et al., 2004a; 2004b) has also applied homogenization theory as a means to analyze and design hierarchical tissue engineering scaffolds.

Homogenization theory assumes that a microstructured material is periodic, that is the complete material volume can be made by repeating a basic unit mathematical cell in 3D space. The unit cell may be different at different locations in the material, allowing for heterogeneity. Homogenization theory addresses multiscale physics problems by first taking an asymptotic expansion of the relevant field variable as:

$$u_i^\varepsilon = u_i^0 + \varepsilon u_i^1 + \cdots \tag{6.1}$$

where u_i^ε is the total value of the field variable, u_i^0 is the average value at the global level, u_i^1 is the perturbation of the average value due to the presence of a microstructure, and ε is the ratio of the microstructure scale to the global scale, which is much less than 1. The gradient of the field variable is taken as:

$$\frac{\partial}{\partial x_i^\varepsilon} = \frac{\partial}{\partial x_i^0} + \frac{1}{\varepsilon}\frac{\partial}{\partial x_i^1} \tag{6.2}$$

where x_i^0 denotes the global coordinate scale, x_i^1 denotes the microstructural coordinate level, $\partial/\partial x_i^\varepsilon$ is the total gradient of the field variable, $\partial/\partial x_i^0$ is the gradient of the field variable with respect to the global coordinates, and $\partial/\partial x_i^1$ is the gradient of the field variable with respect to the microstructural coordinates.

By substituting the relevant field variable expansion and gradient into the governing equilibrium equation we obtain a microstructural equilibrium equation and a global equilibrium equation. These two scale equilibrium equations are linked by the effective physical property. Thus, for elasticity by expanding the displacements u_i we obtain the governing microstructural elasticity equation;

$$\frac{\partial}{\partial x_j^1} E_{ijkl} \frac{\partial \chi_k^{pq}}{\partial x_l^1} = \frac{\partial}{\partial x_j^1} E_{ijpq} \tag{6.3}$$

For which χ_k^{pq} represent nine different microstructural displacements due to nine global unit strains and E_{ijkl} are the elastic properties of the base scaffold material. We can then directly calculate the effective elastic properties E_{ijkl}^{eff} as a function of base material elastic properties E_{ijkl} and the architectural arrangement of materials and voids in 3D represented by $\delta_{kp}\delta_{lq} - \frac{\partial \chi_p^{kl}}{\partial x_q^1}$ through the following integration over the microstructure:

$$E_{ijkl}^{eff} = \frac{1}{|V_{micro}|} \int_{V_{micro}} E_{ijpq} \left(\delta_{kp}\delta_{lq} - \frac{\partial \chi_p^{kl}}{\partial x_q^1} \right) dV_{micro} \qquad (6.4)$$

Equation 6.4 thus gives us a direct mathematical way to compute the intuitive concept that effective scaffold elastic properties are determined by the base scaffold material elastic properties and the way these materials and pores are arranged in 3D space.

Similarly, for diffusion and electrical conductivity (which have the exact same form) we expand species concentration (for diffusion) or voltage (electrical conductivity) to obtain the following governing equation:

$$\frac{\partial}{\partial x_i^1} D_{ij} \frac{\partial \chi^p}{\partial x_j^1} = \frac{\partial}{\partial x_i^1} D_{ip} \qquad (6.5)$$

where χ^p represents local species concentration or voltage under three unit global concentration or voltage gradients. Again, we can directly calculate the effective diffusivity or electrical conductivity D_{ij}^{eff} by integrating the base scaffold diffusivity or conductivity D_{ij} times the quantity $\delta_{ip} - \frac{\partial \chi^p}{\partial x_j^1}$ that gives the architectural influence of material and pores arrangement in 3D space:

$$D_{ij}^{eff} = \frac{1}{|V_{micro}|} \int_{V_{micro}} D_{ip} \left(\delta_{jp} - \frac{\partial \chi^j}{\partial x_p^1} \right) dV_{micro} \qquad (6.6)$$

Again, Eq. 6.6 provides a mathematical computation for the concept that effective diffusivity (conductivity) depends on base scaffold material diffusivities (conductivities) and the way these materials and pores are arranged in 3D space.

The final physical characteristic of interest is permeability. Permeability is defined by how easy or difficult it is for fluid to flow through a porous matrix. Thus, since we expect creeping flow, we solve Stokes flow at the microstructural level by expanding both the fluid pressure and velocity. This leads to the following equation that is solved to determine scaffold level pore fluid velocity under three unit pressure gradients:

$$\frac{\partial p^{1k}}{\partial x_i^1} - \frac{\partial}{\partial x_j^1}\left(\frac{\partial v_i^{0k}}{\partial x_j^1}\right) = e_i^k \tag{6.7}$$

where p^{1k} is the internal pore pressure, v_i^{0k} is the average pore fluid velocity, and e_i^k represents three unit externally applied pressure gradients. The effective scaffold permeability K_{ik}^{eff} is then computed directly from the average pore fluid velocity as:

$$K_{ik}^{eff} = \frac{1}{\mu |V_{micro}^{pore}|} \int_{V_{micro}^{pore}} v_i^{0k} dV_{micro}^{pore} \tag{6.8}$$

where μ is the fluid viscosity and V_{micro}^{pore} represents the pore volume occupied by fluid. It is interesting to note that permeability is only determined by the average pore fluid velocity and thus the 3D pore arrangement. Assuming the basic scaffold material is impermeable, effective scaffold permeability does not depend upon scaffold material.

Homogenization theory gives us a means to compute effective mechanical (elasticity, Eq. 6.4), electrical (electrical conductivity, Eq. 6.6), and mass transport properties (diffusion Eq. 6.6; permeability Eq. 6.8) for porous scaffold architectures. Although homogenization theory was originally developed for periodic microstructures, it gives reasonable effective property estimates for non-periodic microstructure (Hollister and Kikuchi, 1992). We can thus apply homogenization theory to estimate effective properties for scaffolds with designed architecture made by solid free-form fabrication (SFF) techniques, or scaffolds with more random microstructures made by techniques like porogen leaching. Homogenization theory provides a rigorous mathematical technique to estimate effective properties for scaffold microstructures designed using either IBD or CAD techniques. For diffusivity and elasticity, finite element techniques are used to solve the governing microstructural equilibrium equations, while either finite element or finite difference techniques may be used to solve the Stokes flow equations for permeability. Figure 6.2 shows an example of effective elastic, permeability, and diffusivity results for a computationally designed interconnecting cylindrical void unit cell.

Figure 6.2 demonstrates that achieving a desired functional scaffold property often comes at the expense of a desired mass transport property. Ideally, we would like to achieve the best or optimal balance between function and mass transport. Homogenization theory can also be used in optimization algorithms to achieve that desired balance. This optimization approach, termed topology optimization, seeks to automatically design a material layout to achieve a balance of target effective properties (Sigmund, 1994; Lin et al., 2004a; Guest and Prevost, 2006; Hollister and Lin, in press). The optimization problem may be written as:

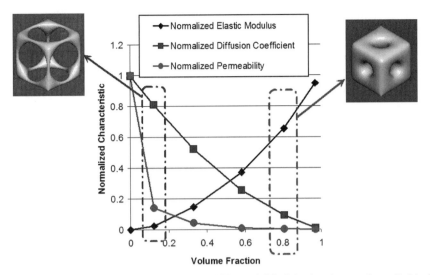

Fig. 6.2 Change in normalized elasticity, permeability, and diffusivity for a intersecting cylindrical pore design as a function of volume fraction. Unit cells on left and right of graph are 15% (dashed rust oval) and 80% (dashed lavender oval) volume fraction structures, respectively. Results show intuitive increase in elasticity and concomitant decrease in permeability and diffusivity with increasing volume fraction

$$\underset{\rho}{Min} \sum_{i=1}^{N} \left(A_{ij(kl)} - A_{ij(kl)}^{target} \right)^2 \qquad (6.9)$$

$Subject\ to:$

$$0 \leq \rho \leq 1$$

$$\frac{1}{V_{micro}} \int_{V_{micro}} \rho dV_{micro} \leq Vol^{target}$$

$$B_{ij(kl)} \geq B_{ij(kl)}^{target}$$

where ρ denotes volume fraction ranging from 0 to 1, $A_{ij(kl)}$ denotes a physical characteristic either diffusivity or permeability (a 2nd order tensor with two indices) or elasticity (a 4th order tensor with four indices, hence the parentheses), $A_{ij(kl)}^{target}$ denotes the target value of the same physical characteristic, V_{micro} is the volume of the microstructure, Vol^{target} is a constraint on the total amount of material that may be used, $B_{ij(kl)}^{target}$ is an upper or lower bound on the physical characteristic $B_{ij(kl)}$ that may or may not be the same as the physical characteristic in the objective. Equation 6.9 is solved using a nonlinear programming technique combined with a homogenization solver. Our group (Lin et al., 2004a) utilized this technique to optimize structures for target elastic constants with a constraint on volume fraction as well as to maximize permeability while matching desired elastic constants. Once the desired architecture design is achieved using IBD, it can readily be integrated into an image data-base defining the external scaffold shape.

6.4 Simulation/Design of Time Dependent Scaffold Performance

Although designing initial scaffold characteristics is critical, scaffolds experience dynamic change in performance as material degrades and tissue is regenerated. Therefore, even if we can design a scaffold for given initial physical characteristics, we ideally would like to simulate how the scaffold degrades and tissue regenerates as a function of the designed material and architecture. Ultimately, we would like to include time dependent processes in initial scaffold design or design/optimization.

6.4.1 Simulation/Design for Material Degradation

Perhaps the less difficult time dependent process to model, although still quite complex in its own right, is material degradation. There are only a small number of computational models of biopolymer degradation. One of the first models of biopolymer degradation was proposed by Göpferich and Langer (1993) and Göpferich (1997). Göpferich noted that degradation was a complex process influenced by crystallinity, hydrophobicity, polymer chain length, and water diffusivity. The Göpferich model therefore assumed that degradation was stochastic and could be modeled as a Poisson process. Material lifetime within the scaffold was assumed to follow a first-order Erlan probability density function:

$$\varepsilon(t) = \lambda e^{-\lambda t} \qquad (6.10)$$

Where $\varepsilon(t)$ is the probability that a piece of material degrades at a time t, λ is a degradation rate constant (a material property), and t is a random variable that designates a material lifetime. The time it takes a piece of material, located at a voxel indexed by i, j, and k in 3D space is denoted as:

$$t_{i,j,k} = \frac{1}{\lambda \ln\left(n_i n_j n_k\right)} \ln(1-\mu) \qquad (6.11)$$

Where $t_{i,j,k}$ is the time it takes material at voxel location i,j,k to degrade, $n_i n_j n_k$ are the number of voxels in the i,j,k direction for a voxel representation of material, λ is the material degradation constant, and μ is a random variable. Thus, for each material location within the scaffold, a lifetime until degradation is given by $t_{i,j,k}$. If the total time elapsed for the scaffold is greater than $t_{i,j,k}$, then that piece of material has degraded, and loses molecular weight. However, recognizing that mass loss often lags molecular weight loss, Göpferich specified that a material voxel would not experience mass loss until all surrounding voxels has loss mass. Using this model, Göpferich (1997) was able to predict degradation of Poly(D,L-lactic acid-co-glycolic acid).

A second model of polymer degradation by hydrolysis was proposed by Adachi et al. (2006). This model focused on water diffusivity as the major factor affecting degradation, and therefore assumed that molecular weight loss was correlated to water diffusion, and elastic modulus was correlated to molecular weight loss. Water diffusion is modeled by a first order diffusion equation:

$$\frac{dc}{dt} = \frac{\partial}{\partial x_i} D_{ij} \frac{\partial c}{\partial x_j} \tag{6.12}$$

where c denotes normalized water concentration in the scaffold material and D_{ij} are the local base scaffold material diffusion coefficients. The material molecular weight loss is assumed to depend on the local water content c as:

$$\frac{dW}{dt} = -\beta c \ (0 \leq c \leq 1) \tag{6.13}$$

where β is a material constant governing degradation rate and W represents molecular weight. The base scaffold material elastic properties are assumed to depend linearly on molecular weight as:

$$E_{ijkl}(t) = \frac{W(t)}{W(t=0)} E_{ijkl}(t=0) \tag{6.14}$$

where $E_{ijkl}(t)$ are the base scaffold elastic properties at a given time t, $E_{ijkl}(t=0)$ are the initial base scaffold elastic properties, $W(t)$ is the molecular weight at a given time t, and $W(t=0)$ is the initial molecular weight. By solving Eqs. 6.12–6.14, and knowing β from experiment for a given material, the rate of change of scaffold elastic properties over time can be predicted.

Beyond predicting elastic property degradation, we ultimately would like to include degradation effects into the initial scaffold design, especially if we seek to prolong scaffold load bearing capability. Essentially, this amounts to predicting the best location for material reinforcement within the scaffold to prolong load bearing without significantly decreasing mass transport characteristics.

Lin et al. (2004c) developed a topology optimization method to account for time dependent material degradation. In this method, the topology optimization method given in Eq. 6.9 is solved for discrete points during the degradation period, with the difference that the degraded scaffold base modulus is used as input for the optimization. An optimized density distribution $\rho(t_d)$ is calculated to achieve a constant target elasticity tensor $E_{ijkl}^{\text{eff target}}$ is at discrete time points t_d ranging from $t=0$ to the desired final time point. Obviously as the elastic properties are reduced, more material must be added to achieve the constant target effective elastic property. The final density is then calculated by summing the predicted densities at each time point, weighted by the total time of scaffold life minus the time point and the ratio of degraded to initial elastic properties at that time point:

$$\rho(x_i) = \sum_{p=1}^{\text{Ntime points}} \rho(t_p) \frac{(t^{\text{total}} - t_p)}{t^{\text{total}}} \frac{\sqrt{E_{ijkl}^{\text{base-degraded}} E_{ijkl}^{\text{base-degraded}}}}{\sqrt{E_{ijkl}^{\text{base-initial}} E_{ijkl}^{\text{base-initial}}}} \quad (6.15)$$

where $\rho(x_i)$ is the weighted reinforced material density, $\rho(t_p)$ is the density predicted at a given time point t_p, t^{total} represents the total degradation time, and the ratio of elastic properties is described by the Frobenius norm $\sqrt{E_{ijkl}^{\text{base-degraded(initial)}} E_{ijkl}^{\text{base-degraded(initial)}}}$ (Bendsoe et al., 1996) of the degraded and initial elasticity tensor. Thus, as time passes, the addition of material reinforcement becomes less. Furthermore, this formula will reduce the influence of time points where severe degradation occurs, as the scaffold will be predicted to become completely solid as the material goes away. The baseline scaffold architecture will be that predicted at the initial time $t_p = 0$, with degradation kinetics controlling where material should be added to bolster elastic properties at the expense of reduced permeability and diffusion. Lin et al. (2004c) combined IBD with degradation topology optimization on both the global and microstructural levels to design a degradable spinal fusion cage. Figure 6.3 illustrates both time dependent and initial condition only topology optimization of the cage within the vertebral body, demonstrating areas of material reinforcement as white areas on design image slices.

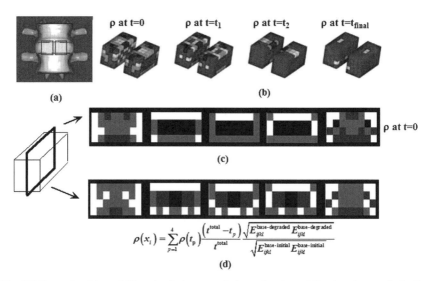

Fig. 6.3 Example of degradable spine fusion cage design using degradation topology optimization from Lin et al. (2004c). (a) initial design domain (red box) in intervertebral space. (b) predicted density to achieve objective at selected points during material degradation. Red indicates high density, blue low density. As material degrades, higher density is needed to achieve the same objective. (c) Density sections through cage design for predicted density at $t = 0$ only, where white is high density material and black is void. (d) Reinforced design achieved by weighting four time points using given equation to account for degradation

6.4.2 Simulation/Design for Tissue Regeneration on Scaffolds

In addition to scaffold degradation, tissue regeneration is the second critical time dependent process that must be modeled. This capability would represent a tremendous advance in tissue engineering therapy, dramatically reducing the time and cost to bring such therapies to clinical use. This capability is also by far the most difficult to achieve, given the need to account for a myriad of factors and the relatively small number of experiments in which a majority of these factors are controlled.

Computational models of tissue regeneration must address cell migration, differentiation and matrix synthesis, in addition to the influence of chemotactic (for example, cytokines) and haptotactic (extracellular and synthetic matrix) factors on cells. Outside of tissue engineering, researchers in tumor-induced angiogenesis have developed inclusive models addressing all these issues. For example, Anderson and Chaplain (1998) developed a model predicting blood vessel growth near tumors in which the path of vessel growth was equated with endothelial cell movement. This movement was assumed to be a random walk modeled as a diffusion process influenced by a chemotactic factor, VEGF produced by the tumor, and a haptotactic factor, fibrin, produced by migrating endothelial cells. The governing equations for this model are:

$$\frac{dn}{dt} = \underbrace{\frac{\partial}{\partial x_i} D^n_{ij} \frac{\partial n}{\partial x_j}}_{\text{Random Cell Motion}} - \underbrace{\chi_0 \frac{\partial}{\partial x_i} n \frac{\partial c}{\partial x_j}}_{\text{chemotactic influence}} - \underbrace{\rho_0 \frac{\partial}{\partial x_i} n \frac{\partial f}{\partial x_j}}_{\text{haptotactic influence}} \quad (6.16)$$

$$\frac{df}{dt} = \underbrace{\omega n}_{\text{fibronectin production}} - \underbrace{\mu n f}_{\text{fibronectin loss due to cell binding / degradation}}$$

$$\frac{df}{dt} = \underbrace{-\lambda n c}_{\text{cell uptake of VEGF}}$$

where n is the population of endothelial cells, D is the diffusion coefficient describing random cell motion, c is the concentration of the chemotactic factor like VEGF, f is the concentration of the haptotactic factor like fibronectin, χ_0 is a constant describing the influence of the chemotactic factor on cell motion, ρ_0 is a constant describing the influence of haptotactic factor on cell motion, and ω, μ and λ are constants to be determined by experiment. Models such as these could form the basis for predicting cell infiltration and tissue production for any scaffold based tissue regeneration. The key of course, will be determining the matrix and cytokine production parameters as well as diffusion of these parameters from the scaffold to validate the model.

Kelly and Prendergast (2005; 2006) used a model very similar to the angiogensis diffusion models, with the exception that cell proliferation and death were assumed to be influenced by mechanical stimulus rather than chemotactic and haptotactic factors. They used the following equations in their model:

$$\frac{dn^k}{dt} = \frac{\partial}{\partial x_i} D_{ij}^k \frac{\partial n^k}{\partial x_j} + P^k(S) n^k - A^k(S) n^k \qquad (6.17)$$

$$P^k(S) n^k - A^k(S) n^k = a^k + b^k S_0 + c^k S_0^2$$

$$S = \frac{S_0}{\alpha} + \frac{v}{\beta}$$

where n^k is the number of a particular cell population (osteocyte, chondrocytes, fibroblast), D_{ij}^k is the random cell motility coefficient, $P^k(S)$ is a proliferation constant that depends on mechanical stimulus S, $A^k(S)$ is an apoptosis constant that also depends on mechanical stimulus, So is the octahedral shear strain, v is fluid velocity in the tissue, and a^k, b^k, c^k, α, β are empirical coefficients that must be estimated by experiment. Kelly and Prendergast used this simulation to predict optimal properties of a biphasic cartilage scaffold, estimating that a Young's modulus of 5MPa and a permeability of 1.0^{-15} m^4/Ns would lead to the most cartilage formation.

Adachi et al. (2006) combined their model of scaffold degradation with a mechanobiologic model of bone regeneration. In the bone regeneration model, it was assumed that trabecular bone seeks to produce uniform stress under applied load. Rate of bone gain or loss is determined by how far the bone stress deviates from uniformity. The governing equations for this model are:

$$\Gamma = \ln \left(\frac{\sigma_c \int_A w(l) \, dA}{\int_A \sigma_r w(l) \, dA} \right) \qquad (6.18)$$

$$\dot{M} = \begin{cases} < 0 \; if \; \Gamma < \Gamma_l \\ = 0 \; if \; \Gamma_l \leq \Gamma \leq \Gamma_l \\ > 0 \; if \; \Gamma > \Gamma_u \end{cases}$$

where σ_c is the stress a point on the bone surface, σ_r is the stress at a neighboring point, Γ is a parameter characterizing stress uniformity, Γ_L is a lower bound on stimulus below which bone is resorbed, Γ_U is the upper bound above which bone is added, and M describes the position of the bone surface, with $\dot{M}$ describing the velocity of surface motion in the direction normal to the surface. Using this combined model, Adachi et al. (2006) predicted simultaneously scaffold degradation and bone regeneration on a unit cell model (Fig. 6.4)

Adachi et al. (2006) suggested a scaffold design criteria where the total strain energy of the scaffold should remain as stable as possible through scaffold degradation and bone regeneration. Their results suggested that 60 to 64% porosity was optimal to minimize changes in mechanical strain energy throughout scaffold degradation and bone regeneration (Fig. 6.5).

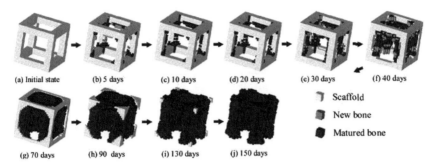

Fig. 6.4 Simultaneous prediction of scaffold degradation and bone formation on a unit cell of scaffold architecture over 150 days by Adachi et al. (2006). Grey voxels are scaffold which starts to degrade at 40 days and is gone by 130 days. Black represents bone formation which begins at 5 days and is finished by 150 days (Reprinted with permission [1])

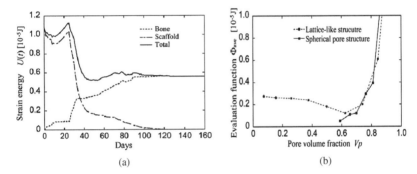

Fig. 6.5 (a) Mapping of overall construct strain energy density (black line), scaffold strain energy density (dashed line) and bone strain energy density (dotted line) for architecture in Fig. 6.4. (b) Overall variation in construct strain energy density over 150 days for a lattice scaffold architecture as shown in Fig. 6.4 and a spherical pore architecture. If the change in time weighted strain energy is to be minimized, then the computational model suggests that a scaffold volume fraction of 60% is optimal (Reprinted with permission [1])

Results of Adachi et al. (2006) demonstrate that scaffold performance in terms of load bearing and tissue regeneration can be designed for time dependent goals in addition to initial trade-offs between function and mass transport.

6.5 Summary

Scaffolds play a critical role in tissue regeneration. The ability to effectively design scaffolds and simulate their performance over time would greatly enhance tissue engineering therapies. To date, significant progress has been made in designing scaffold architectures to achieve desired physical characteristics. To a lesser degree, models have been introduced to simulate time dependent scaffold processes, either scaffold degradation or tissue growth, and in one case (Adachi et al., 2006) coupled scaffold degradation and tissue regeneration. Since functional animal trials and human clinical trials are very expensive, the ability to computationally assess how scaffold/biologics combinations influence tissue regeneration would make a substantial contribution to the field.

However, a significant amount of research remains to be done. First, models such as those reviewed here for designing initial scaffold characteristics and simulating time dependent performance must be more tightly integrated. For example, one could foresee designing a scaffold for a given application to achieve desired functional and mass transport characteristics, and then simulating how the scaffold degrades and the tissue grows over time. Thus, before this scaffold is even implanted in an animal, the computational design/simulation would provide insight into how the scaffold would perform.

Second, computationally designed/simulated scaffolds must be fabricated with high fidelity to the design and tested in a variety of in vitro and in vivo models under controlled conditions. Such experiments require the tight integration of solid free-form fabrication (SFF) techniques with computational design to build scaffolds having complex 3D architectures from a wide range of biomaterials (Hollister et al., 2002; Hollister 2005; Hutmacher et al., 2004). These experiments would serve to provide unknown parameters, verify and thereby help improve the computational models, which in turn would stimulate further experiments and help bring tissue engineering therapies into the clinic.

Acknowledgment Some work presented in this chapter was supported by NIH R01 DE 13608.

References

Adachi T, Osako Y, Tanaka M, Hojo M, Hollister SJ (2006) Framework for optimal design of porous scaffold microstructure by computational simulation of bone regeneration. *Biomaterials* 27:3964–3972

Anderson ARA, Chaplain MAJ (1998) Continous and discrete mathematical models of tumor-induced angiogenesis. *Bull Math Biology* 60:857–900

Aoubiza B, Crolet JM, Meunier A (1996) On the mechanical characterization of compact bone structure using the homogenization theory. *J Biomech* 29:1539–1547

Bendsoe M, Guedes JM, Plaxton S, Taylor JE (1996) Optimization of structure and material properties composed of softening material. *Int J Solids Structures* 33:1799–1813

Göpferich A (1997) Polymer bulk erosion.*Macromolecules* 26:2598–2604

Göpferich A, Langer R (1993) Modeling of polymer erosion. *Macromolecules* 26:4105–4112

Guedes JM, Kikuchi N (1990) Preprocessing and postprocessing for materials based on the homogenization method with adaptive finite-element methods. *Com Meth App Mech Eng* 83:143–198

Guest JK, Prevost JH (2006) Optimizing multifunctional materials: design of microstructures for maximized stiffness and fluid permeability. *Int J Solids Structures* 43:7028–7047

Hollister SJ, Fyhrie DP, Jepsen KJ, Goldstein SA (1991) Application of homogenization theory to the study of trabecular bone mechanics. *J Biomech* 24:825–839

Hollister SJ, Kikuchi N (1992) A comparison of homogenization and standard mechanics analyses for periodic porous composites. *Comput Mech* 10:73–95

Hollister SJ, Brennan JM, Kikuchi N (1994) A homogenization sampling procedure for calculating trabecular bone effective stiffness and tissue level stress. *J Biomech* 27:433–444

Hollister SJ, Levy RA, Chu TM, Halloran JW, Feinberg SE (2000) An image-based approach for designing and manufacturing craniofacial scaffolds. *Int J Oral Maxillofac Surg* 29:67–71

Hollister SJ, Maddox RD, Taboas JM (2002) Optimal design and fabrication of scaffolds to mimic tissue properties and satisfy biological constraints. *Biomaterials* 23:4095–4103

Hollister SJ (2005) Porous scaffold design for tissue engineering. *Nat Mater* 4:518–524

Hollister SJ, Lin CY (2007) Computational design of tissue engineering scaffolds. *Com. Meth App Mech Eng* 196: 2991–2998

Hutmacher DW, Sittinger M, Risbud MV (2004) Scaffold-based tissue engineering: rationale for computer-aided design and solid free-form fabrication systems. *Trends Biotechnol* 22:354–362

Kelly DJ, Prendergast PJ (2005) Mechano-regulation of stem cell differentiation and tissue regeneration in osteochondral defects. *J Biomech* 38:1413–1422

Kelly DJ, Prendergast PJ (2006) Prediction of the optimal mechanical properties for a scaffold used in osteochondral defect repair. *Tissue Eng* 12:2509–2519

Lin CY, Kikuchi N, Hollister SJ (2004a) A novel method for biomaterial scaffold internal architecture design to match bone elastic properties with desired porosity. *J Biomech* 37:623–636

Lin CY, Hsiao CC, Chen PQ, Hollister SJ (2004b) Interbody fusion cage design using integrated global layout and local microstructure topology optimization. *Spine* 29:1747–1754

Lin CY, Lin CY, Hollister, SJ (2004c) A New Approach For Designing Biodegradable Bone Tissue Augmentation Devices By Using Degradation Topology Optimization, *Proceedings of the 8th World Multiconference on Systemics Cybernetics and Informatics*

Sanchez-Palencia E (1980) *Non-homogeneous media and vibration theory.* Springer, Berlin

Sigmund O (1994) Construction of materials with prescribed constitutive parameters: an inverse homogenization problem. *Int J Solids Structures* 31:2313–2329

Sun W, Darling A, Starly B, Nam J (2004a) Computer-aided tissue engineering: overview, scope and challenges. *Biotechnol Appl Biochem* 39:29–47

Sun W, Starly B, Darling A, Gomez C. (2004b) Computer-aided tissue engineering: application to biomimetic modelling and design of tissue scaffolds. *Biotechnol Appl Biochem* 39:49–58

Wettergreen MA, Bucklen BS, Sun W, Liebschner MA (2005) Computer-aided tissue engineering of a human vertebral body. *Ann Biomed Eng* 33:1333–1343

Yin L, Elliot DM (2005) A homogenization model of the annulus. *J Biomech* 38:1674–1684

Chapter 7
Virtual Prototyping of Biomanufacturing in Medical Applications

Conventional manufacturing processes for three-dimensional scaffolds

Y. S. Morsi, C. S. Wong and S. S. Patel

7.1 Introduction

Tissue engineering (TE) is a multidisciplinary field that combines the principles of engineering and biological sciences to contribute to the development of tissues and organs for the purpose of regeneration, repair or replacement. Although TE was a term that was used and gained popularity in 1987 at the National Science Foundation, the concept of TE had been explored as early as the 1930s whereby living cells and tissues were maintained in a bioreactor by Carrel and Lindbergh (Nerem 2006). Langer and Vacanti, eminent researchers of TE, described a technique using bioabsorbable synthetic polymer as matrices for cell transplantation in 1988 and this formed the basis for the concept of TE (Vacanti et al. 1988). Since isolated cells cannot form new tissue on their own, three-dimensional (3D) scaffolds are required to provide architectural support for the cells and define the anatomical shape of the tissue (Yang et al. 2001).

The techniques employed for manufacturing 3D scaffold for TE has undergone substantial changes in the last decade and these techniques are continuously evolving to accommodate the specific requirements of individual organ or tissue of interest such as pore size and interconnectivity (Buckley and O'Kelly 2004). The techniques for 3D scaffold fabrication can be divided into two broad categories; "conventional methods" such as solvent-casting/particulate-leaching and newer manufacturing technologies such as rapid prototyping. Continuous research into both approaches will lead, no doubt, to dramatically new methodologies, techniques and processes. A critical step in generating an optimum design of scaffolds will be the development of an underlying technical foundation through research by industry, academia and government institutions. This must be guided by

---

Y.S. Morsi
Biomechanics and Tissue Engineering Group, FETS, Swinburne University of Technology, Hawthorn, Victoria, Australia.

understanding the fundamental challenges, which need to be fulfilled in order for the vision to be realised.

In this chapter, we will present and discuss the progress of "conventional methods" that have been used to construct 3D porous scaffolds for TE applications. The use of rapid prototyping techniques which utilises the integration of medical imaging modalities (CT or MRI) for the acquisition of anatomic structural data and 3D CAD modelling for designing and creating the digital scaffold models are beyond the scope of this chapter.

7.2 The Ideal Scaffold and Materials

The selection of techniques for manufacturing 3D scaffolds is dependent on the optimal scaffold required for the application in hand. In general, the ideal scaffold should possess these characteristics:

- **rate of biodegradation** – the rate of scaffold degradation need to be similar to the rate of tissue growth so that the end product will only contain native tissue
- **good surface chemistry** – the surface should be conducive to cell growth, promoting cellular adhesion, migration and differentiation so that the resultant tissue is comparable to the surrounding native tissue with respect to structure and functionality
- **correct pore size and interconnectivity** – this is essential for encouraging cell growth in the correct spatial orientation, vascularisation and nerve innovation. Additionally, it enables the delivery of nutrients and waste removal.
- **optimal structure and volume** – the structure should have a high surface area to volume ratio so that a high density of cells can be embedded. This is because a large amount of cells is required to generate a small volume of tissue (approximately 20 million cells to create 1 cm^3)
- **ease of processing** – the scaffold should be easily processed and cost effective to manufacture

The design of an ideal scaffold has to be complemented by the selection of suitable materials. The types of materials generally used in TE include natural polymers such as collagen, fibrin and alginate, synthetic polymers such as polylactic acid (PLA) and polyglycolic acid (PGA), ceramics such as hydroxyapatite, and metals such as stainless steel and titanium-based alloys. The physical characteristics of these materials vary from one type to another and have specific features that lend themselves to particular applications. There may be a need to combine natural and synthetic materials (biosynthetic or semi-synthetic) so that it can be tailored to a specific application or improve a specific function of the scaffold such as mechanical integrity. Currently, research into biosynthetic materials is an emerging field and a common approach is to crosslink biological components such as fibronectin and extracellular matrix proteins to polyethylene glycol (PEG) (Rosso et al. 2005). However, irrespective of the type of material used and application, the chosen materials should have the following general characteristics:

- **biocompatibility** – prevents scaffold rejection and inflammatory response
- **biodegradability** – the material should degrade or resorbed completely so that only the newly generated tissue remains
- **ease of modification** – permits modification to the material such as incorporation of biological molecules such as growth factors, peptides and drugs. This may facilitate the cellular interaction between the implanted scaffolds and the surrounding host tissue, enhancing the remodelling and/or regeneration process
- **structural stability** – allows the construct to be handled safely and easily during implantation as well as ensuring that while implanted, toxic by-products are not produced
- **malleability** – enables the complex anatomy of interested organ to be reproduced, resulting in better outcomes (functional and aesthetic) and possibly reducing the number of surgical procedures
- **versatility** – enables the material to be applicable for various medical purposes
- **efficiency** – the material should be easily accessed and available at a low cost

The conventional techniques that have been developed for producing porous scaffolds are presented and discussed below. These conventional methods are defined herein as processes that create scaffolds having a continuous, uninterrupted pore structure that lacks any long-range channelling micro-architecture.

7.3 Solvent-Casting Particulate-Leaching

The technique of solvent-casting particulate-leaching (SC/PL) is based on a simple principle. Polymer such as poly-L-lactide (PLLA) or polylactic-co-glycolic acid (PLGA) is dissolved in a solvent, for example, chloroform or methylene chloride in a petri dish, and is then mixed with the porogen (Fig. 7.1). The porogen used

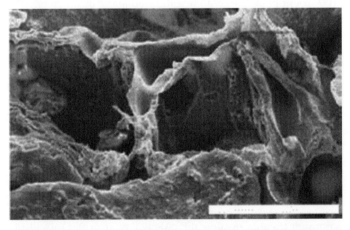

Fig. 7.1 Poly DL-lactide scaffold created by solvent casting/particulate leaching technique. The size of the salt crystals range from 200–400 µm. Scale bar = 300 µm. *Reproduced with kind permission from Springer Science and Business Media, Draghi L et al., Microspheres leaching for scaffold porosity control, J Mater Sci Mater Med (2005); 16(12):1093–97*

is usually salt particles of a calibrated size such as sodium chloride, tartrate and citrate. The solvent is then left to completely evaporate from the polymeric mixture. The porogen can be removed from the composite via immersion in water for 2 days. After this process, the porosity of the scaffold can be easily determined by the amount and size of the salt crystals (Mikos et al. 1994). The scaffolds fabricated using SC/PL have been extensively used with various cell types and does not have any adverse effects on new tissue formation (Freed et al. 1993; Mikos et al. 1993b; Ishaug et al. 1997; Ishaug-Riley et al. 1998; Goldstein et al. 1999, Shastri et al. 2000; Kim et al. 2003; Lee et al. 2003; McGlohorn et al. 2004; Draghi et al. 2005).

Although this process is simple, cheap and capable of producing scaffolds of a wide range of pore sizes, it suffers from poor control over internal structure of scaffold and has limited interconnectivity. Moreover, the side of scaffold "foam" that is exposed to air produces a rougher surface characteristic than that exposed to the petri dish. However, further research on modifications to this technique is being conducted. Goldstein et al. (1999) proposed a modified technique whereby pieces of the polymer/salt composite (PLLA/PLGA) were moulded into cylindrical form at temperatures just above the melting temperature of PLLA or glass transition temperature of PLGA. The cylinder was cut into discs of desired thickness before undergoing leaching in water. As a result, a more precise control of scaffold thickness and uniformity of the foam surface can be achieved. However, the limitation of this modification was the effect of the thermal degradation of the polymer during the compression-moulding step. Shastri et al. (2000) offered an alternative fabrication method for PLLA and PLGA scaffolds by using waxy hydrocarbons as porogens. The porogen and polymer, dissolved in methylene chloride or chloroform, was mixed to form a paste. This composite was then packed in a Teflon mold and immersed in a hydrocarbon solvent (pentane or hexane) to remove the wax without dissolving the PLLA/PLGA. The remaining foam was vacuum-dried for several days to extract any solvents. Samples having thickness up to 2.5 cm with interconnected pores could be produced using this technique. Porosity up to 87% and pore size well over 100 μm in diameter could be produced using this method. The strength or electrical conductivity of the final structure can be increased by the addition of particulate phase to the paste. It was observed that when blended with PEG and seeded with bovine chondrocytes for 4 weeks, there was formation of cartilage-like tissue in these foams, demonstrating biocompatibility (Shastri et al. 2000).

In all of the above procedures, organic solvents are used and thus exclude the possibility of adding pharmacological agents to the scaffold during fabrication. The scaffold preparation time for water-soluble porogens is significantly increased by the leaching step.

7.4 Gas Foaming

The gas foaming technique requires the polymer such as PLGA and PLLA to be initially melted with an extrusion machine (Mooney et al. 1996). After that, gas such as carbon dioxide (CO_2), also known as the blowing agent, is added at high pressure

Fig. 7.2 Scanning electron micrographs of hydroxyapatite scaffolds produced by gas foaming technique. *Reprinted from Biomaterials, 25(17), Almirall A et al., Fabrication of low temperature macroporous hydroxyapatite scaffolds by foaming and hydrolysis of an a-TCP paste, 3671–3680, 2004, with permission from Elsevier*

(approximately 800 psi) to saturate the material (Lips et al. 2005). The high pressure CO_2 is then brought back to atmospheric level so that the solubility of the gas in the polymer can be decreased rapidly. Expansion occurs when the polymer/gas mixture is exposed to atmospheric pressure (Biron 2005) or high temperature (Lips et al. 2005). As the solubility of the gas decreases, the mixture is quenched so that the physical state of the material can be set. This produces nucleation and gas bubbles of various sizes ranging from 100–500 μm within the polymer. The nucleation of the material can be controlled by varying the rate of gas diffusion into the polymeric mixture before quenching occurs. Other factors that play a role include temperature, degree of saturation, hydrostatic pressure, interfacial energy and visco-elastic properties of the material/gas mixture (Lips et al. 2005). The stiffness of the material, on the other hand, can be manipulated by the foaming temperature (Lips et al. 2005). The gas foaming technique can create scaffolds with pore sizes of 200–300 μm (Fig. 7.2) and porosities of up to 93% (Mikos and Temenoff 2000). In addition, this method does not have a leaching step and uses no harsh chemical solvents. However, the pores on the surface of these foams are largely unconnected. Also the use of high temperature prevents the incorporation of cells or bioactive molecules.

7.5 Gas Foaming and Particle Leaching (GF/PL)

Particulate leaching with high pressure gas foaming (GF/PL) has made it possible to fabricate uniform, interconnected pore matrices. This combined method proposed by Harris et al. (1998) yields a highly porous, closed pore matrix. As discussed previously, the mechanism of pore formation with gas foaming involve the formation of the polymer gas solution, the creation of pore nucleation, as well as pore growth and density reduction. The polymer/gas solution is formed by subjecting the polymer to high pressure CO_2 gas and allowing it to saturate with gas in 2 days. Thermodynamic instability is achieved by bringing the gas pressure to ambient. Diffusion of gas in areas of the polymer adjacent to the nucleation sites causes the growth of the pores and the gas phase separates from the polymer through pore nucleation. The pore growth in turn reduces the polymer density. However, this leads to a predominantly closed-pore structure (Harris et al. 1998). The combined technique of GF/PL uses sodium chloride (NaCl) as the particulate and as the polymer/gas solution expands during foaming, the polymer particles in the compressed polymer/NaCl disc fuse together, leading to a continuous polymer matrix. The subsequent leaching of the salt particles produces an interconnected pore structure. The ratio of NaCl and the polymer, for example, PLGA determines the porosity of the matrix and the size of the NaCl particles determines the pore size. This approach provides structurally stable matrices with very high porosities of up to 97%. Matrices created by GF/PL exhibit more uniform pore structure and enhanced mechanical properties than those created using SC/PL. This is due to the polymer particles foaming uniformly throughout the salt bed when the GF/PL technique is employed whereas an uneven distribution of polymer and NaCl is produced by the SC/PL technique. Although matrices fabricated by GF/PL are stiffer than those by SC/PL, GF/PL matrices have a lower elongation at break. Unlike SC/PL matrices there is no residual chloroform and thus the polymer here is less ductile (Harris et al. 1998).

Scaffolds produced by GF/PL have been shown to support cellular growth such as smooth muscle cells, which adhered to the matrices and proliferated within the 3D porous structures (Harris et al. 1998). The in vitro study showed that cell growth was concentrated on the periphery of the scaffold. This may be due to a reduction in the transport of nutrient to the center of the matrices. While matrices for a variety of cell such as hepatocytes, chondrocytes and osteoblasts can be achieved by other techniques (SC/PL and phase separation), GF/PL fabricated matrices will probably produce better outcomes due to their enhanced mechanical properties (Harris et al. 1998). A study using rat liver cells demonstrated 40% viability after 1 week in culture (Nam et al. 2000), indicating the biocompatibility of the scaffold. The putty-like consistency of the polymer-salt mixture can also be used to mould constructs for surgery. However, scaffolds created by GF/PL may not be presently suitable for TE applications as the use of organic solvents and long-term effects of residues of ammonium bicarbonate on cells poses safety concerns (Mikos and Temenoff 2000).

7.6 Phase Separation/Emulsification

The techniques used for fabrication of porous scaffolds in this section are based on the concepts of phase separation rather than the incorporation of a porogen (Fig. 7.3). These techniques include emulsification/freeze-drying (Whang et al. 1995) and liquid-liquid phase separation (Lo et al. 1995; Lo et al. 1996; Schugens et al. 1996; Nam and Park 1999a; 1999b) which are briefly discussed below.

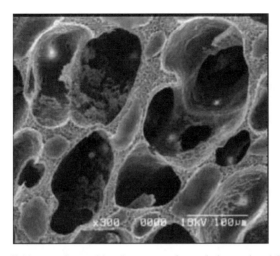

Fig. 7.3 PLLA scaffolds manufactured by phase separation technique using 4.5%wt PLLA solution. *Reprinted from Biomaterials, 25(12), Kim H. et al., Effect of PEG–PLLA diblock copolymer on macroporous PLLA scaffolds by thermally induced phase separation, 2319–2329, 2004, with permission from Elsevier*

7.6.1 Phase Separation

The phase separation technique can be used to incorporate biological components into a scaffold. This involves dissolving biodegradable synthetic polymer in molten phenol or naphthalene, followed by the addition of bioactive molecules such as alkaline phosphatase (Lo et al. 1995). A liquid-liquid phase separation is formed by reducing the temperature and then quenched to produce a two-phase solid. Once the solvent is removed via sublimation, a porous scaffold consisting of bioactive molecules within its structure will be produced.

7.6.2 Emulsification/Freeze-drying

The process of emulsification/freeze-drying requires the addition of ultrapure water to a solvent/polymer solution. Based on a method by (Whang et al. 1995) an emulsion is produced by dissolving polymer such as PLGA in methylene chloride and mixing the solution with distilled water. The polymer/water mixture is then cast into a mould and quenched in liquid nitrogen. Freeze-drying the scaffolds at $-55\,°C$ resulted in the removal of dispersed water and polymer solvents.

This technique has been used to fabricate scaffolds with porosities up to 95% and pore sizes of 13–35 µm. These characteristics are dependent on parameters such as the ratio of polymer solution to water and viscosity of the emulsion as they affect the stability of the emulsion prior to quenching (Whang et al. 1995). Therefore, the pore size can be altered by modifying the before mentioned factors. The emulsification/freeze-drying technique is simplified as an extra washing/leaching step has been omitted. However, as with the GF/PL technique, the concern of using organic solvents as part of the manufacturing process is warranted (Mikos and Temenoff 2000). A modification of the technique such as types of solvent, solvent/nonsolvent ratio, types of polymer and thermal quenching strategy may eventually produce a porous scaffold that can be applicable in the field of tissue engineering (Yang et al. 2001).

7.6.3 Freeze-drying

Scaffolds produced by freeze-drying are obtained by initially dissolving synthetic polymers in glacial acetic acid or benzene. The polymeric solution is then frozen and freeze-dried, resulting in a porous matrix, as shown in Fig. 7.4. The technique of freeze-drying can also be used to create scaffolds made from natural polymers such as collagen (Yannas et al. 1980). The freezing of the collagen solution causes ice crystals to form and this in turn forces the collagen molecules to aggregate into the available interstitial spaces. The frozen collagen is then dehydrated by ethanol and ice crystals are removed by freeze-drying, producing pores within the scaffold. The pore size is determined by the freezing rate and pH whereby a high freezing rate produces smaller pores (Dagalakis et al. 1980; Doillon et al. 1986; Sachlos and Czernuszka 2003). A homogenous 3D-pore structure can be created using unidirectional solidification (Schoof et al. 2000; 2001). Other natural polymers such as chitin (Madihally and Matthew 1999) and alginate (Glicklis et al. 2000) have also been fabricated into scaffolds using freezing-drying (Sachlos and Czernuszka 2003). These natural polymeric scaffolds can be crosslinked to reduce their antigenicity, solubility and degradability. Crosslinking is achieved either by physical

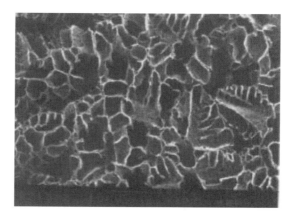

Fig. 7.4 Scanning electron micrographs of PLLA scaffold prepared using freeze-drying method. *Reprinted from Biomaterials, 25(1), Ho M. et al., Preparation of porous scaffolds by using freeze-extraction and freeze-gelation methods, 129–138, 2004, with permission from Elsevier*

or chemical means. Physical crosslinking utilises ultraviolet (Miyata et al. 1971), gamma irradiation (Miyata et al. 1980) or dehydrothermal treatment (Weadock et al. 1983–84; Thompson and Czernuszka 1995) while chemical crosslinking is usually achieved by using glutaraldehyde (Ruijgrok et al. 1994), hexamethylene diisocyanate (Olde Damink et al. 1995) or by carboxyl group activation with carbodiimides (Weadock et al. 1983–84).

7.6.4 Liquid-liquid Phase Separation

The liquid-liquid phase separation method is based on the thermodynamic principles of creating polymer-rich and polymer-poor phases within a polymer solution. The removal of the polymer poor phase creates a highly porous polymer network. PLLA and PLGA scaffolds have been produced with this technique (Lo et al. 1995; 1996; Schugens et al. 1996; Nam and Park 1999a; 1999b).

This fabrication technique utilises solvents that have low melting point and are easy to sublime such as naphthalene, phenol or 1,4 dioxane to dissolve polymers. A small amount of water may be required to act as a non-solvent to induce phase separation (Schugens et al. 1996; Nam and Park 1999a; 1999b). The polymer solution is cooled below the melting point of the solvent (polymer poor phase) and vacuum dried for several days so that the solution undergoes complete solvent sublimation. The cooling parameters for the solution are important as they determine the morphology of the resultant scaffold. Temperatures just below the critical temperature (or cloud point for polydisperse polymers) cause phase separation to occur via a nucleation and growth mechanism. Temperatures beneath the spinodal curve of a phase diagram cause the separation to occur via spinodal decomposition. The separation mechanism influences the domain formation in that nucleation and growth mechanism leads to spheroidal domains and spinodal decomposition causes the formation of interconnected cylinders. Spinodal decomposition may be the preferred mechanism for TE applications as it produces scaffold with good interconnectivity within the network (Nam and Park 1999a). Domains formed by either mechanism can be enlarged by annealing (Sperling 1992). There are various parameters that influence the phase diagram of the system. These factors include polymer concentration, solvent/non-solvent ratio, cooling method and time and the presence of surfactants. Surfactants have been known to reduce the interfacial tension between phases and increase pore size and interconnectivity (Nam and Park 1999a; 1999b). This method has produced scaffolds with porosities of up to 90% and pores of approximately 100 µm. Such technique has been used to create scaffolds which have the potential to support nerve tissue engineering (Fig. 7.5). Although the use of harsh organic solvents is again a disadvantage, this technique warrants further investigation because modifications to some of the parameters may lead to the creation of scaffolds with desired pore size and porosity, which can be beneficial in biomedical applications. A better understanding and characterization of the phase diagrams of the polymer-solvent systems should also be gained with respect to tissue-engineered constructs, (Mikos and Temenoff 2000).

Fig. 7.5 Scaffolds produced by liquid-liquid phase separation using varying concentrations of poly L-lactide solutions at (a) 3% (w/v), (b) 7% (w/v) and (c) 9% (w/v). *Reprinted from Biomaterials, 25(10), Yang F et al., Fabrication of nano-structured porous PLLA scaffold intended for nerve tissue engineering, 1891–1900, 2004, with permission from Elsevier*

7.7 Melt Molding

The meld moulding technique utilises a teflon mould and fills it with polymer such as PLGA in the form of powder and gelatin microspheres of specific diameter. The mould is then heated above the glass-transition temperature of PLGA while applying pressure to the mixture (Thomson et al. 1995a). This treatment enables the PLGA particles to bond together. The gelatin component is leached out by immersing in water and the scaffold is then dried. A porous scaffold is produced (Fig. 7.6) with its shape determined by the shape of the mold.

Melt moulding can also be utilised to incorporate carbohydrate polymers such as hyaluronic acid (HA) into PLGA scaffolds. The process needs to be modified as a uniform distribution of HA fibers throughout the PLGA scaffold is required and can only be achieved by the solvent-casting technique. This is so that a composite material of HA fibers, PLGA matrix and gelatin or salt porogen can be produced and used in the melt molding process (Thomson et al. 1995; Sachlos and Czernuszka

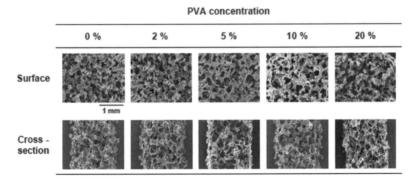

Fig. 7.6 PLGA scaffolds with varying concentrations of polyvinyl alcohol produced using melt molding particulate leaching method. *Reprinted from Biomaterials, 24(22), Oh S et al., Fabrication and characterization of hydrophilic poly(lactic-coglycolic acid)/poly(vinyl alcohol) blend cell scaffolds by melt-molding particulate-leaching method, 4011–4021, 2003, with permission from Elsevier*

2003). The pore size is determined by the microsphere diameter and the general porosity changes with the polymer/gelatin ratio. Other polymeric scaffolds can also be manufactured by substituting PLGA with L-PLA and PGA. However, some parameters will need to be adjusted as higher temperature is required to melt these semicrystalline polymers (Yang et al. 2001). Although melt molding enables macro shape control and independent control of porosity and pore size, high temperatures are required and the scaffold produced contains residual porogens. Additionally, this method does not allow scaffold with tailored specifications to be made as a computer-aided design (CAD) model cannot be used to produce it.

7.8 Compression Molding–particle Leaching

Compression molding is a technique commonly used to shape plastic products with complicated geometry and particulate leaching is a preferred pore-forming method. Thus porous scaffolds with complicated geometry may be fabricated a combination of these two methods (Wu et al. 2005). Blending together a starch-based polymer (in the powder form) and leachable particles such as salt particles of different sizes is the basis of the compression molding and particle leaching method (Fig. 7.7). This amount should be sufficient to provide a continuous phase of a polymer and a dispersed phase of leachable particles in the blend. The blend is then compress molded into a desired shape and leachable particles can be removed from the scaffold by immersion in distilled water (Hou et al. 2003). Wu et al. (2005) have developed a method to fabricate three-dimensional porous scaffolds with auricle-like shape from PDLLA and PGLA by this technique. Porosity as high as 95% was achieved with satisfactory mechanical properties. Good viability of 3T3 fibroblast cells seeded in these porous scaffolds was also observed suggesting the biocompatibility of these scaffolds (Wu et al. 2005).

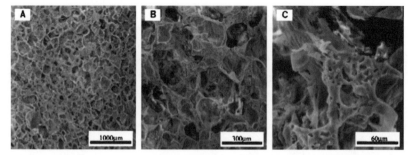

Fig. 7.7 Poly DL-lactide scaffold with pore sizes ranging from 280–450 μm at various magnifications (A–C), produced by compression molding. *Reproduced with kind permission from Wiley Interscience, Jing D et al., Solvent-assisted room-temperature compression molding approach to fabricate porous scaffolds for tissue engineering, Macromol Biosci.(2006); 6(9): 747–57*

Fig. 7.8 An image of polyurethane fibre mesh taken using scanning electron microscopy. Bar = 200 mm. *Reprinted from Biomaterials, 26(7), Sanders J et al. Fibro-porous meshes made from polyurethane micro-fibers: effects of surface charge on tissue response, 813–818, 2005, with permission from Elsevier*

7.9 Fibre Meshes

The textile technology is employed to produce non-woven scaffolds from polymers such as PGA and PLLA (Cima et al. 1991). Fiber meshes are made up of individual fibers, which are either woven or knitted into three-dimensional patterns of variable pore size (Fig. 7.8). Scaffolds produce with this technique presents a large surface area for cell attachment and enables rapid diffusion of nutrients, thus, encouraging cell growth and proliferation. However, fibers meshes may not be suitable for TE applications as these scaffolds lack structural stability, often resulting in significant deformation. The mechanical properties of such scaffolds can be enhanced by fibre bonding technique as described by Mikos et al. (1993b).

7.10 Fibre Bonding

Similar to fibre meshes, the polymer fibres form three-dimensional structure, which encourages cell interaction and growth with the exception that these fibres are bound to produce an interconnected fibre network. There are two different methods for bounding fibres as reported in the literature (Mikos et al. 1993a). The first method immerses polymer fibres, for example, PGA, in a PLLA solution and as the solvent evaporates, a network of PGA fibres embedded in PLLA is produced. The composite is heated above the melting temperature of both polymers and as the PLLA melts, spaces produced by the fibres will be filled. This step resulted in maintaining the spatial arrangement of fibres and preventing the fibre structure from collapsing when the PGA begins to melt. The fibres welded together at the cross-points to minimise interfacial energy and forms a highly porous scaffold. The PLLA component of the scaffold is then removed by methylene chloride. Scaffolds with porosities as high as 81% and pore diameters of up to 500 μm are formed via this fabrication technique.

Hepatocytes cultured in these matrices for 1 week remained viable and were able to form clusters (Mikos and Temenoff 2000).

Another method of fibre bonding utilises atomisation of PLLA or PLGA to coat the fibres. PLLA or PLGA is dissolved in chloroform and sprayed onto the PGA fibres (Mooney et al. 1996). The fibres will not be affected during this process as PGA is only weakly soluble in chloroform. As the solvent evaporates, the fibres adhered to the PLLA or PLGA. Matrices created by this technique produce pore sizes that are similar to those of the aforementioned technique. Tubes manufactured using this method demonstrated fibrous tissue ingrowth in vivo after 17 days of implantation (Mooney et al. 1996). This suggests that the physical properties of such a construct can promote neo-tissue formation.

The fiber bonding techniques produce highly porous scaffolds with interconnected pores that are suitable for tissue regeneration (Freed et al. 1993; Mikos et al. 1993a; Mooney et al. 1996). On the other hand, both methods described in this section uses solvents that may be harmful and toxic to cells, and the porosity and pore sizes can be hard to control. The solvents can be removed by drying the scaffold in vacuum for several hours. The completion of this step enables the scaffold to be used immediately, rendering it to be a favorable option in surgeries. However, aside from toxic chemicals, another limitation of fiber bonding is the extreme temperature used in the first method, which may not be feasible if bioactive molecules such as growth factors are to be included in the scaffold during processing. Additionally, there is little flexibility in using other polymers in this technique as the choice of solvent, immiscibility of the two polymers, and their relative melting temperatures is quite restrictive.

7.11 Current Research

The "conventional" fabrication techniques discussed in this chapter are still being used in the present day for biomedical applications as shown in Table 7.1. Most of the studies were conducted in vitro using a variety of cells such as stem cells, hepatocytes, osteoblasts and smooth muscle cells. Since there are both advantages and disadvantages in manufacturing 3D scaffolds with these methods, some studies have modified the techniques to suit their applications. The modifications include incorporation of extracellular matrix proteins such as collagen and HA, addition of minerals and combining different types of polymer, all with the aim of producing scaffolds that are conducive to cells. While the results from in vitro studies are encouraging, such data are usually not reproduced in vivo. Thus far, there have only been a handful of studies that examined the biocompatibility of scaffolds produced by the discussed methods in vivo. The short-term results appear to be promising whereby PLGA scaffolds with extracellular matrix proteins fabricated by SC/PL techniques has been shown to promote epithelial membrane coverage over conjunctival wounds. However, further investigations will need to be conducted to assess the biosafety of these manufacturing processes such as cell viability and functionality, and toxic by-products.

Table 7.1 Selected tissue engineering applications for 3D scaffolds produced by conventional fabrication techniques

Process	Authors	Application	Findings	Validation
Solvent casting–particulate leaching	McGlohorn et al. (2004)	Rat aortic smooth muscle cells	Initial SMC attachment is better when NaCl is used as compared with glucose but those produced with glucose supported a higher cell population	In vitro
	Lee et al (2003)	PLLA scaffold		
		Corneal epithelial cells and human stromal fibroblasts	Enhanced adhesion and proliferation was observed in treated scaffold as compared to untreated scaffold	In vitro
		PLGA scaffold Modified by collagen, HA*, AM**		
	Lee et al. (2003)	PLGA/ collagen/ HA scaffolds grafted on conjunctival wounds on eyes of rabbit	Wounds were covered with epithelial membranes	In vivo
	Kim et al. (2003)	Mesenchymal stem cells	Amount of mineralisation was significantly higher and osteogenesis of mesenchymal stem cells was increased	In vitro
		PLGA scaffolds containing AsAP$^+$ & Dex^{++}		
	Shastri et al. (2000)	Bovine chondrocytes	Formation of cartilage-like tissue seen after 4 weeks	In vitro
		PLLA/PLGA + PEG		
	Draghi, L. et al. (2005)	Osteosarcoma MG63 cells	Cell viability was unaltered in presence of medium incubated with scaffolds up to 7 days	In vitro
		PLA scaffolds		

Method	Reference	Material/Application	Result	Study
Gas foaming – particulate leaching	Yoo et al. (2005)	Cartilage tissue	Scaffolds excelled in inducing cartilage tissue formation in terms of collagen type II expression and tissue morphological characteristics	In vitro
	Kim et al. (2005)	PLGA coated with HA* (PLGA/HA) Rat calvarial osteoblasts	Bone formation was observed	In vitro
	Nam and Park (1999)	PLGA/HA Rat hepatocytes	40% cell viability at 1 day of cell seeding	In vitro
	Harris, L. et al. (1998)	Smooth muscle tissue adherence and formation of 3 D tissues PLGA scaffolds	78% increase in cell density after 2 weeks	In vitro
	Montjovent et al. (2005)	PLLA scaffolds Fetal and adult bone cells	Cell spreading and proliferation was observed	In vitro
Fibre mesh	Mikos and Temenoff (2000)	Hepatocytes culturing	Cells remained alive for 1 week and began to interact with each other	In vitro
	Mooney et al. (1996)	PGA + PLLA Tubes for implantation in rats	Fibrous tissue ingrowth was observed	In vivo
Thermally induced phase separation (TIPS)	Day et al. (2004)	Tubes were implanted subcutaneously in adult male Lewis rats PLGA tubular foams	Cellular infiltration mainly of fibrovascular tissue was observed	In vivo

*HA = Hyaluronic acid
**AM = Human amniotic membrane
+AsAP = Ascorbate 2-phosphate
++Dex = Dexamethasone

Research into biosynthetic scaffolds is currently an emerging field (Rosso et al. 2005). Although PGLA and similar polymers have great potential, they are rigid or semi-rigid in nature and not 'user friendly' for clinical applications. Most of the scaffolds produced will elicit a foreign body response and surface modification of these scaffolds with biological components such as collagen is required to reduce or eliminate these immunological responses. There is a need for scaffolds to be used in TE to possess both the mechanical characteristics of synthetic polymers and the biofunctionality of natural materials.

A hydrophilic polymer known as hydrogels is attracting considerable interest as a potential material for fabricating scaffolds for TE. This is because hydrogels have similar structure to natural extracellular matrices and as 3D cross-linked structures, they exhibit good interconnected network pore structures with high pore to polymer volume ratio. The hydrophilic polymer chains can absorb a large amount of water, well beyond their own dry weight without dissolution producing similar characteristics to soft tissues in the human body. The ability to incorporate growth factors, enzymes and proteins adds to the appeal of using hydrogels as a scaffold for TE (Ratner and Bryant 2004; Rosso et al. 2005). These characteristics allow tissue structure to be controlled and cell function to be regulated. The high permeability of hydrogels also facilitates exchange of oxygen, nutrients and other water-soluble metabolites. Over the past decade, hydrogels have been widely used as materials for drug delivery (Gulsen and Chauhan 2004), contact lenses (Liu et al. 2006) and corneal implants (Ziegelaar et al. 1999), and more recently for applications of TE (Halstenberg et al. 2002; Lutolf et al. 2003; Almany and Seliktar 2005).

Since hydrogels can be made from various types of polymer, thus, requiring different solvents and chemicals relevant to the chosen polymer, a generalised two-step method has been described here. Functionalised hydrophilic and hydrophobic precursors are dissolved and a crosslinker that reacts with the functional groups of the precursors is added to the solution. An efficient crosslinking process will result in a good liquid-to-solid transition by minimising steric hindrances (Almany and Seliktar 2005). The polymeric solution is then polymerised via UV irradiation at 365 nm with the aid of a photoinitiator solution. This produces a network consisting of hydrophobic and hydrophilic portions that are interconnected. The solvent is then substituted with an aqueous solvent, usually water and this causes the network to swell due to the hydrophilic precursors. The hydrophobic component enhances the rigidity of the scaffold and the combined characteristics leads to a scaffold that is spongy and elastic with good mechanical strength (Almany and Seliktar 2005).

Hydrogels made from PEG is commonly used because it is biocompatible, has good physical characteristics and able to sustain the viability of encapsulated cells (Temenoff et al. 2002; Alhadlaq and Mao 2003). Alginate, agarose and peptides are other sources that can be used to create hydrogels and studies, both in vitro and in vivo, have been shown to support the growth of chondrocytes and bone formation (Kisiday et al. 2002; Alsberg et al. 2003). After 4 weeks of culture in vitro, chondrocytes encapsulated in a peptide hydrogel produced cartilage-like extracellular matrix (Kisiday et al. 2002). The development of bone tissue was observed after a

13-week implantation of alginate hydrogels containing osteoblasts in rats (Alsberg et al. 2003).

7.12 Conclusion

Conventional fabrication techniques have been previously thought to be unsuitable for TE purposes. This is because scaffolds produce by these techniques lack continuous interconnectivity and uniform spatial distribution of pores. In addition, the pore size and geometry cannot be easily or precisely controlled. The 3D scaffolds also lack the mechanical strength required to withstand various stresses and forces encountered in a living system. However, this chapter has shown that these techniques have the potential to produce scaffolds that is applicable to TE through some modifications to the systems or adjustment of the parameters used. In fact, most of the conventional fabrication methods are still being employed in recent studies demonstrating good short-term biocompatibility.

While improvements are observed, the manufacturing process needs to be evaluated so that the scaffolds produced are suitable for TE applications. The use of harsh organic solvents for dissolving polymers and other chemicals required by the fabrication process may produce toxic by-products and/or leaching of residual chemicals from the scaffold. This in turn affects the biocompatibility of the scaffold and raises concerns about safety of patients receiving the implants. It should be appreciated that there is no ideal fabrication technique that is suitable for all TE purposes. The choice of technique implemented is dependent on the application of interest. Continual research into the fabrication process will lead to advancements and breakthroughs in addressing the complex requirements of manufacturing scaffolds for TE.

References

Alhadlaq A, Mao JJ (2003) Tissue-engineered neogenesis of human-shaped mandibular condyle from rat mesenchymal stem cells. J Dent Res 82:951–956

Almany L, Seliktar D (2005) Biosynthetic hydrogel scaffolds made from fibrinogen and polyethylene glycol for 3D cell cultures. Biomaterials 26:2467–2477

Alsberg E, Kong HJ, Hirano Y, Smith MK, Albeiruti A, Mooney DJ (2003) Regulating bone formation via controlled scaffold degradation. J Dent Res 82(11):903–908

Biron M (2005) Foams, structural foams II – Physical blowing agents, Foam property overview. SpecialChem Polymer Additives and Colours article, Oct 3, 2005

Buckley CT, O'Kelly KU (2004) Regular scaffold fabrication techniques for investigations in tissue engineering. In: P. J. Prendergast and P. E. McHugh (eds.) Topics in Bio-Mechanical Engineering. Dublin and Galway: Trinity Centre for Bioengineering & National Centre for Biomedical Engineering Science, Chapter 5, 147–166

Cima LG, Vacanti JP, Vacanti C, Ingber D, Mooney D, Langer R (1991) Tissue engineering by cell transplantation using degradable polymer substrates. J. Biomech. Eng. 113(2):143–151

Dagalakis N, Flink J, Stasikelis P, Burke JF, Yannas IV (1980) Design of an artificial skin. Part III. Control of pore structure. J. Biomed. Mater. Res. 14(4):511–528

Doillon CJ, Whyne CF, Brandwein S, Silver FH (1986) Collagen-based wound dressings: control of the pore structure and morphology. J. Biomed. Res. 20(8):1219–1228

Draghi L, Resta S, Pirozzolo MG, Tanzi MC (2005) Microspheres leaching for scaffold porosity. J. Mater. Sci.: Mater. In Med. 16:1093–1097

Freed LE, Marquis JC, Nohria A, Emmanual J, Mikos AG, Langer R (1993) Neocartilage formation in vitro and in vivo using cells cultured on synthetic biodegradable polymers. Journal of Biomedical Materials Research 27:11–23

Glicklis R, Shaprio L, Agbaria R, Merchuk JC, Cohen S (2000) Hepatocyte behaviour within three-dimensional porous alginate scaffolds. Biotechno. Bioeng. 67(3):344–353

Goldstein AS, Zhu G, Morris GE, Meszlenyi RK, Mikos AG (1999) Effect of Osteoblastic Culture conditions on the structure of poly(DL-lactic-co-glycolic acid) foam scaffolds. Tissue Engineering 5:421–433

Gulsen D, Chauhan A (2004) Ophthalmic drug delivery through contact lenses. Invest Ophthalmol Vis Sci 45:2342–2347

Halstenberg S, Panitch A, Rizzi S, Hall H, Hubbell JA (2002) Biologically engineered protein-graft-poly(ethylene glycol) hydrogels: a cell adhesive and plasmin-degradable biosynthetic material for tissue repair. Biomacromolecules 3:710–723

Harris LD, Kim B, Mooney DJ (1998) Open pore biodegradable matrices formed with gas foaming. J. Biomed. Mater. Res. 42(3):396–402

Hou Q, Grijpma DW, Feijen J (2003) Porous polymeric structures for tissue engineering prepared by a coagulation, compression moulding and salt leaching technique. Biomaterials 24: 937–1947

Ishaug-Riley SL, Crane-Kruger GM, Yaszemski MJ, Mikos AG (1998) Three-dimensional culture of rat calvarial osteoblasts in porous biodegradable polymers. Biomaterials 19:1405–1412

Ishaug SL, Crane GM, Miller MJ, Yasko AW, Yaszemski MJ, Mikos AG (1997) Bone formation by three-dimensional stromal osteoblast culture in biodegradable polymer scaffolds. Journal of Biomedical Materials Research 36:17–28

Kim H, Kim HW, Suh H (2003) Sustained release of ascorbate-2-phosphate and dexamethasone from porous PLGA scaffolds for bone tissue engineering using mesenchymal stem cells. Biomaterials 24(25):4671–4679

Kim SS, Sun Park M, Jeon O, Yong Choi C, Kim BS (2006) Poly(lactide-co-glycolide)/ hydroxyapatite composite scaffolds for bone tissue engineering. Biomaterials 27(8):1399–1409

Kisiday J, Jin M, Kurz B, Hung H, Semino C, Zhang S, Grodzinsky AJ (2002) Self-assembling peptide hydrogel fosters chondrocyte extracellular matrix production and cell division: implications for cartilage tissue repair. Proc Natl Acad Sci U S A 99(15):9996–10001

Lee WK, Ichi T, Ooya T, Yamamoto T, Katoh M, Yui N (2003) Novel poly(ethylene glycol) scaffolds crosslinked by hydrolysable polyrotaxane for cartilage tissue engineering. J. Biomed. Mater. Res. – Part A 67(4):1087–1092

Lips PAM, Velthoen IW, Dijkstra PJ, Wessling M, Feijen J (2005) Gas foaming of segmented poly(ester amide) films. Polymer 46(22):9396–9403

Liu L, Fishman ML, Hicks KB, Kende M, Ruthel G (2006) Pectin/zein beads for potential colon-specific drug delivery: synthesis and in vitro evaluation. Drug Deliv 13:417–423

Lo H, Kadiyala S, Guggino SE, Leong KW (1996) Poly(L-lactic acid) foams with cell seeding and controlled-release capacity. Journal of Biomedical Materials Research 30:475–484

Lo H, Ponticiello MS, Leong KW (1995) Fabrication of controlled release biodegradable foams by phase separation. Tissue Engineering 1:15–28

Lutolf MP, Weber FE, Schmoekel HG, Schense JC, Kohler T, Muller R, Hubbell JA (2003) Repair of bone defects using synthetic mimetics of collagenous extracellular matrices. Nat Biotechnol 21:513–518

Madihally SV, Matthew HW (1999) Porous chitosan scaffolds for tissue engineering. Biomaterils 20(12):1133–1142

McGlohorn JB, Holder WD Jr, Grimes LW, Thomas CB, Burg KJ (2004) Evaluation of smooth muscle response using two types of porous polylactide scaffolds with differing pore topography. Tissue Eng. 10(3-4):505–514

Mikos AG, Thorsen AJ, Czerwonka LA, Bao Y, Langer R, Winslow DN, Vacanti JP (1994) Preparation and characterization of poly(L-lactic acid) foams. Polymer 35: 1068–1077

Mikos AG, Bao Y, Cima LG, Ingber DE, Vacanti JP, Langer R (1993a) Preparation of Poly(glycolic acid) bonded fiber structures for cell attachment and transplantation. Journal of Biomedical Materials Research 27:183–189

Mikos AG, Sarakinos G, Leite SM, Vacanti JP, Langer R (1993b) Laminated three-dimensional biodegradable foams for use in tissue engineering. Biomaterials 14:323–330

Mikos A.G, Sarakinos G, Lyman MD, Ingber DE, Vacanti JP, Langer R (1993c) Prevascularization of porous biodegradable polymers. Biotechnol. Bioeng. 42:716–723

Mikos AG, Temenoff JS (2000) Formation of highly porous biodegradable scaffolds for tissue engineering. Elec. J. Biotech. 3(2):114–119

Miyata T, Sode T, Rubin AL, Stenzel KH (1971) Effects of ultraviolet irradiation on native and telopeptide-poor collagen. Biochim. Biophys. Acta. 229(3):672–680

Miyata T, Arai M, Sakumoto A, Washino M (1980) Effect of 60Co gamma-ray irradiation on dilute acqeous solution of phthalate. Radioisotopes 28(8):479–484

Montjovent MO, Mathieu L, Hinz B, Applegate LL, Bourban PE, Zambelli PY, Manson JA, Pioletti DP (2005) Biocompatibility of bioresorbabale poly(L-lactic acid) composite scaffolds obtained by superficial gas foaming with human fetal bone cells. Tissue Eng. 11(11-12): 1640–1649

Mooney DJ, Mazzoni CL, Breuer C, McNamara K, Hern D, Vacanti JP (1996) Stabilized polyglycolic acid fibre-based tubes for tissue engineering. Biomaterials 17:115–124

Nam YS, Yoon JJ, Park TG (2000) A novel fabrication method of macroporous biodegradable polymer scaffolds using gas foaming salt as a porogen additive. Journal of Biomedical Materials Research (Applied Biomaterials) 53:1–7

Nam YS, Park TG (1999a) Biodegradable polymeric microcellular foams by modified thermally induced phase separation method. Biomaterials 20:1783–1790

Nam YS, Park TG (1999b) Porous biodegradable polymeric scaffolds prepared by thermally induced phase separation. Journal of Biomedical Materials Research 47:8–17

Nerem RM (2006) Tissue engineering: the hope, the hype, and the future. Tissue Eng 12: 1143–1150

Olde Damink LH, Dijkstra PJ, Van Luyn MJ, Van Wachem PB, Nieuwenhuis P, Feijen J (1995) Changes in mechanical properties of dermal sheep collagen during in vitro degradation. J Biomed. Mater. Res. 29(2):139–147

Ratner BD, Bryant SJ (2004) Biomaterials: where we have been and where we are going. Annu Rev Biomed Eng 6:41–75

Rosso F, Marino G, Giordano A, Barbarisi M, Parmeggiani D, Barbarisi A (2005) Smart materials as scaffolds for tissue engineering. J Cell Physiol 203:465–470

Ruijgrok JM, De Wijn JR, Boon ME (1994) Optimizing glutaraldehyde crosslinking of collagen: Effects of time, temperature and concentration as measured by shrinkage temperature. J Mater. Sci.: Mater. In Med. 5(2):80–87

Sachlos E, Czernuszka JT (2003) Making tissue engineering scaffolds work. Review of the application of solid freeform fabrication technology to the production of tissue engineering scaffolds. European Cells and Mater. 5:29–40

Schoof H, Bruns L, Fischer A, Heschel I, Rau G (2000) Dendritic ice morphology in unidirectionally solidified collagen suspensions. J Crystal Growth 209:122–129

Schoof H, Apel J, Heschel I, Rau G (2001) Control of pore structure and size in freeze-dried collagen sponges. J Biomed. Mater. Res. 58(4):352–357

Schugens C, Maquet V, Grandfils C, Jerome R, Teyssie P (1996) Polylactide macroporous biodegradable implants for cell transplantation II. Preparation of polylactide foams for liquid-liquid phase separation. Journal of Biomedical Materials Research 30: 449–461

Shastri VP, Martin I, Langer R (2000) Macroporous polymer foams by hydrocarbon templating. Proceedings of the National Academy of Sciences USA 97:1970–1975

Sperling LH (1992) Introduction to physical polymer science. In: John Wiley and Sons (eds), 2nd edn. Wiley, New York

Temenoff JS, Athanasiou KA, LeBaron RG, Mikos AG (2002) Effect of poly(ethylene glycol) molecular weight on tensile and swelling properties of oligo(poly(ethylene glycol) fumarate) hydrogels for cartilage tissue engineering. J Biomed Mater Res 59:429–437

Thomson RC, Yaszemski MJ, Powers JM, Mikos AG (1995a) Hydroxyapatite fiber reinforced poly(a-hydroxy ester) foams for bone regeneration. Biomaterials 19:1935–1943

Thomson RC, Wake MC, Yaszemski MJ, Mikos AG (1995b) Biodegradable polymer scaffolds to regenerate organs. Advances in Polymer Science 122:245–274

Thompson JI, Czernuszka JT (1995) The effect of two types of cross-linking on some mechanical properties of collagen. Biomed. Mater. Eng. 5(1):37–48

Vacanti JP, Morse MA, Saltzman WM, Domb AJ, Perez-Atayde A, Langer R (1988) Selective cell transplantation using bioabsorbable artificial polymers as matrices. J Pediatr Surg 23:3–9

Weadock K, Olson RM, Silver FH (1983-84) Evaluation of collagen crosslinking techniques. Biomater. Med. Devices Artif. Organs 11(4):293–318

Whang K, Thomas CH, Healy KE, Nuber G (1995) A novel method to fabricate bioabsorbable scaffolds. Polymer 36:837–842

Wu L, Zhang H, Zhang J, Ding J (2005) Fabrication of three-dimensional porous scaffolds of complicated shape for tissue engineering. I. Compression molding based on flexible-rigid combined mold. Tissue Eng. 11(7-8):1105–1114

Yang S, Leong K, Du Z, Chua C (2001) The Design of Scaffolds for Use in Tissue Engineering. Part I. Traditional Factors. Tissue Engineering 7(6):679–689

Yannas IV, Burke JF, Gordon PL, Huang C, Rubenstein RH (1980) Design of an artificial skin. II. Control of chemical composition. J. Biomed. Mater. Res. 14(2):107–132

Yoo HS, Lee EA, Yoon JJ, Park TG (2005) Hyaluronic acid modified biodegradable scaffolds for cartilage tissue engineering. Biomaterials 26(14):1925–1933

Ziegelaar BW, Fitton JH, Clayton AB, Platten ST, Maley MA, Chirila TV (1999) The modulation of corneal keratocyte and epithelial cell responses to poly(2-hydroxyethyl methacrylate) hydrogel surfaces: phosphorylation decreases collagenase production in vitro. Biomaterials 20:1979–1988

Chapter 8
Advanced Processes to Fabricate Scaffolds for Tissue Engineering

Paulo J. Bártolo, Henrique A. Almeida, Rodrigo A. Rezende, Tahar Laoui and Bopaya Bidanda

8.1 Introduction

Tissue engineering is an interdisciplinary field that necessitates the combined effort of cell biologists, engineers, material scientists, mathematicians, geneticists, and clinicians toward the development of biological substitutes that restore, maintain, or improve tissue function (Fig. 8.1). It comprises tissue regeneration and organ substitution (Table 8.1). The first definition of tissue engineering was provided by Skalak and Fox (1988) who stated it to be "the application of principles and methods of engineering and life sciences toward the fundamental understanding of structure-function relationships in normal and phatological mammalian tissues and the development of biological substitutes to restore, maintain, or improve tissue function". An historical overview of this field can be found in a recent report published by National Science Foundation, USA (2003).

Three strategies have been explored for the creation of a new tissue (Fuchs et al. 2001; Langer, 1997; Langer and Vacanti, 1993):

- The use of isolated cells or cell substitutes. This strategy avoids potential surgical complications but has the disadvantages of possible rejection or loss of function.
- Tissue-induced substances. The success of this strategy depends on the growth factors and controlled released systems
- Cells placed on or within constructs. This is the most common strategy and involves either a closed or an open system. In a closed system, cells are isolated from the body by a permeable membrane that allows exchange of nutrients and wastes and protects cell from the immune response of the body. An open system begins with the in vitro culture of cell, which are then seeded onto a scaffold. The cells-matrix construct is then implanted into the body.

Cells used in tissue engineering may be allogenic, xenogenic, syngeneic or autologous (Fuchs et al. 2001). They should be nonimmunogenic, highly proliferate, easy to harvest and with high capacity to differentiate into a variety of cell types with specialized functions (Fuchs et al. 2001; Marler et al. 1998). Cell attachment to materials is correlated to many factors, such as the stiffness and attachment area. Skeletal muscle satellite cells, cardiomyocytes and endothelial cells have been used in many tissue engineering applications.

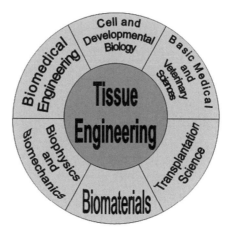

Fig. 8.1 Multidisciplinary nature of the tissue engineering field

Table 8.1 Tissue engineering main areas (Tabata, 2001)

	Purpose	Techniques/methodology
Tissue regeneration	In vitro production of tissue constructs	Cell scaffolding, bioreactor, microgravity
	In vivo natural healing process	Cell scaffolding, controlled release, physical barrier
	Ischemia therapy	Angiogenesis
Organ substitution	Immunoisolation	Biological barrier
	Nutrition and oxygen supply	Angiogenesis
	Temporary assistance for organ function	Extracorporeal system

Scaffolds provide an initial biochemical substrate for the novel tissue until cells can produce their own extra-cellular matrix (ECM). Therefore scaffolds not only define the 3D space for the formation of new tissues, but also serve to provide tissues with appropriate functions. These scaffolds are often critical, both ex vivo as well as in vivo, as they serve some of the following purposes (Gomes and Reis, 2004; Leong et al. 2002):

- Allow cell attachment, proliferation and differentiation
- Deliver and retain cells and growth factors
- Enable diffusion of cell nutrients and oxygen
- Enable an appropriate mechanical and biological environment for tissue regeneration in an organised way

To achieve these goals an ideal scaffold must satisfy some biological and mechanical requirements (Table 8.2):

a) Biological requirements:

Table 8.2 Relationship between scaffold characteristics and the corresponding biological effect (Mahajan, 2005)

Scaffold characteristics	Biological effect
Biocompatibility	Cell viability and tissue response
Biodegradability	Aids tissue remodelling
Porosity	Cell migration inside the scaffold Vascularisation
Chemical properties of the material	Aids in cell attachment and signiling in cell environment
	Allows release of bioactive substances
Mechanical properties	Affects cell growth and proliferation response
	In-vivo load bearing capacity

- Biocompatibility – the scaffold material must be non-toxic and allow cell attachment, proliferation and differentiation
- Biodegradability – the scaffold material must degrade into non-toxic products
- Controlled degradation rate – the degradation rate of the scaffold must be adjustable in order to match the rate of tissue regeneration
- Appropriate porosity macro- and microstructure of the pores and shape, highly interconnected pore structure and large surface area to allow high seeded cells and to promote neovascularisation. Large number of pores may be able to enhance vascularisation, while smaller diameter of pore is preferable to provide large surface per volume ratio. Typical desirable porosity are around 90% with pore size in the range of 20–250 μm (Freyman et al. 2001; Whang et al. 1995). Optimum pore sizes of 20 μm have been reported for fibroblast ingrowth, between 20 and 125 μm for regeneration of adult skin and 100–250 μm for the regeneration of bone. Figure 8.2 shows the effect of pore size on the percentage of cells attached onto collagen-glycosaminoglycan (CG) scaffolds.
- Should encourage the formation of ECM by promoting cellular functions
- Ability to carry biomolecular signals such as growth factors. Numerous growth factors have been identified such as fibroblast growth factor (FGF), platelet-derived growth factor, bone morphogenic protein (BMP), insulin growth

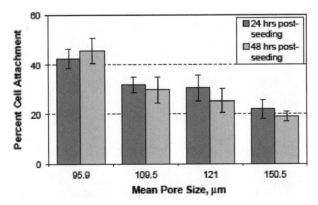

Fig. 8.2 Percentage of cells attached to the collagen-glycosaminoglycan scaffolds with different pore sizes (O'Brien et al. 2005)

Table 8.3 Biomaterials commonly used in tissue engineering

Biopolymers	
Natural polymers	**Synthetic polymers**
Alginate	Poly(glycolic acid), poly(lactic acid) and their co-polymers
Collagen	Poly(ε-caprolactone)
Chitosan	Poly(dioxanone)
Hyaluronic acid	Polyethylene oxide/polybutylene teraphthalate co-polymersPoly
Poly(hydroxybutyrate)	(propylene fumarate)
	Polyanhydride
Bioceramics	
Hydroxyapatite and other types of calcium phosphate like fluorapatite or tricalcium phosphate	
Biphasic hydroxyapatite/ tricalcium phosphate ceramics	
Bioactive glass ceramics	

factor, transforming growth factor-β, epidermal growth factor, vascular endothelial growth factor, etc. (Nathan and Sporn, 1991; Wei et al. 2007)

b) Mechanical and physical requirements

- Sufficient strength and stiffness to withstand stresses in the host tissue environment
- Adequate surface finish guaranteeing that a good biomechanical coupling is achieved between the scaffold and the tissue. New efforts to encourage cell attachment focusing on mimicking the surface chemistry of autogenous ECM has been also reported (Hynes, 1992). Surface properties such as surface charge and surface topography can influence biocompatibility
- Easily sterilised either by exposure to high temperatures or by immersing in a sterilisation agent remaining unaffected by either of these processes.

A variety of biodegradable materials have already been used for tissue scaffolds, including ceramics and polymers (Gomes and Reis, 2004). The primary use of ceramics has been in bone tissue engineering, presenting long degradation times often on the order of years. Polymeric scaffolds are used in the form of fibrous meshes, porous sponges, or hydrogels showing short degradation times, often on the order of days or months. The most common materials used for scaffolds are indicated in Table 8.3.

8.2 Conventional Fabrication Techniques

Conventional methods to fabricate scaffolds include (Gomes and Reis, 2004; Ho et al. 2004; Leong et al. 2002; Reignier and Huneault, 2006; Whang et al. 1995):

- Solvent casting/salt leaching: involves mixing solid impurities, such as sieved sodium chloride particles, into a polymer solvent solution, and casting the dispersion to produce a membrane of polymer and salt particles. The salt particles are then leached out with water to yield a porous membrane. Porosity and pore

size have been shown to be dependent on salt weight fraction and particle size. Pore diameters of 100–500 μm and porosities of 87–91% have been reported.
- Phase separation: involves dissolving a polymer in a suitable solvent, placing it in a mould, then cooling the mould rapidly until the solvent is frozen. The solvent is removed by freeze-drying, leaving behind the polymer as a foam with pore sizes of 1–20 μm in diameter.
- Foaming: is carried out by dissolving a gas, usually CO_2, at elevated pressure or by incorporating a chemical blowing agent that yields gaseous decomposition products. This process generally leads to pore structures that are not fully interconnected and produces a skin-core structure.
- Gas saturation: this technique uses high pressure carbon dioxide to produce macroporous sponges at room temperature. Polymeric sponges with large pores (~100 μm) and porosities up to 93% have been reported (Mooney et al. 1996)
- Textile meshes: these processes include all technologies successfully employed to fabricate non-woven meshes of different polymers. Major limitations are due to difficulties in obtaining high porosity and regular pore size.

Each of these techniques presents several limitations as they usually do not enable to properly control pore size, pore geometry and spatial distribution of pores, besides being almost unable to construct internal channels within the scaffold. Beyond these limitations, these techniques usually involve the use of toxic organic solvents, long fabrication times on top of being labour-intensive processes. Therefore, rapid prototyping technology (also called Solid Freeform Fabrication) is considered a viable alternative to fabricate scaffolds for tissue engineering.

8.3 Rapid Prototyping and Manufacturing Techniques for Tissue Engineering

Rapid prototyping and manufacturing (RP&M) represents a new group of non-conventional fabrication techniques recently introduced in the medical field. The main advantages of RP&M are both the capacity to rapidly produce very complex 3D models and the ability to use various raw materials. In the tissue engineering field, RP&M have been used to produce scaffolds with customised external shape and predefined internal morphology, allowing good control of pore size and pore distribution (Bártolo, 2006).

Figure 8.3 provides a general overview of the necessary steps to produce rapid prototyping scaffolds for tissue engineering. The first step is the generation of the corresponding computer solid model through one of the currently available medical imaging techniques such as computer tomography, magnetic resonance imaging, etc. These imaging methods produce continuous volumetric data (voxel-based data), which provide the input data for the digital model generation (Bártolo, 2006). The model is then tessellated as an STL file, which is currently the standard file for facetted models in RP&M. Finally, the STL model is mathematically sliced into thin layers (sliced model). RP&M technologies are similar to 2D printing and

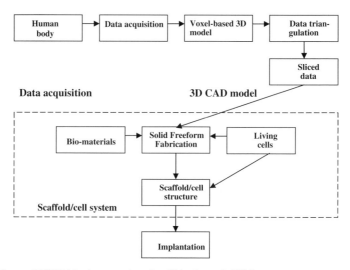

Fig. 8.3 Steps of RP&M in tissue engineering (Bártolo et al. 2004)

plotting technologies using both vector-based and raster-based imaging techniques. The various RP&M technologies for tissue engineering, described in the following sections, include stereolithographic processes, laser sintering, extrusion and three dimensional printing.

8.3.1 Stereolithographic Processes

Stereolithographic processes produce three-dimensional solid objects in a multi-layer procedure through the selective photo-initiated cure reaction of a polymer (Bártolo and Mitchell, 2003). These processes usually employ two distinct methods of irradiation. The first method is the mask-based method in which an image is transferred to a liquid polymer by irradiating through a patterned mask. The irradiated part of the liquid polymer is then solidified. In the second method, a direct writing process using a focused UV beam produces polymer structures (Fig. 8.4).

The direct or laser writing approach consists of a vat containing a photosensitive polymer, a moveable platform on which the model is built, a laser to irradiate and cure the polymer and a dynamic mirror system to direct the laser beam over the polymer surface "writing" each layer. After drawing a layer, the platform dips into the polymer vat, leaving a thin film from which the next layer will be formed.

Mask-based writing systems build models by shining a flood lamp through a mask, which lets light pass through it. These systems generally require the generation of a lot of masks with precise mask alignments. One solution for this problem is the use of a liquid crystal display (LCD) or a digital processing projection system as a flexible mask.

8 Advanced Processes to Fabricate Scaffolds for Tissue Engineering

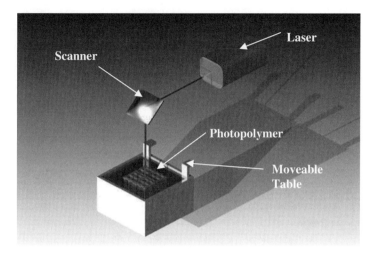

Fig. 8.4 Stereolithography system

Microstereolithography is a relatively recent development, similar to conventional stereolithography. However, to get a better resolution, the beam is focused more precisely in order to reduce the spot size to a few micrometers of diameter. Several strategies have been proposed (Bertsch et al. 2003): constrained surface techniques, free surface techniques, and integral processes. Integral microstereolithography represents the most recent advancement in this field, enabling the solidification of each layer in one irradiation step by projecting the corresponding image onto the surface of the photo-polymerisable resin through either a liquid crystal display or a digital micro mirror device. MicroTEC (Germany) is one of the few companies commercialising a microstereolithography. Their propriety technology is known as Rapid Micro Product Development (RMPD) and uses an excimer laser as a light source that works on a vector-by-vector basis.

All of the abovementioned stereolithographic approaches are based on a single-photon initiated polymerisation procedure. Two-photon-initiated polymerisation curing processes represent a useful stereolithographic alternative strategy to produce micro/nanoscale structures by using femtosecond infrared laser without photomasks (Kowata and Sun 2003; Lemercier et al. 2005; Tormen et al. 2004,). In this process, the molecule simultaneously absorbs two photons instead of one, being excited to higher singlet state. The use of two-photon-initiated polymerisation allows a submicron 3D resolution, on top of enabling both 3D fabrication at greater depth and an ultra-fast fabrication.

Many groups have used and developed stereolithographic processes for tissue engineering. Levy et al. (1997) used a direct irradiation stereolithographic process to produce hydroxyapatite (HA) ceramic scaffolds for orbital floor prosthesis. A suspension of fine HA powder into a UV-photocurable resin was formulated and used as building material. The photo-cured resin acts as a binder to hold the HA particles together. The resin is then burnt out and the HA powder assembly sintered for consolidation. A similar approach was used by Griffith and Halloran (1996) that

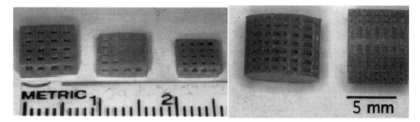

Fig. 8.5 Sintered HA scaffolds produced by a lost-mould technique (Chu et al. 2001) elective laser sintering process

produced ceramic scaffolds using suspensions of alumina, silicon nitride and silica particles with a photo-curable resin. The binder was removed by pyrolysis and the ceramic structures sintered.

Chu et al. (2001) developed a lost-mould technique to produce implants with designed channels and connection pattern (Fig. 8.5). Stereolithography was used to create epoxy moulds designed from negative image of implants. A highly loaded HA-acrylate suspension was cast into the mould. The mould and the acrylic binder were removed by pyrolysis and the HA green scaffold submitted to a sintering process. The finest channel size achieved was about 366 μm and the range of implant porosity between 26 and 52%.

In another study, Cooke et al. (2002) used a biodegradable resin mixture of diethyl fumarate, poly(propylene fumarate) and bisacylphosphine oxide as photoinitiator to produce scaffolds for bone ingrowth. Similarly, Matsuda and Mizutani (2002) developed a photopolymer containing biodegradable copolymer of trimethylene carbonate and ε-caprolactone. UV light can also be used to fabricate hydrogel polymer scaffolds. The main difficulty is the development of water-soluble components that are both functional and photolabile (Fischer et al. 2001).

8.3.2 Laser Sintering

Selective laser sintering (SLS) uses a laser emitting infrared radiation, to selectively heat powder material just beyond its melting point (Fig. 8.6). The laser traces the shape of each cross-section of the model to be built, sintering powder in a thin layer. It also supplies energy that not only fuses neighbouring powder particles, but also bonds each new layer to those previously sintered. For polymeric powders, the sintering process takes place in a sealed heated chamber at a temperature near the melting point filled with nitrogen or argon. After each layer is solidified, the piston over the model retracts to a new position and a new layer of powder is supplied using a mechanical roller. The powder that remains unaffected by the laser acts as a natural support for the model and remains in place until the model is complete.

Materials most commonly used in tissue engineering scaffolds through laser sintering are biocompatible polymers such as polycaprolactone (PCL) and poly lactic acid (PLA) and biocompatible ceramics. PCL is a bioresorbable polymer used for

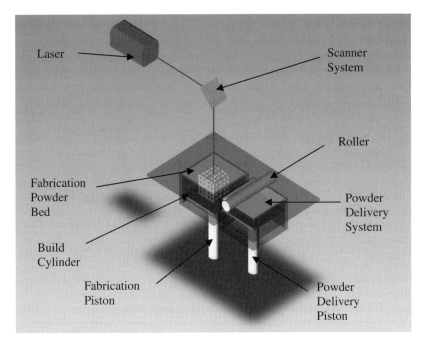

Fig. 8.6 Selective laser sintering process

bone and cartilage repair. It is more stable at ambient conditions that PCL and is also less expensive and readily available (Williams et al. 2005).

The potential of SLS to produce PCL scaffolds for replacement of skeletal tissues was shown by Williams et al. (2005). The scaffolds were seeded with bone morphogenetic protein-7 (BMP-7) transduced fibroblasts. In vivo results show that these scaffolds enhance tissue in-growth, on top of possessing mechanical properties within the lower range of trabecular bone. Compressive modulus (52 to 67 MPa) and yield strength (2.0 to 3.2 MPa) were in the lower range of properties reported for human trabecular bone.

Lee and Barlow (1996) coated calcium phosphate powder with polymer by spray drying slurry of particulate and emulsion binder. The coated powder was then sintered to fabricate calcium phosphate bone implants. Afterwards, these structures were infiltrated with calcium phosphate solution or phosphoric acid-based inorganic cement.

Popov and co-authors (2004) proposed the concept of Surface Selective Laser Sintering (SSLS) technique that enables to extend the range of polymers that can be used to extend the range of polymers that can be used for scaffold fabrication. Unlike conventional selective laser sintering, where polymer has a strong absorption at the laser wavelength, the SSLS process is based on melting the particle, which are transparent for laser radiation, due to the laser beam absorption by a small amount (<0.1 wt%) of biocompatible carbon black homogeneously distributed along the polymer surface. This process allows preventing significant overheating of the particles internal domains that can lead to properties changes and degradation.

8.3.3 Extrusion-based Processes

The extrusion-based rapid prototyping technique, commercially known as Fused Deposition Modelling (FDM), was developed by Crump (1989). By this process, thin thermoplastic filaments are melted by heating and guided by a robotic device (extruder) controlled by a computer, to form the three-dimensional object (Fig. 8.7). The material leaves the extruder in a liquid form and hardens immediately. The previously formed layer, which is the substrate for the next layer, must be maintained at a temperature just below the solidification point of the thermoplastic material to assure good interlayer adhesion.

Extrusion-based processes have been used to successfully produce scaffolds in PCL, PP-TCP, PCL-HA, PCL-TCP with resolution of 250 µm. Some of the major limitations of FDM are due to the use of filament-based materials and the high heat effect on raw material. In order to solve some limitations of the FDM process, such as the requirement of precursor filaments or high processing temperatures, some alternative processes have been proposed.

Hutmacher et al. (2001) optimised the FDM processing parameters for the production of PCL honeycomb-like scaffolds. Similar work was conducted by Zein et al. (2002) that produced PCL scaffolds with a range of channel size 160–700 µm, filament diameter 260–370 µm, porosity 48–77% and regular honeycomb pores. The compressive stiffness ranged from 4 to 77 MPa, yield strength from 0.4 to 3.6 MPa, and yield strain from 4 to 28%.

Koh et al. (2006) exploited the fact that when a warm PCL-HA/acetone solution is extruded into a reservoir containing ethanol, the extruded filament rapidly solidifies via solvent extraction producing a continuous rigid filament, to fabricate macro-channelled scaffolds. The diameter and morphology of the filament were controlled by adjusting the deposition speed and volume flow rate.

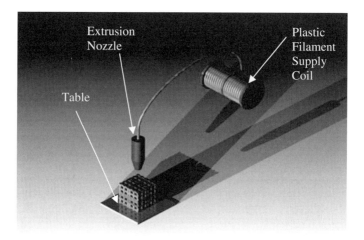

Fig. 8.7 Fused Deposition Modelling process

Woodfield et al. (2004) used a FDM-like technique, called 3D Fiber Deposition, to produce poly(ethylene glycol)-terephthalate-poly(butylenes terephthalate) (PEGT/PBT) block co-polymer scaffolds with a 100% interconnecting pore network for engineering of articular cartilage (Fig. 8.8). By varying the co-polymer composition, porosity and pore geometry, scaffolds were produced with a range of mechanical properties close to articular cartilage. The scaffolds seeded with bovine chondroccytes supported a homogeneous cell distribution and subsequent cartilage-like tissue formation.

Recently, Tellis et al. (2007) used micro CT to create biomimetic tissue engineering scaffolds. CAD models were exported to a FDM machine, producing polybutylene terephthalate (PBT) trabecular scaffolds. The scaffolds were compression tested at two different load rates (49 and 294N/s). Some scaffolds were soaked in a 25 °C saline solution for 7 days before compression. When compressed at 49 N/s the dry trabecular scaffolds had a compressive stiffness ranging from 2.46±0.55 MPa

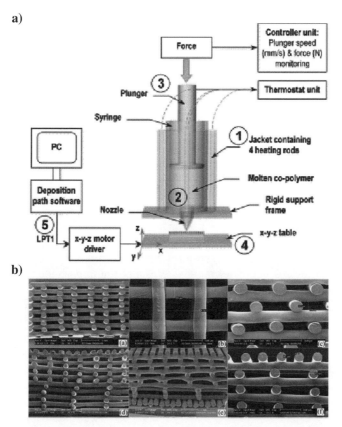

Fig. 8.8 (a) The 3D Fiber Deposition system. (b) SEM sections of 3D deposited scaffolds with varying deposition geometries (Woodfield et al. 2004)

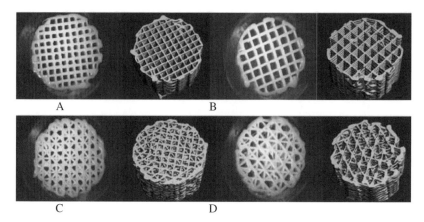

Fig. 8.9 Digital photographs and micro CT 3D segmentations of PBT scaffolds (Tellis et al. 2007)

for the complex interconnected pore structure (case E in Fig. 8.9) to 5.11±1.89 MPa for the simple linear structure (case A in Fig. 8.9). At 294 N/S, the compressive stiffness values roughly doubled. It was also observed that soaking the scaffolds in saline solution had an insignificant effect on stiffness and that compressive stiffness decreased as pore size increased. Compressive trabecular scaffolds matched bone samples in porosity. However, physiologic connectivity density and trabecular separation requires optimisation of scaffold processing.

Drexel University developed a variation of FDM called precision extruding deposition (PED) for fabrication of bone tissue scaffolds. In this process, material in pellet or granule form is fed into a chamber where it is liquefied. Pressure from a rotating screw forces the material down a chamber and out through a nozzle tip. This process was used by Wang et al. (2004) to directly fabricate PCL scaffolds with controlled pore size of 250 µm and designed structural orientations (0°/90°, 0°/120° or combined 0°/120° and 0°/90° patterns). Proliferation studies were performed using cardiomyoblasts, fibroblasts and smooth muscle cells. Similarly, Xiong et al. (2001) proposed the concept of precise extrusion manufacturing (PEM) to fabricate PLLA scaffolds for bone tissue engineering with controlled porous architectures from 200 to 500 µm. The sprayer of this system is equipped with a built-in heating unit to melt the feedstock. Compressed air is used as a piston to push the melted material through the nozzle.

In order to eliminate the elevated temperatures required by the extrusion-based processes, Tsinghua University developed a process called Low-temperature Deposition Manufacturing (LDM) to produce scaffolds at a low temperature environment under 0°C (Xiong et al. 2005). The LDM system comprises a multi-nozzle extrusion process and a thermally induced phase separation process (Fig. 8.10). Scaffolds having a macroporous structure larger than 100 µm in diameter and a microporous structure smaller than 100 µm have been reported. The LDM process was used to produce poly(L-lactide) (PLLA) and TCP composite scaffolds with BMP growth factor. The scaffolds were implanted into rabbit radius and canine radius large-segmental defects. After 12 weeks it was possible to observe that the rabbit radius

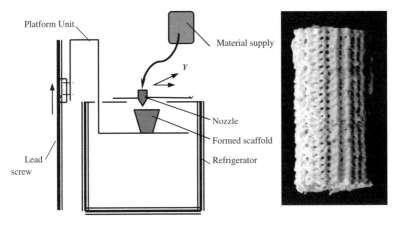

Fig. 8.10 (a) Schematic illustration of the LDM system. (b) Example of a porous PLLA/TCP composite scaffold produced by LDM process (Yan et al. 2003)

defect was successfully repaired and the regenerated bone had properties similar to the healthy bone. For the canine radius it was observed similar results after 24 weeks (Yan et al. 2003a).

Bioplotting (Fig. 8.11) is a technique developed by the Freiburg Materials Research Center and Envisiontec, Germany, that uses a pressure-controlled dispenser to deposit material into a reactive liquid medium of comparable density. This balance of media densities, which allows scaffolds to be created without the need of support structures, is a key characteristic of this process. Buoyancy can be provided to the plotted material, making strands of material remain in the correct position instead of sacking due to gravitational effects. The use of materials such as melts of poly(lactides), poly(lactide-co-glycolide), poly(hydroxybutyrate-co-valeriate), poly(caprolactone), poly(butylenes terephthalate-block-oligoethylene oxide), solutions of agar and gelatine, collagen and reactive biosystems involving

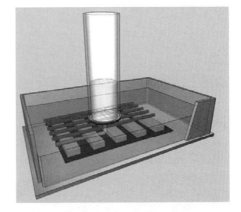

Fig. 8.11 (a) The 3D-Bioplotter system. (b) Process building of the first layer (Carvalho et al. 2005)

fibrin formation and polyelectrolyte complexation have been reported (Pfister et al. 2004). Fig. 8.12 illustrates some examples of scaffold structures made by the 3D-Bioplotter system.

Moroni et al. (2006) reported a novel strategy to create hollow fibers with controlled shell thickness and lumen diameter, organizing them into 3D scaffolds. Hollow fibers (Fig. 8.13), are made by extrusion of a blend of poly(butylmethacrylatemethylmethacrylate) (P(BMA/MMA) and poly(ethylene oxideterephtalate)-co -poly(butylene terephtalate) (PEOT/PBT) using the Bioplotter system. During the flow through the nozzle of the extruder and due to viscosity differences, the polymer with lower viscosity tends to shift towards the walls. The consequent separation of the polymers produces a stratification effect. Hollow fibers are produced by removing the core polymer by selective dissolution. It was also observed that bovine primary articular chondrocytes grow and form ECM not only in the scaffold macropores but also inside the hollow cavities (Fig. 8.14). The use of these hollow matrices for selective drugs release is being investigated.

An alternative process is the pressure assisted microsyringe (PAM) that involves the deposition of polymer dissolved in solvent through a syringe (Vozzi et al. 2003). The thickness of the polymer stream can be varied by changing the syringe pressure, solution viscosity, syringe tip diameter and motor speed. Resolution as low as 10 µm on a 2D structure was achieved.

Robocasting (Fig. 8.15), also known as direct-write assembly, consists on the robotic deposition of highly concentrated colloidal suspensions capable of fully supporting their own weight during assembly due to their viscoelastic properties (Miranda et al. 2006). This technique have been used to β-tricalcium phosphate (β-TCP) scaffolds (Miranda et al. 2006; Saiz et al. 2007).

Ang et al. (2002) developed a rapid prototyping robotic dispersing (RPBOD) system using the same principle as the 3D bio-plotting system, which was used to produce chitosan-HA scaffolds. Solutions of chitosan-HA were extruded into a sodium hydroxide and ethanol medium to induce the precipitation of chitosan. The scaffolds were then hydrated, frozen and freeze-dried.

Fig. 8.12 (a) Hydroxyapatite scaffold. (b) PLGA scaffold (Carvalho et al. 2005)

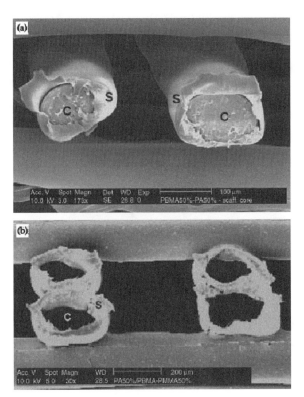

Fig. 8.13 SEM micrographs of a scaffold before (a) and after (b) leaching out the core material (Moroni et al. 2006)

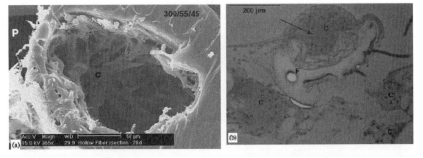

Fig. 8.14 SEM (a) and optical microscope (b) micrographs showing chondrocytes and ECM formation inside and outside the hollow fibers (Moroni et al. 2006). P = pore; F = fiber; C = chondrocytes

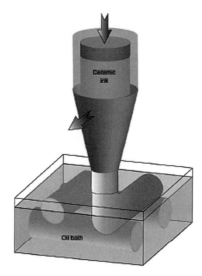

Fig. 8.15 Illustration of the robocasting fabrication process (Miranda et al. 2006)

8.3.4 Three-dimensional Printing

Three-dimensional printing (3DP) was developed at the Massachusetts Institute of Technology (USA) by Sachs et al. (1989). The process deposits a stream of microparticles of a binder material over the surface of a powder bed, joining particles together where the object is to be formed (Fig. 8.16). A piston lowers the powder bed so that a new layer of powder can be spread over the surface of the previous

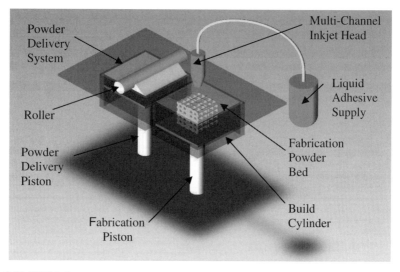

Fig. 8.16 3D Printing process

layer and then selectively joined to it. Therics Incorporated applied the 3DP process to tissue engineering and developed the *TheriForm* process to fabricate drug delivery devices and scaffolds.

Kim et al. (1998) employed 3DP with particulate leaching to create porous scaffolds, using polylactide-coglycolide (PLGA) powder mixed with salt particles and a suitable organic solvent. The salt particles were leached using distilled water. Cylindrical scaffolds measuring 8 mm (diameter) by 7 mm (height) with pore sizes of 45–150 µm and 60% porosity were fabricated. Hepatocytes were successful attached to the scaffolds.

The influence of pore size and porosity on cell adhesion and proliferation were investigated by Zeltinger et al. (2001). Disc shaped poly(L-lactic acid) (L-PLA) scaffolds measuring 10 mm (diameter) by 2 mm (height) were produced through both 3DP and salt and leaching methods. The scaffolds were produced with two different porosities (75 and 90%) and four different pore size distributions (<38, 38–63, 63–106 and 106–150 µm), and tested with cell culture using canine dermal fibroblasts, vascular smooth muscle cells and microvascular epithelial cells.

Lam et al. (2002) developed a blend of starch-based powder containing cornstarch (50%), dextran (30%) and gelatine (20%), bounded by printing distilled water. Cylindrical scaffolds were produced measuring 12.5 mm (diameter) by 12.5 mm (height) and infiltrated with different amounts of a copolymer solution consisting of 75% L-PLA and 25% polycaprolactone in dichloromethane to improve their mechanical properties.

Leukers et al. (2005) produced HA scaffolds with complex internal structures and high resolution. MC3T3-E1 cells were seeded on the scaffolds and cultivated under static and dynamic setups. Dynamic cultivation was performed in perfusion containers. A flow rate of 18 µl/min. Histological evaluation was carried out to characterise the cell ingrowth process. It was observed that the dynamic cultivation method lead to a stronger population compared to the static cultivation method. Static cells culture led to multiple cell layers located on the surface of HA granules. Dynamic cells culture tends to grow in between cavities of the granules. Additionally, it was found that cells proliferated deep into the structure forming close contact to HA granules.

Sachlos et al. (2003) used an indirect approach to produce collagen scaffolds with complex internal morphology and macroscopic shape by using a 3DP sacrificial mould. A dispersion of collagen was cast into the mould and frozen. The mould was then dissolved with ethanol and the collagen scaffold was critical point dried with liquid carbon dioxide. Other research works, like the ones of Taboas et al. (2003), Limpanuphap and Derby (2002) and Park et al. (1998), have also exploited the capabilities of 3DP for tissue engineering.

Ink-jet printing systems have been used to print both aqueous solutions onto supports and cell within a scaffold (Pardo et al. 2003; Saunders et al. 2004). During the droplet formation process the liquid material experiences shear rates close to 10^4 s^{-1} and similar strains occurs during the impact (Saunders et al. 2004). Therefore, cells in suspension are subjected to large stresses and deformation. Nevertheless, ink-jet printing has been reported to be a viable method for cell deposition and patterning (Saunders et al. 2004). Boland et al. (2006), explored a cell and organ printing fabrication strategy to print cells and proteins within 3D hydrogel structures. Several examples of printed tissues such as contractile cardiac hybrids have been considered. Alginate hydrogels were used as support structures. As indicated in Fig. 8.17 endothelial cell attachment was observed. Filopodia can be seen at the leading edge of the cell and lamellapodia at the trailing edge suggesting cell migration into pores. It was postulated that local variation in the mechanical compliance of alginate structures causes cells to attach to the areas with greatest stiffness or highest stress. Endothelial cells are also known to grow well on surfaces, so it is not surprising to see cell attachment on the inner surfaces of the alginate pores. Nevertheless, the exact mechanism of cell attachment to alginate structures is still unknown and requires further research.

Similar procedures have been used by Mironov et al. (2003) and Yan et al. (2003), which developed the concept of cell printing. This process prints gels, single cells and cell aggregates offering a possible solution for organ printing. An analogous process, called alginate-based rapid prototyping, has been developed at the Polytechnic Institute of Leiria. This process produces alginate solid structures, by extruding a solution of sodium alginate, mixed with a solution of calcium chloride, providing a temporary support for the seeded cells in culture (Bártolo, 2006; Bártolo et al. 2004). Alginate is an anionic copolymer composed of homopolymeric regions of 1,4-linked β-D-mannuronic (M blocks) and α-L-guluronic acid (G blocks), interspersed with regions of alternating structure. Gelation occurs when

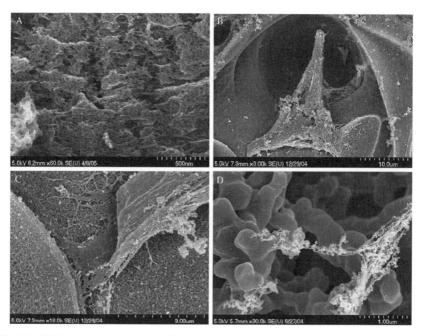

Fig. 8.17 SEM micrographs of endothelial cells attached to alginate structure. (A) Wall with nanosize pores. (B) An endothelial cell attach inside an alginate structure. (C) Filopodia and lamellapodia interacting with the alginate material. (D) Interactions between fibrous secretions and alginate (shellCell-jet printing equipment and some cell-gel mixture printing structures (Boland et al. 2006)

divalent ions take part in the interchain ionic binding between G-blocks in the polymer chain giving rise to a three dimensional network. Such binding zones between the G-blocks are often referred to as "egg boxes". These ions act as cross-linkers that stabilise alginate chains forming a gel structure, which contains cross-linked chains interspersed with more freely movable chains that bind and entrap large quantities of water. The gelification process is characterised by a re-organisation of the gel network accompanied by the expulsion of water. Gels made of M-rich alginate are softer and more fragile, and may also have lower porosity. This is due to the lower binding strength between the polymer chains and to the higher flexibilities of the molecules. The gelification process is highly dependent upon diffusion of gelification ions into the polymer network. Trasmittancy, swelling and viscoelasticity of alginate structures are highly affected by the M/G ratio.

8.4 Conclusions

RP&M technologies have a great potential for tissue engineering. These technologies offer a high degree of freedom for tissue engineering either for the design of scaffolds (pore size, pore geometry, orientation, interconnectivity, etc.) or for its fabrication. Several materials can also be used enabling the production of both soft and hard scaffolds. These characteristics can enhance the fabrication of biomimetic scaffolds and scaffolds for complex biomechanical applications. Future developments will possibly lead to establishing rapid prototyping as a key tool for tissue reconstruction and regeneration. The main advantages and limitations of rapid prototyping scaffolds for tissue engineering are listed in Table 8.4.

Table 8.4 Characteristics of rapid prototyping scaffolds for tissue engineering

Rapid prototyping	Advantages	Limitations
Stereolithoigraphic processes	Relatively easy to achieve small feature	Limited by the development of photo-polymerisable, biocompatible and biodegradable materials; currently limited to reactive and mostly toxic resins
SLS	Relatively higher scaffold strength; solvent free	Powder material trapped in small inner holes can be difficult to remove; high temperatures in the chamber
FDM	No materials trapped in the scaffold; solvent free	High heat effect on raw material; limited geometrical complexity
3DP	Low heat effect on raw powder; easy process; low cost	Materials trapped in small inner holes; low mechanical properties
Bio-plotting	Large variety of materials for both soft and hard tissues	Low geometrical complexity

References

Ang, T.H., Sultana, F.S.A., Hutmacher, D.W., Wong, Y.S., Fuh, J.Y.H., Mo, X.M., Loh, H.T., Burdet, E. and Teoh, S.H., Fabrication of 3D chitosan-hydroxyapatite scaffolds using a robotic dispersing system, *Mater Sci Eng*, C20, 35–42, 2002

Bártolo, P., Mendes, A. and Jardini, A., Bio-prototyping, in Design and Nature II, Edited by M.W. Collins and C.A. Brebbia, WIT Press, Southampton, UK, 535–543, 2004

Bártolo, P.J. and Mitchell, G., Stereo-thermal-lithography, *Rapid Prototyping J*, 9(3), 150–156, 2003

Bártolo, P.J., State of the art of solid freeform fabrication for soft and hard tissue engineering, in Design and nature III: comparing design in nature with science and engineering, 233–243, 2006.

Bertsch, A., Jiguet, S., Bernhards, P. and Renaud, P., Microstereolithography: a review. *Mat Res Soc Symp Proc*, 758: LL.1.1.1–13, 2003.

Boland, T., Tao, X., Damon, B.J., Manley, B. and Kesari, P., Drop-on-demand printing of cells and materials for designer tissue constructs, *Mater Sci Eng*: C 27(3), 372–376, 2007.

Carvalho, C., Landers, R., Mulhaupt, R., Hubner, U. and Schmelzeisen, R., Fabrication of soft and hard biocompatible scaffolds using 3D-Bioplotting, in Virtual Modelling and Rapid Manufacturing – Advanced Research in Virtual and Rapid Prototyping, Edited by P.H. Bártolo et al. Taylor & Francis, London, UK, 2005.

Chu, T.-M.G., Halloran, J.W., Hollister, S.J. and Feinberg, S.E., Hydroxyapatite implants with designed internal architecture, *J Mater Sci: Mater Med*, 12, 471–478, 2001.

Cooke, M.N., Fisher, J.P., Dean, D., Rimnac, C., Mikos, A.G., Use of stereolithography to manufacture critical-sized 3D biodegradable scaffolds for bone ingrowth, *J Biomed Mater Res Part B: Appl Biomater*, 64B, 65–69, 2002

Crump, S.S., Apparatus and method for creating three-dimensional objects, US Pat. 5121329, 1989

Fischer, J.P., Dean, D., Engel, P.S., Mikos, A., Photoinitiated polymerization of biomaterials, *Annu Rev Mater Res*, 31, 171–181, 2001.

Freyman, T.M., Yannas, I.V. and Gibson, L.J., Cellular materials as porous scaffolds for tissue engineering, *Progress in Mate Sci*, 46, 273–282, 2001.

Fuchs, J.R., Nasseri, B.A. and Vacanti, J.P., Tissue engineering: a 21st century solution to surgical reconstruction, *Ann Thorac Surg*, 72, 577–581, 2001.

Gomes, M.E. and Reis, R.L., Biodegradable polymers and composites in biomedical applications: from catgut to tissue engineering. Part 2 Systems for temporary replacement and advanced tissue regeneration, *Int Materials Rev*, 49, 274–285, 2004.

Griffith, M.L. and Halloran, J.W., Freeform fabrication of ceramics via stereolithography, *J Am Ceram Soc*, 79, 2601–2608, 1996.

Ho, M.H., Kuo, P.Y., Hsieh, H.J., Hsien, T.Y., Hou, L.T., Lai, J.Y. and Wang, D.W., Preparation of porous scaffolds by using freeze-extraction and freeze-gelation methods, *Biomaterials*, 25, 129–138, 2004.

Hutmacher, D.W., Schantz, T., Zein, I., Ng, K.W., Teoh, S.H. and Tan, K.C., Mechanical properties and cell cultural response of polycaprolactone scaffolds designed and fabricated via fused deposition modelling, *J Biomed Mater Res*, 55, 203–216, 2001.

Hynes, R.O., Integrins, versatility, modulation, and signalling in cell adhesion, *Cell*, 69, 11–25, 1992.

Kim, S.S., Utsunomiya, H., Koski, J.A., Wu, B.M., Cima, M.J., Sohn, J., Mukai, K., Griffith, L.G. and Vacanti, J.P., Survival and function of hepatocytes o a novel three-dimensional synthetic biodegradable polymer scaffolds with an intrinsic network of channels, *Ann Surg*, 228, 8–13, 1998.

Koh, Y.-H, Jun, I.-K. and Kim, H.-E., Fabrication of poly(ε-caprolactone)/hydroxyapatite scaffold using rapid direct deposition, *Materials Letters*, 60, 1184–1187, 2006.

Kowata, S. and Sun, H.B., Two-photon photopolymerization as a tool for making micro-devices, *Appl Surf Sci* 208/209, 153–158, 2003.

Lam, C.X.F., Mo, X.M., Teoh, S.H. and Hutmacher, D.W., Scaffold development using 3D printing with a starch-based polymer, *Mater Sci Eng*, 20, 49–56, 2002.

Langer, R., Tissue engineering: a new field and its challenges, *Pharm Res*, 14, 840–841, 1997.
Langer, R. and Vacanti, J.P., Tissue engineering, *Science*, 260, 920–926, 1993.
Lee, G. and Barlow, J.W., Selective laser sintering of bioceramic materials for implants, *Proceedings of the '96 SFF Symposium*, Austin, TX, August 12–14, 1996.
Lemercier, G., Mulatier, J.C., Martineau, C., Anémian, R., Andraud, C., Wang, I., Stéphan, O., Amari, N. and Baldeck, P., Two-photon absorption: from optical limiting to 3D microfabrication, *Comptes Rendus Chimie*, 8, 1308–1316, 2005.
Leong, K.F., Cheah, C.M. and Chua, C.K., Solid freeform fabrication of the three-dimensional scaffolds for engineering replacement tissues and organs, *Biomaterials*, 24, 2363–2378, 2003
Leukers, B., Gulkan, H., Irsen, S.H., Milz, S., Tille, C., Schieker, M. and Seitz, H., Hydroxyapatite scaffolds for bone tissue engineering made by 3D printing, *J Mater Sci Mater Med*, 16, 1121–1124, 2005.
Levy, R.A., Chu, T.G.M., Holloran, J.W., Feinberg, S.E. and Hollister, S., CT-generated porous hydroxyapatite orbital floor prosthesis as a prototype bioimplant, *Am J Neuroradiol*, 18, 1522–1525, 1997.
Limpanuphap, S. and Derby, B., Manufacture of biomaterials by a novel printing process, *J Mater Sci Mater Med*, 13, 1163–1166, 2002.
Mahajan, H. P., Evaluation of chitosan gelatine complex scaffolds for articular cartilage tissue engineering, MSc Thesis, Mississipi State University, USA, 2005.
Marler, J.J., Upton, J., Langer, R. and Vacanti, J.P., Transplantation of cells in matrices for tissue regeneration, *Adv Drug Del Rev*, 33, 165–182, 1998.
Matsuda, T. and Mizutani, M., Liquid acrylate-endcapped poly(e-caprolactone-co-trimethylene carbonate). II. Computer-aided stereolithographic microarchitectural surface photoconstructs, *J Biomed Mater Res*, 62, 395–403, 2002.
Miranda, P., Saiz, E., Gryn, K. and Tomsia, A.P., Sintering and robocasting of β-tricalcium phosphate scaffolds for orthopaedic applications, *Acta Biomaterialia*, 2, 457–466, 2006.
Mironov, V., Boland, T., Trusk, T., Forgacs, G. and Markwald, R.R., Organ printing: computer-aided jet-based 3D tissue engineering, *Trends Biotechnol*, 21, 157–161, 2003.
Mooney, D.J., Baldwin, D.F., Suh, N.P., Vacanti, J.P. and Langer, R., Novel approach to fabricate porous sponges of poly(D,L-lactic-co-glycolic acid) without the use of organic solvents, *Biomaterials*, 17, 1417–1422, 1996.
Moroni, L., Schotel, R., Sohier, J., Wijn, J.R. and Blitterswijk, C.A., Polymer hollow fiber three-dimensional matrices with controllable cavity and shell thickness, *Biomaterials*, 27, 5918–5926, 2006.
Nathan, C. and Sporn, M., Cytokines in context, *J Cell Biol*, 113, 981–986, 1991.
O'Brien, F.J., Harley, B.A., Llanas, I.V. and Gibson, L.J., The effect of pore size on cell adhesion in collagen-GAG scaffolds, *Biomaterials*, 26, 433–441, 2005.
Pardo, L., Wilson, W.C. and Boland, T., Characterization of patterned self-assembled monolayers and protein arrays generated by the ink-jet method, *Langmuir*, 19, 1462–1466, 2003.
Park, A., Wu, B. and Griffith, L.G., Integration of surface modification and 3D fabrication techniques to prepare patterned poly(L-lactide) substrates allowing regionally selective cell adhesion, *J Biomater Sci-Polym E*, 9, 89–110, 1998.
Pfister, A., Landers, R., Laib, A., Hübner, U., Schmelzeisen, R., Mülhaupt, Biofunctional rapid prototyping for tissue-engineering applications: 3D bioplotting versus 3D printing, *Journal of Applied Polymer Science Part A: Polymer Chemistry*, 42, 624–638, 2004.
Popov, V.K., Antonov, E.N., Bagratashvili, B.N., Konovalov, A.N. and Howdle, S.M., Selective laser sintering of 3-D biodegradable scaffolds for tissue engineering, *Mat Res Soc Symp Proc.*, EXS-1, F5.4.1–F.5.4.3, 2004.
Reignier, J. and Huneault, M.A., Preparation of interconnected poly(ε-caprolactone) porous scaffolds by a combination of polymer and salt particulate leaching, *Polymer*, 47, 4703–4717, 2006.
Sachlos, E., Reis, N., Ainsley, C., Derby, B. and Czernuszka, J.T., Novel collagen scaffolds with predefined internal morphology made by solid freeform fabrication, *Biomaterials*, 24, 1487–1497, 2003.
Sachs, E.M., Haggerty, J.S, Cima, M.S. and Williams, P.A., Three-dimensional printing techniques, US Pat. 5204055, 1989.

Saiz, E., Gremillard, L., Menendez, G., Miranda, P., Gryn, K and Tomsia, A.P., Preparation of porous hydroxyapatite scaffolds, Materials Science and Engineering C, 27, 546–550, 2007.

Saunders, R., Derby, B., Gough, J. and Reis, N., Ink-jet printing of human cells, Mat Res Soc Symp Proc, EXS-1, F.6.3.1–F.6.3.3, 2004.

Skalak, R. and Fox, C.F., Tissue engineering, Alan R. Liss, New York, 1988.

Tabata, Y., Recent progress in tissue engineering, *Drug Discov Today*, 6, 483–487, 2001.

Taboas, J.M., Maddox, R.D., Krebsbach, P.H. and Hollister, S.J., Indirect solid free form fabrication of local and global porous biomimetic and composite 3D polymer-ceramic scaffolds, *Biomaterials*, 24, 181–194, 2003.

Tellis, B.C., Szivek, J.A., Bliss, C.L., Margolis, D.S., Vaidyanathan, R.K. and Calvert, P., Trabecular scaffolds created using micro CT guided fused deposition modeling, Materials Science and Engineering C (2007).

Tormen, M., Businaro, L., Altissimo, M., Romanato, F., Cabrini, S., Perennes, F., Proitti, R., Sun, H.B., Kawata, S., Fabrizio, E.D., 3D patterning by means of nanoimprinting, X-ray and two-photon lithography, *Microelectronic Engineering*, 73/74, 535–541 2004.

Vozzi, G., Flaim, C., Ahluwalia, A. and Bhatia, S., Fabrication of PLGA scaffolds using soft lithography and microsyringe deposition, *Biomaterials*, 24, 2533–2540, 2003.

Wang, F., Shor, L., Darling, A., Khalil, S., Güçeri, S. and Lau, A., Precision deposition and characterization of cellular poly-e-caprolactone tissue scaffolds, *Rapid Prototyping J*, 10, 42–49, 2004.

Wei, G., Jin, Q., Giannobile, W.V. and Ma, P.X., The enhancement of osteogenesis by nano-fibrous scaffolds incorporating rhBMP-7 nanospheres, *Biomaterials*, 2087–2096, 2007.

Whang, K., Thomas, C.H., Healy, K.E. and Nuber, G., A novel method to fabricate bioabsorbable scaffolds, *Polymer*, 36, 837–842, 1995.

Williams, J.M., Adewunmi, A., Schek, R.M., Flanagan, C.L., Krebsbach, P.H., Feinberg, S.E., Hollister, S.J. and Das, S., Bone tissue engineering using polycaprolactone scaffolds fabricated via selective laser sintering, *Biomaterials*, 26, 4817–4827, 2005.

Woodfield, T.B.F., Malda, J., de Wijn, J., Péters, F., Riesle, J. and van Blitterswijk, C.A., Design of porous scaffolds for cartilage tissue engineering using a three-dimensional fiber-deposition technique, *Biomaterials*, 25, 4149–4161, 2004.

Xiong, Z., Yan, Y. and Zhang, R. and Sun, L., Fabrication of porous poly(L-lactide acid) scaffolds for bone tissue engineering via precise extrusion, *Scripta Materialia*, 45, 773–779, 2001.

Xiong, Z., Yan, Y., Zhang, R. and Wang, X., Organism manufacturing engineering based on rapid prototyping principles, *Rapid Prototyping J* 11(3), 160–166, 2005.

Yan, Y., Wu, R. and Zhang, R, Biomaterial forming research using RP technology, *Rapid Prototyping J*, 9, 142–149, 2003a.

Yan, Y., Zhang, R. and Lin, F., Research and applications on bio-manufacturing, Proceedings of the 1st International Conference on Advanced Research in Virtual and Rapid Prototyping, Leiria, Portugal, 2003.

Zein, I., Hutmacher, D.W., Tan, K.C. and Teoh, S.H., Fused deposition modeling of novel scaffolds architectures for tissue engineering applications, *Biomaterials*, 23, 1169–1185, 2002.

Zeltinger, J., Sheerwood, J.K., Graham, D.M., Mueller, R. and Griffith, L.G., Effects of pore size and void fraction on cellular adhesion, proliferation, and matrix deposition, Tissue Engineering, 7(5), 557–572, 2001.

Chapter 9
Rapid Prototyping to Produce POROUS SCAFFOLDS WITH CONTROLLED ARCHITECTURE for Possible use in Bone Tissue Engineering

Alexander Woesz

In this chapter we review the various attempts to use Rapid Prototyping (RP) methods to directly or indirectly produce scaffolds with a defined architecture from various materials for bone replacement and summarize our own investigations in that field.

We start with an introduction on the porous bone replacement materials used today and an explanation why it is beneficial to have the possibility to adjust the pore size, pore size distribution, pore shape and pore interconnectivity of such a porous bone replacement material by the usage of Rapid Prototyping methods from the mechanical as well as the biological point of view. Mechanical properties strongly depend not only on the apparent density and the bulk material's properties, but also on the architecture, the distribution of the material in space, and speed and depth of cell ingrowth into such a porous implant also depend on this architecture.

The second part describes and compares the published work in the field of porous implant production utilising RP, which includes the production of structures from polymers and ceramic materials as well as the usage of various composites from either groups. Among those are direct (meaning a one step process) procedures like fused deposition modelling (FDM), selective laser sintering (SLS), three-dimensional printing (3DP) and layered object manufacturing (LOM). For indirect, two step processes, where the Rapid Prototyping is just used to fabricate a mould for the future bone replacement material, nearly all RP methods can be used. Reports in literature can be found on the usage of stereolithography (SLA), three-dimensional printing (3DP), ballistic particle manufacturing (BPM) and others. Advantages and drawbacks of all methods are discussed and possibilities for future applications of the various methods are mentioned.

The last part is a summary of the results of the usage of the RP methods introduced before for the production of scaffolds with controlled architecture as bone replacement materials and a description of methods to investigate the applicability of the products is given.

9.1 Introduction

Bone is a highly complex, hierarchically structured material with remarkable mechanical performance. On a macroscopic scale, most bones consist of a dense outer shell (compact bone, corticalis) and a spongy interior (cancellous bone, spongiosa). Clearly the bones geometry, the distribution of bone material in space, is adapted to the loading conditions. Besides its load bearing assignment, bone is part of a living being, and as such involved in a permanent cycle of remodelling, which is not only necessary to remove and replace damaged material, but also to adapt the bones geometry to new loading conditions.

Both aspects, the load bearing assignment and the fact that bone is a living material, have to be taken into consideration when bone replacement materials are developed.

Implants have been used for thousands of years [1, 2], but the achievements of modern times were necessary to allow us to produce a wide variety of materials usable as bone replacement materials in various application sites. Among these are materials as different from each other as titanium alloys [3] and polymethyl methacrylate bone cements [4]. Both are advantageous compared to others, depending on the application.

In Doremus et al., an orthopaedic surgeon described the ideal implant material as a material performing its function without toxicity and foreign body reaction that completely resorbs and is replaced by new tissue [2]. Anyhow, implants bearing heavy weight, as an artificial hip, cannot be manufactured from a resorbable porous calcium phosphate. Nevertheless, in other fields of application, the porous material is the best solution. In this chapter we present several different porous bone replacement materials and compare them to dense (non-porous) implants. Later on we present approaches found in literature to produce porous bone replacement materials with a designed pore size and architecture using Rapid Prototyping (RP) methods.

9.1.1 Currently Used Porous Implants

The best replacement for diseased bone material is healthy bone material. That's what happens in humans and animals through all their lifetime – osteoclasts resorb material by dissolution of the hydroxyapatite platelets and degradation of the organic matrix [5], which is soon replaced by new collagen produced by the osteoblasts. The new collagen is mineralised within several weeks. This regeneration process is probably controlled by the osteocytes, cells entrapped within the bone material, which are supposed to work as strain sensors. It is necessary to remove damaged material, e.g. material with micro-cracks induced by high mechanical loading, or to adapt the bones to new, higher impact [6].

Sometimes, the body is not able to produce new bone material, or to produce enough material, or to produce it in sufficient time. For these problems surgery knows several solutions, and again, in an case the best would be healthy bone material, if there were no side effects. **Autografts**, bone parts harvested from the patient himself, are osteoconductive, meaning they have the possibility to induce bone ingrowth and bone formation. Furthermore, they are osteoinductive, they form bone when they are implanted in non-osseous sites, which seems for this material self-evident, but is for synthetic bone replacement materials the final goal [7]. Besides these properties, what makes the natural bone material superior compared to other materials? On the one hand, other biologically relevant properties as the pore size and the pore size distribution, the interconnectivity of the pores and, very important, the resorbability, are, easy comprehensible, optimised by evolution. On the other hand, the mechanical properties of both cancellous and cortical bone grafts are, although not highest, yet optimal when combined with the surrounding natural bone material. All these arguments make surgery call the autografts a "gold standard".

Alas, when using autografts one has to cope with some disadvantages, too. The amount of autografts to be harvested from the iliac crest or other donor sites is limited, and every extraction comes along with an additional trauma, increasing the probability of morbidity [7–9].

Another possibility to use natural bone materials to fill supercritical three-dimensional cavities in bones (supercritical means that the bone cannot fill the defect itself) is to use **allografts**, bone parts harvested from donators of the same species as the acceptor. Advantageous is the avoidance of the morbidity associated with autograft harvest and the availability of the allografts [10]. Anyhow, the use of allografts is accompanied by several disadvantages, too. Their mechanical properties are inferior compared to autografts, the probability of immuno-rejection and inflammation is quite high [11, 12] and, even worse, a residual risk of disease transmission has to be admitted [13]. Furthermore, they are quite expensive.

A third option to use natural materials, are the so called **xenografts**. These are bone parts harvested from animals, e.g. cows or horses, whose organic material has carefully been removed. These materials are offered as powder, granules or blocks [14]. Some of them have impressive mechanical properties. They have been used for years and revealed good results, but, again, they bear a residual risk of disease transmission and immuno-rejection.

A near-nature possibility to replace bone material is the use of **coral derived hydroxyapatite/calcium carbonate composites**, which are used quite frequently. Coral skeletons consist of calcium carbonate in aragonite or calcite form. This material is easily dissolvable and thus unsuitable for most implant purposes. The material is therefore partly (on the surface) converted into hydroxyapatite in a hydrothermal exchange process. Due to its porous structure with a pore size between 150 and 500 µm, which is in the range of the pore size of human cancellous bone, and the fact, that it forms chemical bonds with bone and soft tissue, it is an auspicious candidate for bone substitution. The composition from durable hydroxyapatite and highly soluble calcite causing unknown degradation properties are, together with the low mechanical loading capacity, the limiting factors of the applicability of this material [10, 14, 15].

An advantage of the materials mentioned above is their resorbability. Nevertheless, materials which are stable when implanted gain some advantages when used in porous form, too. Porous **titanium, nickel-titanium alloy** or **hydroxyapatite coated titanium** is proposed as bone graft substitute in weight bearing sites

[3, 16–18]. The stiffness of these materials, which is close to the stiffness of human bone and their strength make them an alternative worth considering.

Rapid bonding to bone, bonding to soft connective tissue and a thick bonding gel layer minimising interfacial stress between implant and tissue are reported to be responsible for the high success rate of **biocompatible glass** (consisting of SiO_2, CaO, Na_2O and P_2O_5, and, sometimes Al_2O_3 [19]) in more than a decade of application [20, 21]. In most applications, this material is used in powder or granular form, but usage in block form was reported, too. Yuan et al reported of a dog model study, were Bioglass® (45S5, U.S. Biomaterials) turned out to be osteoinductive [22]. The effect of the geometry on the bone formation was investigated in [19], comparing bioglass fibre bundles and fibre balls.

Other bone replacement materials which are partly used in porous form are the synthetic calcium phosphates. The first successful repair of a bone defect with calcium phosphate materials was reported in 1920 [23]. Among the materials used today, there are grafts made of hydroxyapatite, tricalcium phosphate and biphasic calcium phosphate.

Hydroxyapatite (HA) is prepared by precipitation and sintering and has a Ca to P ratio of 1.67. It is commercially available in powder, granular and block form, with and without macro pores. The difference between the synthetic material ($Ca_{10}(PO_4)_6(OH)_2$) and the HA in bones is carbonate ions replacing PO_4, and other impurities [1]. The Ca to P ratio in the Carbonatehydroxyapatite ($(Ca, Mg, Na)_{10}(PO_4, HPO_4, CO_3)_6(OH)_2$) in bones, dentin and enamel is somewhere around 1.67. One of the properties strongly influenced by this difference is the solubility. Artificial hydroxyapatite is hardly soluble in water or simulated body fluid and its osteoclast resorbability is low, too. Nevertheless, other properties advise an application as bone graft: Hydroxyapatite develops a strong bond with bone material and there are no fibrous particles encapsulated on the bone/ceramic interface [2]. Dense, pure hydroxyapatite has, due to the absence of micro pores, relatively high strength and an excellent fatigue resistance [2]. However, the intrinsic brittleness of all ceramics limits the applicability of hydroxyapatite as a weight bearing bone graft. **Tricalcium phosphate** is prepared by precipitation and sintering or by solid state reactions at high temperatures. It has a Ca to P ratio of 1.5 and has been used as bone replacement material, although its solubility is very high, which is a strong restriction for most applications [20]. **Biphasic calcium phosphate** is a composite of tricalcium phosphate and hydroxyapatite with varying composition and thus Ca to P ratio varying between 1.5 and 1.67. It is prepared by sintering precipitated calcium deficient apatite and has been used in macro porous form for more than 10 years in dental applications and for bone substitution, although its mechanical stability is low in porous form [24].

Frayssinet et al assessed the osteointegration of pure hydroxyapatite in comparison to several biphasic calcium phosphates in vivo. The higher the amount of β-tricalcium phosphate, the higher the bone ingrowth was [25]. On the other hand, Kasten et al. found in vitro, that the bone ingrowth performance of hydroxyapatite was higher than the one of β-tricalcium phosphate [26]. Yuan et al found a higher probability of bone formation in hydroxyapatite and biphasic calcium phosphates compared to tricalcium phosphate [27]. In [28] the influence of bone morphogenetic protein (BMP) on the formation of bone is investigated. The authors observed that on solid particles of hydroxyapatite no bone formation took place, whereas on porous hydroxyapatite new bone was formed. In [29], the osteogenesis induced by BMP was studied, using porous hydroxyapatite samples with different pore sizes between 100 and 500 μm as substrates. The samples with a pore size between 300 and 400 μm turned out to induce to highest bone formation. The osteoconductive properties of two commercially available hydroxyapatite bone grafts are compared in [30]. Both are macro porous, nonetheless the authors found one material to be osteoconductive, the other not. They suggest the microstructure to be responsible for their findings.

These examples show that though the mechanisms behind osteointegration are complex, bone ingrowth was observed with several calcium phosphate materials in several different forms. These materials combine good biological properties (osteoconductivity, even osteoinductivity and solubility/resorbability) with partially good mechanical stability.

9.1.2 Why is it Beneficial to have the Possibility to Control the Architecture of a Porous Implant

The first and most important benefit gained by using porous bone grafts instead of dense materials is the possibility for the surrounding natural bone to grow into the graft and thus connect it mechanically. The ingrowth of bone into a porous implant can actually stabilise and strengthen the material itself, as described in [31]. The optimal pore size for bone cell ingrowth is mentioned in many publications, but not all authors get the same results: In [32], 50 μm and higher are mentioned, a minimum of 200 μm is advised by Klawitter in [33], Kuboki et al. mentioned 300–400 μm as an optimum [28] and in [34], 565 μm is found to be better than 300 μm. With an optimal pore size alone, bone ingrowth and proliferation is not assured, pore interconnectivity plays an important role, too [35].

When a dense metallic implant is driven into a bone, the surrounding bone material is dilated and exerts pressure onto the implant and thus locates it. If the implant is loosened once, the mechanical connection gets lost. For porous materials, strain-induced fixation is not absolutely necessary, as the bone ingrowth fixes the graft. Furthermore, if a metallic implant within a bone, with a stiffness that is much higher then the stiffness of the bone, is loaded in bending, the bone bears higher strain than the implant, a mechanism leading to disconnection between implant and bone (Fig. 9.1).

Hence, for materials with high stiffness, as metal alloys, the loss in strength and stiffness associated with increasing porosity can be advantageous. On the contrary, materials with a strength and stiffness below natural bone cannot be used in porous form in weight bearing applications. For a material with optimised architecture, the strength and stiffness scales linearly with the apparent density, but for random open cell foam structures, the strength and stiffness decrease with higher powers of

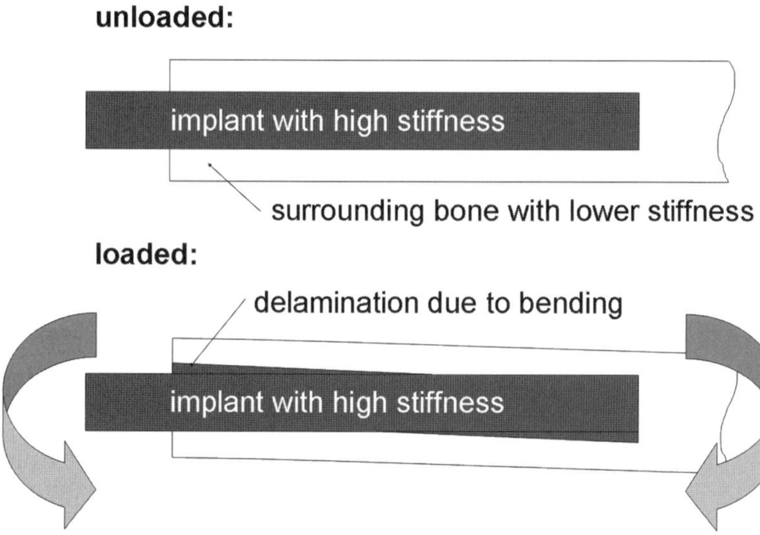

unloaded:

[implant with high stiffness]

surrounding bone with lower stiffness

loaded:

delamination due to bending

[implant with high stiffness]

Fig. 9.1 The mechanism leading to disconnection between implant with high stiffness and bone with lower stiffness when loaded in bending

the density [36]. Anyway, besides the apparent density and the materials properties itself, the architecture controls the mechanical properties of porous materials [37].

Another advantage accompanying porosity is resorbability. The osteoclast cells attach on the surface of the material to be resorbed and dissolve it by producing an acidic environment locally [5]. The resulting resorption lacunae have a depth of approximately 50–100 μm. This suggests a feature thickness of 200 μm (twice the depth of the lacunae) as maximum for resorbable, but not soluble materials, if one wants the graft to be removed and replaced by natural bone material.

Besides the adjustment of the internal architecture, the usage of some Rapid Prototyping methods allows to design the outer shape of an implant according to the needs of the patients. As human bodies are fairly different from one another, the need for custom products is much higher then in any other engineering field. Several reports of the successful application of Rapid Prototyping for the production of parts which are used at or in the human body can be found, a good example being dentistry, where brackets for the correction of tooth position are produced by RP in high quantities [38].

9.2 Rapid Prototyping Methods used to Produce Porous Bone Replacement Materials with Controlled Architecture

So far several Rapid Prototyping methods were utilised for the production of porous bone replacement materials. In general it makes sense to distinguish between direct and indirect methods. The former directly yield the product, whereas the indirect methods are used to produce an intermediate product, usually a mould, which is in a second step moulded into the final product, the porous bone replacement material, in the following called scaffold. In case there is a post-processing step like a heat-treatment necessary, the method is still considered as being direct.

RP methods were originally, developed to fasten the process of developing a new product. In the meantime they have, due to achievements in the field of accuracy and materials quality, reached the status of methods also used for tool production and small batch production (Rapid Tooling and Rapid Manufacturing). They can be used to manufacture parts from polymers, ceramics, metals and composites of those, in the μm range up to the meter range. What they do have in common is the fact that the information transferred into the actual product in the RP process stems from a virtual source. This can be a virtual 3-dimensional drawing performed usually with solid modelling computer aided design (CAD) software or it can be the outcome of a 3-dimensional scanning procedure (reverse engineering). It can also be the virtual 3-dimensional result of a computer tomography or magnetic resonance investigation as commonly used in medicine.

A second feature characterising most RP methods (all commercially available RP methods) is the layer by layer production process. RP machines do not, as with many engineering methods, remove material until the desired shape is obtained (although sometimes CNC machining is considered a RP method), but they add material in small portions, in a layer by layer fashion.

9.2.1 Direct Methods

9.2.1.1 Fused Deposition Modelling, FDM

A movable nozzle deposits a thread of molten material onto a substrate. The build material is heated slightly above its melting temperature so that it solidifies right after it has left the nozzle and was placed at its position. It is necessary that a steady nozzle speed and extrusion rate is chosen and support material has to be used for overhangs or island features (Fig. 9.2).

Most recent systems use two nozzles, one for the build material and one for the support material, which can be broken off after the part is finished, according to the manufacturer without harming the surface of the part. Another system uses water-soluble support material.

Materials used for conventional applications are thermoplastics, acrylonitrile butadiene styrene (ABS), high impact grade of ABS, investment casting wax and elastomers. Feeding of the material into the heated nozzle is performed by a system which takes the thread-like material from a spool. New threads of material are not laid down onto the surface, but squeezed onto the surface. The distance between the nozzle and the surface is about half the diameter of the tread coming from the nozzle. Usual nozzle diameters are around 0.3 mm. Advantages of FDM are that the materials used are cheap, non-toxic, not smelly and environmentally safe, thus it is possible to use the machines in an office surrounding. Drawbacks are the relatively

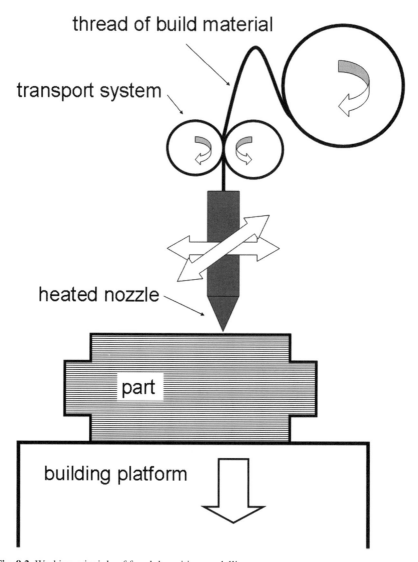

Fig. 9.2 Working principle of fused deposition modelling

bad surface finish and the fact that just one nozzle is utilised, making the usage of a relatively thick nozzle necessary to achieve reasonable building times. This limits the resolution of these systems [38–41]. A technique similar to FDM utilises a feedstock from a polymer/ceramic composite, which is deposited as usual. In the following temperature treatment the polymer is burnt off and a ceramic structure is obtained. This technique was named fused deposition of ceramics, **FDC** [42]

The first task to be faced when FDM should be used to process polymers or polymer/ceramic composites for bone replacement is the fabrication of a filament with an adequately constant diameter, which allows the material to be fed into the nozzle by the machine. The filament can be fabricated by melting the material in a piston and pressing it through a nozzle, but this process is already the first possibility to degrade the biopolymers by heat, a process which has carefully to be avoided when the desired molecular size and crystallinity and the according mechanical and biological properties should be preserved. This process is less probable during the actual FDM process, because the time during which the polymer is in the molten state is much shorter in the depositing head of the FDM machine. The geometry of the scaffold produced with FDM is influenced by the following parameters to be adjusted at the machine: the nozzle diameter, the filament speed, the head temperature, the distance between two adjacent ropes, the layer thickness and the pattern with which the ropes are deposited. Restrictions in the geometry of the scaffolds to be built are quite manifold in the FDM process as compared to other RP methods. Given that a strut of the scaffold consists of one, not several ropes from the nozzle, the system does not offer arbitrary three-dimensional fabrication, because the feeding of the material cannot be switched off and on during the fabrication of one layer. It is thus not possible to fabricate, for example, a strut oriented in the z-direction. Additionally, it is necessary to start and end the rope somewhat outside the scaffold to be built, requiring post-processing after finishing of the building process. Finally, as the usage of support structures which have to be removed afterwards is not possible in a scaffold for bone replacement (with the exception of water-soluble support), the pore size is restricted by the maximum gap between the ropes deposited in the first layer that can be spanned by the next rope without unacceptable sag [43–47].

9.2.1.2 Selective Laser Sintering, SLS

SLS was developed by the University of Austin in Texas and commercialised in 1987. Powder is spread on a building platform with a roller and sintered by a powerful CO_2 laser. The whole chamber is heated to a temperature just below the melting point of the powder, so that the laser has to heat the material just a little more to melt it, thus causing the particles to fuse together. The hot building chamber is also necessary to prevent thermal distortion and to facilitate fusion with the previously built layer. After the layer is finished, the powder bed is moved down a layer thickness, new powder is applied by moving the powder-feed chamber up and spreading the powder with the roller. The sintered material is the part, the unsintered powder remains in place and acts as a support for overhangs and island features. The unsintered powder is brushed off after the part is finished, it can be reused after sieving (Fig. 9.3).

Materials used have to be produced in powder form and need to be sintered, so materials which decompose before melting cannot be used. Apart from this restriction, the variety of materials which have been used up to now comprise nylon, nylon composites, polystyrenes and polycarbonates, materials which are quite cheap, non-toxic, safe and can be sintered with relatively low laser power (10–20 W). But the method in principle allows to process metals and ceramics, too. Metal powder and sand, coated with a suitable binder layer can also be sintered with the equipment used for polymers. They are used for the direct production of metal tooling inserts

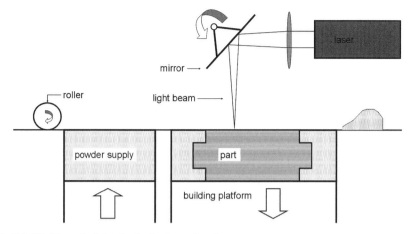

Fig. 9.3 Working principle of selective laser sintering

and sand cores for metal casting. Special powders have been developed for applications like parts with especially high accuracy or parts for investment casting of parts with high flexibility.

SLS has been used for the fabrication of porous scaffolds from polymers as well as polymer/ceramic composites. The first challenge to be faced is the production of a powder with appropriate particle shape, size and size distribution. When aiming at composites, the mixing of the particles has to be addressed, too. It can be done by physical mixing or, for example, spray drying of a suspension of the ceramic particles in a solution of the polymer [48]. Usually the building chamber of the machine has to be kept at a temperature slightly beneath the melting temperature of the material processed, a fact which does not allow to process polymers which tend to degrade at this temperature.

The resolution of the SLS system is restricted by the powder particle size, on the one hand, and by the size of the focal point of the laser, on the other hand. Additionally, heat transfer to adjacent regions of the focal point causes unwanted particles to stick to the surface of the scaffold, giving the surface a fur-like appearance. Removal of the unsintered material within the pores of the scaffold can become quite difficult, if the pore size becomes too small or the size of the scaffold becomes to large [48–53].

The usage of SLS to fabricate structures from pure ceramic or metal powders is possible, if high energy lasers and machines with high heat resistance are used [54, 55]. Another, more plain method is the usage of particles covered with a binder, which is fused in the SLS process. During a subsequent temperature treatment the binder is burnt off and the ceramic or metallic particles are sintered [56, 57]. Of course the obtained polymer/ceramic composite can also be used directly, as suggested in [58].

9.2.1.3 Three-dimensional Printing, 3DP

This method was originally developed at Massachusetts Institute of Technology (MIT) and is based on the selective fusing of particles by a binder, which is printed into a powder bed by inkjet print heads. These print heads act either on a thermal basis, where a pulse of current heats a tiny chamber filled with a water based ink (or binder), thus heating it and causing a small bubble of ink to be spread onto the substrate, or with a piezo pump, where the current makes a piezo crystal bend and force a small droplet of ink to leave the nozzle (Fig. 9.4).

The first step in the production process in 3DP is the spreading of powder onto a platform with a roller, followed by the inkjet print head printing a two-dimensional pattern, equalling the cross-section of the part to be built, onto the powder layer. The ink used is a polymeric binder which bonds the powder particles together. Then the next powder layer is spread and the process is repeated until the part is finished. The unused powder acts, as in the SLS process, as support for the part and is brushed or blown off afterwards.

Materials used in commercialised machines are starch or gypsum based powders and some metal powders and sand for the core production for casting. The main drawback of these systems is the fact that the parts are relatively fragile after production and usually have to be infiltrated before using. As with SLS, the removal of unattached powder from small cavities can be difficult. The resolution of the print heads is 600 dpi, but capillary forces make the binder penetrate adjacent regions, so that the actual resolution is lower. Main advantages are the usage of cheap and non-toxic materials, the high build speed of the systems and, of interest in some applications, the possibility to print in colour.

The 3DP process is applicable for the production of metallic or ceramic scaffolds by binding particles from such materials, followed by a temperature treatment to burn the binder off and a final sintering step [59, 60]. The problems are similar to those faced when using the SLS process. There it is heat transported from the focal point by heat transfer, here it is binder liquid transported into adjacent regions by capillary forces, which causes unwanted material to stick to the surface and give the

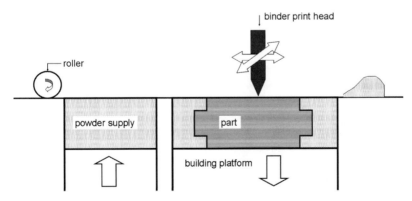

Fig. 9.4 Working principle of 3-dimensional printing

surface a fur-like appearance. This spreading of the binder also limits the resolution of the 3DP process, which is, apart from this, defined by the resolution of the print heads and the powder particle size.

9.2.1.4 Layered Object Manufacturing, LOM

LOM is used to manufacture 3-dimensional models by stacking individually cut sheets of paper or other materials, which are covered by a thermally activated adhesive. Cutting takes place either with a laser or with a knife, the latter is sometimes called PLT, paper laminated technology. Another variation of the process bonds foils of a polymer by curing them with UV light (Fig. 9.5).

Usually the build material is applied to the part from a roll, then bonded to the previous layers using a hot roller which activates the heat sensitive adhesive. Then the contour of the layer is cut with a laser which is carefully modulated in order not to cut more then one layer thickness deep. The surrounding waste material is cut into small portions to facilitate the later removal, but it remains in place to act as a support for overhangs and island features. The sheet material from the roll is wider then the part, so that the edges remain intact and the material can be advanced just by winding the excess material onto a second roller until a fresh area lies over the part.

The laser is a 25–50 W CO_2 laser. Advantages of LOM, generally speaking, are the wide range of cheap materials available, which can be paper, plastic or even fibre-reinforced glass ceramic and the fact that large parts can be manufactured at relative high speed, since the laser does not have to meet the whole cross-sectional

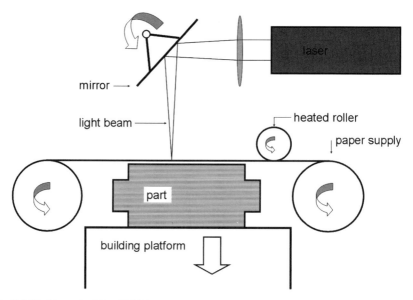

Fig. 9.5 Working principle of LOM

area of the part but just the contour. Major drawbacks are the need to remove the excess material after the building process is finished, a procedure which can even harm the surface of the part, the difficulties when producing hollow parts, the large amount of scrap and the bad mechanical properties and high anisotropy of the part [38–41, 61].

The usage of this method was reported for the production of bone replacements from hydroxyapatite particles bonded together by a calcium phosphate glass phase, but as the removal of small sections cut from the sheets is not possible, it does not allow to produce porous scaffolds with a defined porosity [62].

9.2.1.5 Others

For the production of porous metal scaffolds as bone replacement materials, the **LENS** process, **laser engineering net shaping**, should be mentioned. In this process metal powders are fed through a nozzle where they are fused with a laser. It is capable of forming parts from metals like tooling steel, stainless steel and also titanium alloys e.g. Ti-6-4 directly and the mechanical properties are equal to the ones achieved with conventional processing techniques. Although currently overhangs and island structures need support, a fact which limits the applicability of the system for the production of scaffolds for medical applications, the manufacturers are developing a fully flexible system which will enable most complex geometries without the usage of support structures.

Stereolithography, SLA, described in detail in the following, was used for the direct production of poly(ethylene glycol)dimethacrylate constructs for the study of cell behaviour in 3-dimensional surroundings. The monomer was mixed with photo-initiators and processed (photo-polymerised) in the SLA machine. Scaffolds with defined architecture were obtained [63].

In recent years, a method similar to FDM, named "bioplotting" was developed. It utilises a material tank which extrudes the plotting material into a plotting medium, which causes the dispensed material to solidify and stabilises the structure by buoyancy. Solidification can take place either by a chemical reaction between dispensed material and plotting medium or by cooling the hot dispensed material below its melting point. Although the resolution of the Bioplotter is not very high, it has a huge advantage: the whole plotting process can take place under sterile conditions and the plotted material can contain living cells [38, 64–67].

9.2.2 Indirect Methods

As indirect methods for the production of bone replacement materials nearly all RP methods are conceivable. Only those methods already used and mentioned in literature should be presented and discussed here.

9.2.2.1 Stereolithography, SLA

SLA is the first commercially available Rapid Prototyping method and was developed by teams in France, USA and Japan, independently from each other. It is the most widely used Rapid Prototyping system today and is based on the selective polymerisation of a photosensitive resin with light, usually a laser (Fig. 9.6).

A building platform is immersed in a tank filled with the photosensitive resin. The building platform is positioned (in z-direction) such that just a very thin layer of resin covers it, or, later in the building process, covers the previously built layers of the part. A mirror guides the laser beam across the surface of the resin where the future part is located. The laser light causes the resin to polymerise and thus to solidify.

After the complete cross-sectional area of one layer of the part has been met by the laser and solidified, the building platform is moved down a layer thickness, the coating system applies a new layer of resin on top of the previously built layer, and this procedure is repeated until all layers of the part have been built. Then the remaining liquid resin is washed off and the part is post-cured with UV light to fully polymerise the resin. Layer thicknesses are usually in the range between 0.02 and 0.2 mm, the laser exposes first the contour and afterwards the inner region. Advantages of SLA are that with this method or at least with similar methods known as µ-SLA, the highest resolutions of all RP methods can be achieved. Feature size below 1 µm is possible. Furthermore, some resins yield parts with mechanical properties comparable to those of many engineering polymers. Drawbacks are that

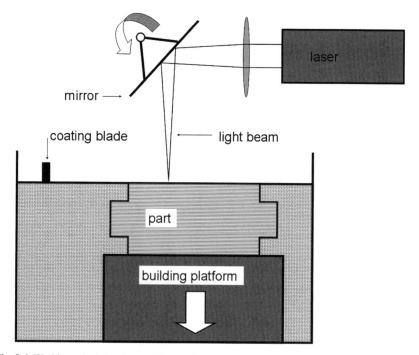

Fig. 9.6 Working principle of stereolithography

undercuts and island features have to be supported, causing damage to the parts, surface when removed, the shrinkage of the polymer during polymerisation, which causes internal stresses or warpage, and the restrictions as far as the available materials are concerned. Photosensitive resins and powder filled resins, both being toxic, are the only materials which can be directly used in SLA.

Advantages of SLA for the production of moulds for bone replacement materials are the above mentioned high resolution and small minimum feature size and the high achievable surface quality. The main disadvantage is the limitation in materials which can be processed with this system. The photosensitive resins usually need high temperatures to be removed after moulding, thus restricting the scaffold materials to be processed to ceramics and metals [68–70]. Additionally, the high strength of the resins can be a disadvantage, too. The higher the strength of the mould, the higher the risk of crack development in the unsintered green part due to thermal expansion mismatch. Biopolymers and polymer composites are damaged or destroyed before the mould is burnt off. Recent development of water-soluble photosensitive resins could give rise to the usage of SLS for the production of scaffolds from polymers and polymer composites [71].

Miniaturised methods similar to SLA are **Digital Light Projection, DLP**, a method based on digital mirror devices or dynamic mask generators, and **Two Photon Lithography**. The basic principle of DLP comprises the illumination of a digital mirror device or a dynamic mask with non-coherent light. The pixels of the mirror or mask can be individually turned off and on. Projecting the image of a layer of the part to be built onto the photosensitive resin allows one to fabricate one layer at once, a fact responsible for the high building speed achieved with these machines. Two-Photon-Lithography, by contrast, is a method with extremely low build speed, but the highest achievable resolution of all RP methods. It either uses a focused laser beam or two crossing laser beams such that just in the point of focus or in the area where the two laser beam cross the light intensity is high enough to trigger photo-polymerisation. (Fig. 9.7)

In principle, the DLP system has the same advantages and disadvantages as the SLA method. General advantages of DLP over SLS in the context of bone replacement production are the fact that DLP does not use a laser, which reduces the system costs significantly, and has a higher build speed due to the exposure of one layer at once. Disadvantages compared to SLS are the larger minimum feature size and lower resolution [72].

9.2.2.2 Ballistic Particle Manufacturing, BPM

A stream of molten materials is ejected from a nozzle, separates into droplets which hit the substrate where they immediately weld to form the part. The stream can be "drop on demand" or a continuous jet. By controlling the current, which activates the piezo-pump, the size, speed and the amount of drops per time unit can be adjusted. The proper combination of temperature, speed and distance has to be chosen to make the drops weld to the substrate (high enough temperature to partly melt the substrate), but not to deform too much (low enough temperature to have sufficiently

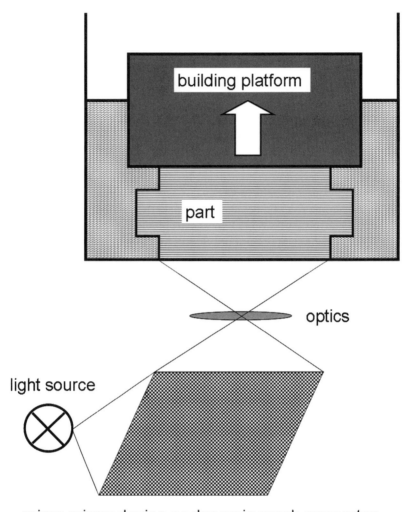

Fig. 9.7 Working principle of digital light projection

high viscosity). In systems like the Solidscape Modelmaker, two piezo-pumps are utilised to deposit two materials, a build material and a support material. Other manufacturers offer systems using just one kind of material, which requires the usage of support structures for overhangs and island features. Other systems also use just one material, but many nozzles, thus increasing the build speed drastically. Sometimes these systems are referred to as **Multi Jet Modelling, MJM** (Fig. 9.8).

BPM, especially the Model Maker series (Solidscape, Inc., Merrimack, NH, USA), has been used extensively for the production of moulds for bone replacement materials with defined architecture. The main advantage of this system is the usage of wax as mould material. The wax can either be removed after moulding

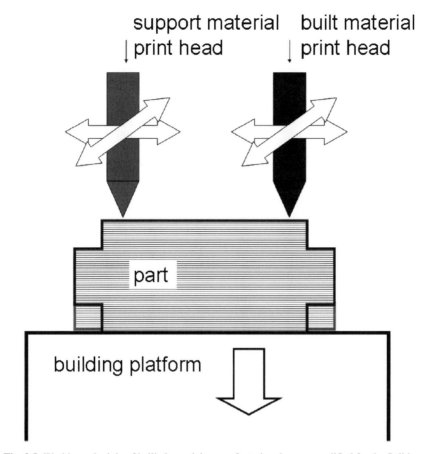

Fig. 9.8 Working principle of ballistic particle manufacturing, here exemplified for the Solidscape Modelmaker, which uses two kinds of materials, a support and a build material

by dissolution in alcohol or by a comparably moderate temperature treatment. The usage of a second material, the support wax, which is dissolved after the build job is finished, allows the production of nearly arbitrary geometries, including overhangs and island features. The resolution of this system is quite good.

Disadvantages are the very low building speed and the high effort to calibrate and prepare the system before every single build job. Additionally, the porosity of the wax can cause problems during moulding, and reactions between the wax and cast material can take place [73–77].

9.2.2.3 Others

The methods presented in the section on direct production routes partly have also been used for the indirect production of scaffolds for bone replacement. In

Table 9.1 RP methods and materials used for the production of porous bone replacement materials

direct methods:	material class	material	
FDM	polymer	polycaprolactone (PCL), poly(l-lactic acid) (PLLA), poly(ethylene glycol)-terephtalate - poly(buthylene terephtalate) copolymer	
	composite	PCL/tricalciumphosphate (TCP), polypropylene (PP)/TCP, chitosan/hydroxyapatite (HA), polyethylene (PE)/HA	
	ceramic	TCP, HA	
SLS	polymer	PCL, poly(D,L-lactic) acid (PLA)	
	composite	high density polyehylene (HDPE)/HA, polyvinylalcohol (PVA)/HA	
	ceramic	TCP, HA, apatite-mullite glass ceramic, HA/phosphate glass composite	
3DP	ceramic	HA	
SLA	polymer	poly(ethylene glycol) dimethacrylate	
	ceramic	HA	
Bioplotter	polymer	agar, fibrin/alginate acid, gelatin, polyurethane	
indirect methods:	mould material class	cast material class	cast material
FDM	polymer	ceramic	calcium aluminates, alumina
3DP	ceramic	polymer	poly(D,L-lactic-co-glycolic acid) (PLGA)
SLA	resin	ceramic	HA
DLP	resin	ceramic	HA
BPM	wax	polymer	collagen, PLGA
		composite	chitosan/HA
		ceramic	TCP, HA

[74], the authors facilitate 3DP for the production of moulds for solvent casting of poly(D,L-lactic-co-glycolic acid). The authors of [78] used FDM to produce wax moulds for ceramic gelcasting of calcium aluminates.

Examples for combinations of methods and materials, which have successfully been used to fabricate scaffolds with defined architecture for bone replacement applications can be found in Table 9.1.

9.3 Scaffolds with Controlled Architecture Produced as Bone Replacement Materials

In the following we summarise the outcome of the various attempts to produce porous scaffolds with defined architecture by RP methods.

9.3.1 Polymeric Scaffolds

Polymers have been and are used to a great extent as porous bone replacement materials, both as permanent and as bioresorbable implants. Especially the resorbable materials suffer from their low strength, not allowing using them in load bearing applications. Nonetheless, their usefulness is without a doubt, as they guide the surrounding bone cells into the space they fill and cause the (otherwise supercritical) defect to be filled with new bone material.

Polymeric scaffolds have been produced directly using FDM, SLS, SLA and the Bioplotter and have been produced indirectly utilising 3DP and BPM. Literature on the usage of direct FDM for the production of porous bone replacement materials from poly(ε-caprolactone) (PCL), poly(L-lactic acid) (PLLA) and poly(ethylene glycol)-terephtatlate-poly(butylene terephtalate) block copolymer (PEGT/PBT) can be found.

PCL is a semi-crystalline polyester which is degraded by hydrolysis of its ester linkages in physiological conditions such as in the human body. It degrades very slowly, thus being especially interesting for long term implant applications. It has an Food and Drug Administration (FDA) approval and is used as drug delivery devices, as sutures and as adhesion barriers [79]. Its mechanical properties are, compared to other biopolymers, quite low, as is its unusually low melting point of around 60 °C and low glass transition temperature of about - 60 °C. Nonetheless it is very temperature resistant, its degradation temperature being above 350 °C. This high temperature stability is one of the huge advantages of this material when processed with FDM. Neither the filament production nor the actual deposition process, which have to take place above the melting point of the material, affect its molecular weight and correlated mechanical properties.

Cao et al. [80] produced scaffolds with a porosity of 60–65%, and a pore size between 300 and 580 µm. Similar scaffolds of two different geometries were fabricated in [81], one having a 90° rotation in the strut orientation between two adjacent layers, the other having a 60° rotation. The authors obtained porosity between 48 and 77% and pore sizes between 160 and 700 µm. Clearly the pore size of horizontally oriented pores is much smaller then the vertical pores, due to the above mentioned restrictions of geometry of the FDM machines when single polymer threads are deposited. Additionally, the scaffolds had to be cut after the building process.

The production procedure is, within the machine limitations, highly flexible and the results of both the investigation of the mechanical properties and of the cell response [44, 82–84] to the scaffolds clearly advises the usage of FDM for the production of PCL scaffolds for low-load bone replacement applications.

PLLA is a biodegradable, thermoplastic, aliphatic polyester derived from lactic acid. Due to the chiral nature of lactic acid several distinct forms of polylactide exist: poly(L-lactide) (PLLA) is the semi-crystalline and relatively hard product resulting from polymerization of lactic acid in the L form. PLLA has a glass transition temperature (T_g) between 50–80 °C and a melting temperature between 173–178 °C. The polymerization of a mixture of both L and D forms of lactic acid leads to the synthesis of poly(DL-lactide) (PDLLA) which is an amorphous, transparent material with a T_g of 50–60 °C. By adjusting the ratio between the two forms of PLA, the crystallinity and the degradation time in the human body can be varied to some extent [85]. PLA is currently used in a number of biomedical applications, such as sutures, dialysis media and drug delivery devices, but it is also evaluated as a material for tissue engineering [86, 87]. It has higher strength and stiffness than PCL (partly because the application temperature in the human body is below its glass transition temperature), but its molecular weight is decreased by a factor of >10 during processing, reducing its mechanical properties to a great extent. The porosity of the scaffolds is around 60%, the pore size was varied between 200 and 500 μm. Although the strength of the material is significantly reduced by the FDM process, their mechanical properties still allow to use them as scaffolds for bone tissue engineering [88].

PEGT/PBT copolymers of varying composition were used in [47] for the production of scaffolds with a porosity of up to 88%, and a pore size between 150 and 1650 μm. Due to the geometrical restrictions of the FDM process, the pore size in horizontal direction was always below 200 μm. The material is biocompatible and bioresorbable, with its degradation time as well as other properties like swelling and mechanical strength being adjustable by varying the composition. Its mechanical strength and stiffness are, compared to other biopolymers, low, thus advising their usage more in the field of cartilage tissue engineering.

The usage of SLS for the production of porous scaffolds with defined architecture from PCL and PLA can be found in literature. In [53] the authors produced scaffolds from PCL in a conventional SLS machine. Again, the reduction in molecular weight correlated with long time high temperature treatment does not occur with PCL, due to its low melting temperature of 60 °C and its high degradation temperature of 350 °C. In this study, the working chamber of the machine was kept at about 50 °C, such that the laser had to heat the material just a little more to melt and sinter it. The above mentioned problem, inevitable in SLS, of sticking of particles onto the surface of the molten material led to significantly lower porosity than designed. Porosity between 35 and 55% was obtained, pore size was designed to be between 1.75 and 2.5 mm, but was not measured after production. Geometrical restrictions of the SLS process are much less then with FDM, overhanging and island features are, because of the unsintered powder acting as support, possible and the outer shape can be designed according to the patients needs. In these studies human [51] and pig [53] mandibular condyle scaffolds were presented as examples of the possibilities of the SLS process. The low mechanical strength and stiffness of the PCL raw material does, again, not allow its usage in high load bearing applications, but its thermal stability nonetheless advises its usage in SLS. The geometrical freedom of the SLS process though makes the combination a promising opportunity for bone and cartilage tissue engineering.

A variation of the conventional SLS process, which avoids the complete melting of PLA particles (and the correlated reduction in molecular weight) during the SLS process, is presented in [49]. The authors covered the surface of the PLA particles with carbon micro-particles which absorb, as opposed to the PLA, the infrared laser light, such that just the surface of the particles is molten and the bulk remains relatively unharmed by the heat. In this process the powder bed is not heated, so not only the molecular weight of the PLA is preserved, but it is also possible to keep temperature-sensitive bioactive substances within the particles active. The authors call the process SSLS, surface-selective laser sintering. In this work 3-dimensional scaffolds with defined architecture were produced, but due to the large particle size used, the accuracy of the process is quite low. Additionally, it would be very difficult to remove unsintered powder if the scaffolds exceeded a certain size.

As mentioned above, the SLA process is limited as far as the direct production of scaffolds from biocompatible materials is concerned, due to the restrictions in the availability of materials, whose polymerisation is initiated by photons, but nonetheless attempts have been undertaken to use this procedure for such purposes. In [63] the authors produced scaffolds from poly(ethylene glycol)dimethacrylate (PEGDMA), which was mixed with a photo-initiator, and in [89] the applicability of several water-soluble polymers for the usage in SLA was investigated. Although the materials used so far do not allow their usage in bone replacement applications, the advantages of the SLA process over other RP methods, especially the high resolution and accuracy, make further investigations in this direction a promising challenge.

In [64] and [65] the authors describe the usage of the Bioplotter for the production of hydrogel scaffolds from agar, gelatine and calcium alginate, all materials with very low strength and stiffness. Although this low stiffness does not allow using such scaffolds directly, they could be used in tissue engineering for the in vitro culture of bone replacement tissue. The main advantage of the Bioplotter is the printing into a liquid, whose density is adjusted such that the buoyancy of the plotted medium compensates the forces of gravity, allowing greater freedom in geometrical design as compared to, for instance, FDM. The restrictions due to the impossibility to stop and start the deposition of material within a layer of the part to be build are, as with FDM, still unsolved. In another work bioplotting of saccharide-based aliphatic polyurethanes, which have good biocompatibility, was compared with the results of an attempt to use such materials in a two-step 3DP process [67]. The scaffolds produced are of good accuracy, with a strut diameter and gap width in between two struts of around 0.7 mm. The mechanical properties of the plotted scaffold are, especially when compared to the 3D-printed material, quite impressive, although they decrease significantly during swelling in water. Nonetheless, bioplotting, which can also be used for the fabrication of scaffolds from the conventional biopolymers like the ones introduced above, is a promising technique in the field of production of porous bone replacement materials with defined architecture.

The main problem with the indirect production of polymeric scaffolds with RP is the removal of the mould after casting. Most materials processed in RP machines are more stable at elevated temperatures or in solvents than the biopolymers with which they are to be filled. The authors of [74] thus utilise 3DP of plaster, which dissolves in water after moulding, for the production of the mould. In this study, Poly(D,L-lactic-co-glycolic acid) (PLGA) was filled into the plaster moulds produced with a regular 3DP machine with a layer thickness of 100 µm, a very small strut width of 150 µm and a large gap region of 850 µm. Purpose of this design is to obtain small channels in the later part, which are possible growth regions for intestinal epithelial cells. In addition to the channels formed by the mould, the formation of a random porosity in the PLGA matrix was achieved by casting of a mixture of the polymer solution (solvent casting) with sucrose particles of defined size, which are later dissolved simultaneously with the mould (particulate leaching). The scaffolds obtained thus have a twofold porosity, a defined one, being the negative of the plaster mould and a random one defined by the amount and size of the sucrose particles. Although the construction of the plaster mould from particles gives the surface of the mould a rather rough appearance, the usage of the quite simple, cheap and fast 3DP process, which has a good resolution and high freedom in 3-dimensional design, for the production of water soluble moulds seems to be quite encouraging.

Another possibility to avoid the problem associated with mould removal is the usage of BPM for the production of moulds from wax, which is, at least the one used in the Modelmaker II by Solidscape (Solidscape Inc., Merrimack, NH, USA) dissolvable in ethanol. Moulds produced with this machine have been used for the production of scaffolds from collagen [90, 91] and PLLA [73]. Although the mechanical properties of collagen are not suitable for load bearing applications, it is, being the major component of the extra-cellular matrix of the human body, considered an ideal candidate for tissue engineering as far as cell attachment, morphology, migration and even differentiation is concerned.

9.3.2 Ceramic Scaffolds

The low mechanical strength and stiffness, being the inherent limitation of the applicability of polymeric scaffolds for bone replacement, is overcome by the usage of ceramic scaffolds, which can also, given the proper choice of material, be bioresorbable. Although the low fracture toughness of ceramics is a major problem, these materials had and have a wide range of use as bone replacement materials. Attempts to design the architecture of bioceramics directly with RP methods like SLS, 3DP and SLA as well as indirectly with methods like SLA, DLP, BPM and FDM can be found in literature.

In [56] the authors report the production of parts from a glass-ceramic based on the $SiO_2 \cdot Al_2O_3 \cdot P_2O_5 \cdot CaO \cdot CaF_2$ system. They ground the glass to powder, mixed it with an acrylic polymer binder, and sintered it in an experimental SLS machine, investigating the optimal combination of laser powder and speed depending on the binder powder content. Too large a binder content is disadvantageous, as the complete removal is necessary for biocompatibility and the mixture becomes difficult to be spread across the building platform, too little binder results in fragile green parts. During the heat treatment following the SLS procedure, the binder is removed and the glass particles sinter together. Infiltration of the green part with a resorbable phosphate glass reduces its porosity and increases the mechanical strength of the sintered parts.

In this study no designed macro-porosity was used, but a remaining micro-porosity was detected. Post-sintering could be used to get rid of this micro-porosity, thus increasing the mechanical strength of the parts, but maybe adversely affecting the bone ingrowth. Nevertheless, the authors describe the process of being capable of producing a designed macro-porosity with a pore size of 1–2 mm.

Another attempt to be found in the literature [57] used a ceramic-filled polymeric precursor which further polymerised and solidifies when heat-treated in the SLS-machine and is turned into a ceramic in a post-treatment step at very high temperatures. Polymethylsilsesquioxan, filled with SiC (40–60 vol %) and mixed with a catalyst is spread on the building platform of a commercial SLS-machine and heated with varying laser power and varying scan speed. Subsequent heating and infiltration with Si yields dense, strong parts, the resolution of the process allows its usage for the production of parts with a designed macro-porosity suitable as bone replacement material. Although this process was not intended to be used for medical applications and the materials processed are not bioresorbable, it could be an interesting start for the production of stable high-strength ceramic implants with designed geometry and architecture.

Bioresorbable ceramics are processed in [54], where the usage of SLS for shaping parts from an apatite-mullite glass ceramic and from a hydroxyapatite/phosphate glass composite is reported. The first is a single phase material. Due to the low built rate and the friability of the parts created, which is mostly based on the very weak bonding between two subsequent layers, the authors consider this attempt to be failed. The latter is a composite, where just one phase, the phosphate glass, is molten during the laser exposure, gluing the HA particles together. Curling of single layers and, again, a weak bonding between the layers and a highly fragile material due to insufficient wetting of the HA by the phosphate glass are problems still to be solved.

Further advanced than the attempts to utilise SLS for the production of ceramic parts with defined architecture for bone replacement is the usage of 3DP. In [60] the usage of an experimental 3DP machine for the production of scaffolds with defined architecture from hydroxyapatite is described. In this process HA powder is glued together with an organic binder, which is pyrolysed during the following sintering step. The authors succeeded in producing cylindrical parts with and without macro-porosity and designed and built a test grid to assess the resolution of the process. The minimum wall thickness printed in this process is about 330 µm, the minimum pore size is restricted by the relatively large powder particle diameter and the resulting difficulties to remove unbound powder from narrow holes. Accordingly, the surface of the parts is quite rough, but has, due to the sintering procedure, no sharp edges. A closed micro-porosity with pore sizes of 10–30 µm is preserved during sintering.

The evaluation of cell culture tests on structures similar to the ones introduced above can be found in [59]. It shows that MC3T3-E1 murine fibroblasts grow well on the scaffolds, and can also be found deeply within the pores.

As the surrounding unbound powder acts as a support for overhanging or island features, 3DP is capable of real full 3-dimensional design. The powder particle size, the unwanted bonding of surrounding powder and the problems associated with the removal of unbound powder after production are the limiting factors of this production procedure. Nonetheless, the first attempts to be found in literature are rather promising, the pore size achieved in these reports is in the range mentioned as being optimal for bone cell ingrowth.

Another attempt for the direct production of ceramic scaffolds with designed architecture utilising SLA is reported in [92]. The authors used a suspension of

HA particles in the photosensitive resin, which is removed during the subsequent heating leaving a ceramic scaffold with micro-pores with pore sizes in the range of tenths of microns as well as macro-pores with a pore size depending on the width of the crosshatching cured by the laser. Unfortunately in this paper more the medical aspects of the procedure are illuminated, details of the architecture and, of special interest, the mechanical loading capacity, are not given.

Thermal removal of the mould is, as opposed to the polymeric scaffolds, not a problem when using an indirect method for the production of ceramic scaffolds for bone replacement. However, usage of moulds with lower melting point or removal of the mould by means other than heating, like dissolution, can be advantageous due to the avoidance of cracking because of thermal expansion mismatch between mould and green part. The same is the reason why moulds with lower stiffness and strength, like wax moulds as compared to resin moulds, can be beneficial as far as intactness of the part after sintering is concerned.

Apart from this, any mould material which can be removed at temperatures below the desired sintering temperature of the ceramic to be processed can be used. Literature provides examples for the usage of FDM, SLA, DLP and BPM for the indirect production of ceramic scaffolds for bone replacement.

In [78] the authors used a commercial FDM machine to produce casting moulds for the production of porous calcium aluminate ceramics for bone-graft applications. The cylindrical moulds, in this case made from wax, were then infiltrated with a ceramic slurry, followed by binder removal and sintering to produce scaffolds with porosities of 29 and 44% and a pore size of 300 µm. Decomposition of calcium carbonate during the sintering procedure causes a residual micro-porosity with a pore size around 1 µm, resulting in an overall porosity, depending on the calcium carbonate fraction in the ceramic powder mixture, between 28 and 42%. Mechanical testing revealed strength of up to 25 MPa, which is above the average strength of trabecular bone [93]. The authors deduce from biological characterisations using an immortalised human osteoblast cell line in vitro that the structures produced were highly biocompatible. This cell line stops proliferation after a confluent cell layer is formed, which happens after a culture period of about 3 weeks.

SLA was used in [68–70] for the production of moulds for casting of a HA slurry prepared from commercially available HA powder. Various mould geometries, yielding interconnected pores in two and three dimensions were designed and moulded with the thermosetting ceramic slurry. After debinding and sintering almost dense (micro-pore free) HA scaffolds with porosities between 26 and 52% and a minimum channel size of 330 µm were obtained. Nice is the bullet-shaped channel geometry, inherited from the shape of the struts within the mould, which is determined by the penetration profile of the laser beam into the photosensitive resin. Mechanical assessment of the scaffolds properties revealed a brittle failure at loads up to 30 MPa and a stiffness of about 1.4 GPa. In vivo (!) performance of the structures was evaluated by implantation into the hemi-mandible of Yucatan minipigs, harvest after 5 and 9 weeks revealed bone formation between the original bone and the porous implant after 5 weeks and bone growth into the scaffolds after 9 weeks. Direct apposition of bone on the HA surface was found in all sections. The fact that less bone formation was observed in the inner part of the scaffold

9 Rapid Prototyping in Bone Tissue Engineering

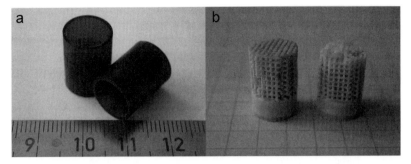

Fig. 9.9 (a) Resin casting moulds produced by DLP, (b) HA scaffolds obtained after demoulding and sintering

strengthens the idea of the importance of vascularisation to maintain nutrition of the bone cells living within the structure.

A cheaper variety of SLA was used in one of our own projects aiming at the indirect production of ceramic scaffolds with defined architecture. We used DLP to produce casting moulds from resin, which were filled with a water-based thermosetting HA slurry by vacuum casting. Casting moulds and scaffolds obtained after thermal treatment can be seen in Fig. 9.9.

The first step in this indirect production method was the design of the structure with computer aided design (CAD) software, Pro/Engineer (PTC, Needham, Massachusetts, USA). The structures consist of layers of parallel struts with quadratic cross-sections. Each layer was turned 90° with respect to the previous one. In this manner, 20 layers were superimposed. The diameter of the whole structure was 10 mm, the side length of one of the struts and the distance between two of them was 500 µm, and hence the height of the whole structure was 10 mm, too. The porosity of the structure was therefore 50 vol %. The RP machine and not the mould filling capacity of the ceramic slurry limited the dimensions of the struts. For the rapid prototyping machine used (Perfactory mini, Envisiontec, Marl, Germany), the obtainable minimum strut size would be about 300 µm, that is in the range of the size of the trabeculae of natural bone material [94, 95].

The second step was to fill the casting mould in vacuum with a thermosetting ceramic slurry, a mixture consisting of water, ceramic powder, water soluble monomers and a dispersion agent. Shortly before casting, a catalyst and an initiator were mixed into the slurry. Thereafter, it was kept for an hour at a temperature of 60 °C, in order to polymerise the monomers and thus stabilise the green body [96–101]. Then the temperature was raised in steps to 1300 °C, which resulted in drying, burning of the resin mould and sintering of the ceramic particles.

Phase composition of the ceramic was assessed after sintering by x-ray diffraction, scanning electron microscopy revealed that we succeeded in the production of an almost micro-pore-free HA material (Fig. 9.10).

In order to assess the biocompatibility of the scaffolds, preosteoblastic MC3T3-E1 cells [102] were seeded onto the scaffolds in a culture medium and kept in culture for 2 weeks. The cells proliferated surprisingly fast on the outer and inner surfaces

Fig. 9.10 Scanning electron microscopy image of the HA surface after sintering. Size of the scale bar is 5 μm

of the scaffolds and embedded themselves into a self-generated tissue like structure consisting of collagen (compare Fig. 9.11) [72, 103, 104].

BPM was used by Wilson et al to produce moulds for HA scaffolds with a designed porosity of about 50% and a square shaped pore size of 400 × 400 μm. As in many other attempts, the Model Maker II (MMII) by Solidscape was used for the mould production from wax. Commercially available HA powder was calcined and mixed into a mixture of water, ammonia and a deflocculent. Shortly before filling the moulds, which took place in a simple vacuum device, a binder was added. The cast parts were dried and subsequently sintered. Biocompatibility was assessed by cultivating goat's bone marrow stromal cells for up to 7 days, partly followed by subcutaneous implantation of the scaffolds into mice. Histological evaluation of the in vitro experiments showed good attachment of the cells onto the scaffolds outer surfaces, soon forming multilayers, and a decreasing amount of cells towards the

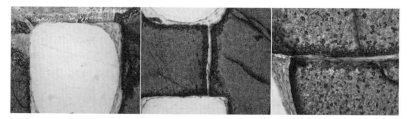

Fig. 9.11 Results of the cell culture on Ha scaffolds (black/grey) after embedding in resin (white/pink), cutting, grinding and polishing. The preosteoblastic MC3T3-E1 cell attached at the HA surfaces proliferated, especially in the edges and into cracks, and embedded themselves into a tissue-like structure consisting of collagen (data not shown)

scaffold's centre. Just a few non-vital cells could be found after 7 days of culture period. Histology of the implanted scaffolds revealed fully tissue-filled scaffolds after 4 weeks, and beginning mineralised bone tissue (MBT) formation at the scaffold's surface, with frequent MBT budding away from the surface after 6 weeks [77].

A similar attempt using the MMII can be found in [75], with a slightly different slurry composition and the usage of TCP instead of HA. Scaffolds with designed pores sized about 460 μm were produced, but no biological evaluation of the scaffolds properties is reported.

9.3.3 Composite Scaffolds

The main disadvantage of bio-resorbable polymeric scaffolds is their low mechanical strength, the one of ceramic scaffolds is their inherent brittleness. A possibility to overcome these problems is to be found in bone itself, which combines the toughness of the biopolymer collagen with the strength and stiffness of its mineral component such that both are preserved to a surprisingly high degree (compare Fig. 9.12).

Bone is a composite. Although engineer's techniques are far from being able to copy the magnificent extent to which nature is able to optimise its structures on

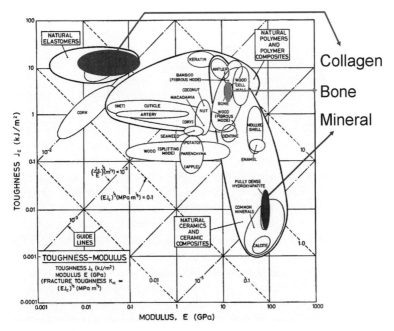

Fig. 9.12 Bone has nearly the toughness of collagen and the stiffness of its mineral component, carbonated apatite. Image adapted after [105]

various orders of magnitude, it has been and is tried to mimic the structure of bone, using a reinforcing stiff mineral phase embedded in a tough polymeric matrix.

Attempts to produce such materials and shape their external and internal architecture directly have been undertaken using FDM and SLS. An indirect method was used, once more utilising BPM.

A report of a study on the usage of FDM to process chitosan and a chitosan/HA mixture to produce scaffolds with defined porosity can be found in [43]. A hydrogel from chitosan, HA and acetic acid was dispensed into a bath of sodium hydroxide and ethanol in order to form a hydrated gel-like precipitate. Composite scaffolds with macro-pores in the range of 200–400 μm were obtained after freeze-drying, sagging of the strands was mostly prevented by the buoyancy of the bath. Human osteoblasts were cultivated on the scaffolds and found to remain viable and proliferate throughout the culture period.

Although chitosan does not provide sufficient strength to be used in load bearing applications, even when reinforced with a stiff mineral phase, this investigations shows a successful attempt to use RP for shaping a composite from a biopolymer reinforced with a bioceramic filler.

Another report utilising FDM can be found in literature, where the authors tried to prepare an extrudable filament from polypropylene (PP) and tricalciumphosphate (TCP) and eventually succeeded for a particle content of 20.5 vol % TCP. Scaffolds with different geometries (as well as scaffolds with a gradient in porosity) were fabricated, porosity calculation revealed porosities of 36, 40 and 52%. Porosimetric plots showed peaks in the range of 150–200 μm and, in addition, a peak in the sub-micron range was detected, which is a result of the evaporation of processing aids during fabrication. The presence of the processing aids is also the reason why composite tensile test specimens (non-porous) had a decreased strength and stiffness as compared to pure PP. Compressive strength of porous samples was found to be in the range of 10 MPa, which is quite similar to the strength of cancellous bone. In vitro characterisation with immortalised human preosteoblastic cells (OPC1) revealed that the matrices were non-toxic and that the cells attached to the surfaces of the scaffolds.

PP is biocompatible, but not biodegradable, which is, besides the fact that incorporation of the ceramic filler and the additives reduces the mechanical properties of the composite as compared to the "non-reinforced" pure polymer, the main disadvantage of the product presented here [45]. Nonetheless, PP is a comparably strong polymer, allowing the authors to produce scaffolds with a porosity of up to 50% with mechanical strength in the range of that of cancellous bone.

A HA particle filled high density polyethylene (HDPE), which is, under the trade name HAPEX, already on the market [106], was used by the authors of [50] and [58] for the production of parts by SLS. HA particles and HDPE (with a HA content of 30 and 40 vol %) were compounded in a twin-screw extruder, pelletised and pulverised using liquid nitrogen cooling. After sieving to obtain a suitable powder particle diameter, the powder was filled into and sintered in an experimental SLS machine, which allowed the adjustment of laser powder and scanning speed within a wide range. A combination of scanning speed and laser powder was found, which allowed the production of coherent single layers of composite, without degrading the polymer. The thickness of this single layer was above 0.5 mm, already indicating the resolution restrictions due to particle size and machine characteristics. Delamination and large pores, possibly because of breakdown of the polymer chains and vaporisation of some of the polymer material, where the problems detected in this study.

Unfortunately, no attempt to produce scaffolds with designed pore architecture is reported, but the study could clearly show that SLS is a promising candidate for such intentions.

A similar attempt is reported in [48], although the authors of this study used a biodegradable polymer, polyvinylalcohol (PVA), instead of HDPE. Polymer coated HA particles were produced by spray-drying of HA suspended in a solution of PVA in water. The high ceramic/polymer ratio did, nonetheless, not allow sintering the obtained powder. Physical blending was chosen to produce powder mixtures with up to 30 w% HA, those were sintered in a commercial SLS machine. The authors report that they have been able to produce disk shaped test specimen with a well defined micro-porosity and good structural integrity. Unfortunately, they do not report any attempt to shape more complicated specimens with their method, and also mechanical properties of the material produced are not mentioned.

An indirect attempt to fabricate composite scaffolds by vacuum casting into wax moulds produced by BPM was undertaken by our group [76]. The Model Maker II by Solidscape was used to produce casting moulds with an internal structure similar to the one introduced above, they can be seen in Fig. 9.13.

A chitosan solution was prepared by dissolving chitosan in a 1% acetic acid solution. HA powder was suspended in this mixture such that the weight ratio of polymer/ceramic equals 60/40. After casting of the mixture into the moulds in vacuum they were immediately frozen at −80 °C, subsequently treated with 10% alcoholic sodium hydroxide solution, which dissolves the wax moulds and ice crystals alike, and finally freeze-dried. The scaffolds retained the shape of the mould, as can be seen in Fig. 9.14, but displayed, due to the freeze-drying procedure, a remarkable micro-porosity (data not shown).

Fig. 9.13 Wax moulds produced by BPM. The inner architecture of these moulds has interconnected pores in all three dimensions, and a porosity of 50%

Fig. 9.14 Chitosan/hydroxyapatite composite scaffolds

Although these scaffolds do not, due to their micro-porosity, exhibit high strength or stiffness, the fact that it was possible to produce composite structures with defined architecture is a promising step towards future resorbable bone replacement material with designed outer and inner shape.

9.4 Conclusions

Many reports of more or less successful attempts to produce porous structures with defined architecture as future bone replacement materials with Rapid Prototyping methods can be found. As conceivable applications in bone surgery are varying to a great extent, so do the solutions for these applications provided by the various Rapid Prototyping methods and the various materials to be utilised. Although many of the results obtained so far are fairly promising, to our knowledge none of the materials has reached readiness for marketing yet. Nonetheless we believe that the great diversity of the Rapid Prototyping methods and especially their possibilities in terms of custom-made solutions will lead to the usage of these methods for the production of bone replacement materials in the near future.

Acknowledgment The author wants to acknowledge the invaluable help from and great possibilities offered by Prof. Peter Fratzl. Furthermore he wants to apologise to the authors of the various papers in the field of porous bone replacement materials which have, due to time- and space-restrictions, not been mentioned in this chapter. It was not on purpose.

References

1. LeGeros, R. Z. (2002). Properties of osteoconductive biomaterials: Calcium phosphates. *Clin Orthop Relat Res*, **395**, 81–98.
2. Doremus, R. H. (1992). Bioceramics. *J Mater Sci*, **27**(2), 285–297
3. Kannan, S., Balamurugan, A., Rajeswari, S. and Subbaiyan, M. (2002). Metallic implants - An approach for long term applications in bone related defects. *Corrosion Reviews*, **20**(4–5), 339–358
4. Kenny, S. M. and Buggy, M. (2003) Bone cements and fillers: a review. *J Mater Sci-Mater Med*, **14**(11), 923–938
5. Teitelbaum, S. L., (2000). Bone resorption by osteoclasts. *Science*, **289**(5484), 1504–1508
6. Currey, J. D. (2002). *Bones structure and mechanics*. Princeton University Press, Princeton, N.J
7. Ben-Nissan, B. (2003). Natural bioceramics: from coral to bone and beyond. *Cur Opin Sol State Mater Sci*, **7**(4–5), 283–288
8. Lane, J. M., Tomin, E. and Bostrom, M. P. G. (1999). Biosynthetic bone grafting. *Clin Orthop Relat Res*, **367**, S107–S117
9. Pelker, R. R. and Friedlaender, G. E. (1987). Biomechanical aspects of bone autografts and allografts. *Orthop Clin North Am*, **18**(2), 235–239
10. Moore, W. R., Graves, S. E. and Bain, G. I. (2001). Synthetic bone graft substitutes. *Aust N Z J of Surg*, 71(6), 354–361
11. Mankin, H. J., Gebhardt, M. C., Jennings, L. C., Springfield, D. S. and Tomford, W. W. (1996). Long-term results of allograft replacement in the management of bone tumors. *Clin Orthop Relat Res*, (324), 86–97
12. Strong, D. M., Friedlaender, G. E., Tomford, W. W., Springfield, D. S., Shives, T. C., Burchardt, H., Enneking, W. F. and Mankin, H. J. (1996). Immunologic responses in human recipients of osseous and osteochondral allografts. *Clin Orthop Relat Res*, (326), 107–114
13. Simonds, R. J., Holmberg, S. D., Hurwitz, R. L., Coleman, T. R., Bottenfield, S., Conley, L. J., Kohlenberg, S. H., Castro, K. G., Dahan, B. A., Schable, C. A., Rayfield, M. A. and Rogers, M. F. (1992). Transmission of human-immunodeficiency-virus type-1 from a seronegative organ and tissue donor. *N Eng J Med*, **326**(11), 726–732
14. Jensen, S., Aarboe, M., Pinholt, E., Hjorting-Hansen, E., Melsen, F. and Ruyter, I. (1996). Tissue reaction and material characteristics of four bone substitues. *Int J Oral Maxillofac Implants*, **11**, 55–66
15. Ben-Nissan, B. (2003). Natural bioceramics: from coral to bone and beyond. *Current Opinion in Solid State and Materials Science*, **7**(4–5), 283–288
16. Kujala, S., Ryhänen, J., Danilov, A. and Tuukkanen, J. (2003). Effect of porosity on the osteointegration and bone ingrowth of a weight-bearing nickel-titanium bone graft substitute. *Biomaterials*, **24**, 4691–4697
17. Fujibayashi, S., Neo, M., Kim, H. M., Kokubo, T. and Nakamura, T. (2004). Osteoinduction of porous bioactive titanium metal. *Biomaterials*, **25**(3), 443–450
18. Oonishi, H., Yamamoto, M., Ishimaru, H., Tsuji, E., Kushitani, S., Aono, M. and Ukon, Y. (1989). The effect of hydroxyapatite coating on bone-growth into porous titanium-alloy implants. *J Bone Joint Surg Br*, **71**(2), 213–216
19. Mahmood, J., Takita, H., Ojima, Y., Kobayashi, M., Kohgo, T. and Kuboki, Y. (2001). Geometric effect of matrix upon cell differentiation: BMP-induced osteogenesis using a new bioglass with a feasible structure. *J Biochem*, **129**(1), 163–171
20. Oonishi, H., Kushitani, S., Yasukawa, E., Iwaki, H., Hench, L. L., Wilson, J., Tsuji, E. I. and Sugihara, T. (1997). Particulate bioglass compared with hydroxyapatite as a bone graft substitute. *Clin Orthop Relat Res*, (334), 316–325
21. Schepers, E., Declercq, M., Ducheyne, P. and Kempeneers, R. (1991). Bioactive glass particulate material as a filler for bone-lesions. *J Oral Rehabil*, 18(5), 439–452
22. Yuan, H. P., de Bruijn, J. D., Zhang, X. D., van Blitterswijk, C. A. and de Groot, K. (2001). Bone induction by porous glass ceramic made from bioglass (R) (45S5). *J Biomed Mater Res*, **58**(3), 270–276

23. Albee, F. H. and Morrison, H. F. (1920). Studies in bone growth: triple CaP as a stimulus to osteogenesis. *Ann Surg*, **71**, 32–39
24. Gauthier, O., Goyenvalle, E., Bouler, J. M., Guicheux, J., Pilet, P., Weiss, P. and Daculsi, G. (2001). Macroporous biphasic calcium phosphate ceramics versus injectable bone substitute: a comparative study 3 and 8 weeks after implantation in rabbit bone. *J Mater Sci-Mater Med*, **12**(5), 385–390
25. Frayssinet, P., Trouillet, J., Rouquet, N., Azimus, E. and Autefage, A. (1993). Osseointegration of macroporous calcium phosphate ceramics having a different chemical composition. *Biomaterials*, **14**(6), 423–429
26. Kasten, P., Luginbühl, R., Griensven, M. v., Barkhausen, T., Krettek, C., Bohner, M. and Bosch, U. (2003). Comparison of human bone marrow stromal cells seeded on calcium-deficient hydroxyapatite, ÿ-tricalcium phosphate and demineralized bone matrix. *Biomaterials*, **24**, 2593–2603
27. Yuan, H. P., Yang, Z., de Bruijn, J. D., de Groot, K. and Zhang, X. D. (2001). Material-dependent bone induction by calcium phosphate ceramics: a 2.5-year study in dog. *Biomaterials*, **22**, 2617–2623
28. Kuboki, Y., Takita, H., Kobayashi, D., Tsuruga, E., Inoue, M., Murata, M., Nagai, N., Dohi, Y. and Ohgushi, H. (1998). BMP-induced osteogenesis on the surface of hydroxyapatite with geometrically feasible and nonfeasible structures: topology of osteogenesis. *J Biomed Mater Res*, **39**(2), 190–199
29. Tsuruga, E., Takita, H., Itoh, H., Wakisaka, Y. and Kuboki, Y. (1997). Pore size of porous hydroxyapatite as the cell-substratum controls BMP-induced osteogenesis. *J Biochem*, **121**(2), 317–324
30. Yuan, H. P., Kurashina, K., de Bruijn, J. D., Li, Y. B., de Groot, K. and Zhang, X. D. (1999). A preliminary study on osteoinduction of two kinds of calcium phosphate ceramics. *Biomaterials*, **20**(19), 1799–1806
31. Vuola, J., Taurio, R., Göransson, H. and Asko-Seljavaara, S. (1998). Compressive strength of calcium carbonate and hydroxyapatite implants after bone-marrow-induced osteogenesis. *Biomaterials*, **19**, 223–227
32. Chang, B. S., Lee, C. K., Hong, K. S., Youn, H. J., Ryu, H. S., Chung, S. S. and Park, K. W. (2000). Osteoconduction at porous hydroxyapatite with various pore configurations. *Biomaterials*, **21**(12), 1291–1298
33. Klawitter, J. (1979). *A basic investigation of bone growth in porous materials*. PhD Thesis, Clemson University, Clemson.
34. Gauthier, O., Bouler, J. M., Aguado, E., Pilet, P. and Daculsi, G. (1998). Macroporous biphasic calcium phosphate ceramics: influence of macropore diameter and macroporosity percentage on bone ingrowth. *Biomaterials*, **19**(1–3), 133–139
35. Ducheyne, P. and Qiu, Q. (1999). Bioactive ceramics: the effect of surface reactivity on bone formation and bone cell function. *Biomaterials*, **20**(23–24), 2287–2303
36. Gibson, L. J. and Ashby, M. F. (1997). *Cellular solids* (2nd ed.). Cambridge University press, Cambridge.
37. Woesz, A., Stampfl, J. and Fratzl, P. (2004). Cellular solids beyond the apparent density - an experimental assessment of mechanical properties. *Advanced Engineering Materials*, **6**(3), 134–138
38. Hopkinson, N. (2005). *Rapid manufacturing technology*. Wiley & Sons, New York.
39. Gebhardt, A. (1996). *Rapid Prototyping. Werkzeuge für die schnelle Produktentwicklung*. Hanser, Fachbuchverlag.
40. Pham, D. T. and Dimov, S. S. (2001). *Rapid manufacturing*. Springer, London.
41. Stampfl, J. (2004). *3D-techiques in material science*. Habilitationsschrift, Wien.
42. Leong, K. F., Cheah, C. M. and Chua, C. K. (2003). Solid freeform fabrication of three-dimensional scaffolds for engineering replacement tissues and organs. *Biomaterials*, **24**(13), 2363–2378
43. Ang, T. H., Sultana, F. S. A., Hutmacher, D. W., Wong, Y. S., Fuh, J. Y. H., Mo, X. M., Loh, H. T., Burdet, E. and Teoh, S. H. (2002). Fabrication of 3D chitosan-hydroxyapatite scaffolds using a robotic dispensing system. *Materials Science & Engineering C-Biomimetic and Supramolecular Systems*, **20**(1–2), 35–42

44. Hutmacher, D. W., Schantz, T., Zein, I., Ng, K. W., Teoh, S. H. and Tan, K. C. (2001). Mechanical properties and cell cultural response of polycaprolactone scaffolds designed and fabricated via fused deposition modeling. *J Biomed Mater Res*, **55**(2), 203–216
45. Kalita, S. J., Bose, S., Hosick, H. L. and Bandyopadhyay, A. (2003). Development of controlled porosity polymer-ceramic composite scaffolds via fused deposition modeling. *Materials Science & Engineering C-Biomimetic and Supramolecular Systems*, **23**(5), 611–620
46. Too, M. H., Leong, K. F., Chua, C. K., Du, Z. H., Yang, S. F., Cheah, C. M. and Ho, S. L. (2002). Investigation of 3D non-random porous structures by fused deposition modelling. *International Journal of Advanced Manufacturing Technology*, **19**(3), 217–223
47. Woodfield, T. B. F., Malda, J., de Wijn, J., Peters, F., Riesle, J. and van Blitterswijk, C. A. (2004). Design of porous scaffolds for cartilage tissue engineering using a three-dimensional fiber-deposition technique. *Biomaterials*, **25**(18), 4149–4161
48. Chua, C. K., Leong, K. F., Tan, K. H., Wiria, F. E. and Cheah, C. M. (2004). Development of tissue scaffolds using selective laser sintering of polyvinyl alcohol/hydroxyapatite biocomposite for craniofacial and joint defects. *J Mater Sci-Mater Med*, **15**(10), 1113–1121
49. Antonov, E. N., Bagratashvili, V. N., Whitaker, M. J., Barry, J. J. A., Shakesheff, K. M., Konovalov, A. N., Popov, V. K. and Howdle, S. M. (2005). Three-dimensional bioactive and biodegradable scaffolds fabricated by surface-selective laser sintering. *Advanced Materials*, **17**(3), 327–330
50. Hao, L., Savalani, M. M., Zhang, Y., Tanner, K. E. and Harris, R. A. (2006). Selective laser sintering of hydroxyapatite reinforced polyethylene composites for bioactive implants and tissue scaffold development. *Proc Inst Mech Eng [H]*, **220**(H4), 521–531
51. Partee, B., Hollister, S. J. and Das, S. (2006). Selective laser sintering process optimization for layered manufacturing of CAPA (R) 6501 polycaprolactone bone tissue engineering scaffolds. *Journal of Manufacturing Science and Engineering-Transactions of the ASME*, **128**(2), 531–540
52. Tan, K. H., Chua, C. K., Leong, K. F., Cheah, C. M., Cheang, P., Abu Bakar, M. S. and Cha, S. W. (2003). Scaffold development using selective laser sintering of polyetheretherketone-hydroxyapatite biocomposite blends. *Biomaterials*, **24**(18), 3115–3123
53. Williams, J. M., Adewunmi, A., Schek, R. M., Flanagan, C. L., Krebsbach, P. H., Feinberg, S. E., Hollister, S. J. and Das, S. (2005). Bone tissue engineering using polycaprolactone scaffolds fabricated via selective laser sintering. *Biomaterials*, **26**(23), 4817–4827
54. Lorrison, J. C., Dalgarno, K. W. and Wood, D. J. (2005). Processing of an apatite-mullite glass-ceramic and an hydroxyapatite/phosphate glass composite by selective laser sintering. *J Mater Sci-Mater Med*, **16**(8), 775–781
55. Hayashi, T., Maekawa, K., Tamura, M. and Hanyu, K. (2005). Selective laser sintering method using titanium powder sheet toward fabrication of porous bone substitutes. *JSME International Journal Series A-Solid Mechanics And Material Engineering*, **48**(4), 369–375
56. Goodridge, R. D., Dalgarno, K. W. and Wood, D. J. (2006). Indirect selective laser sintering of an apatite-mullite glass-ceramic for potential use in bone replacement applications. *Proceedings of the Institution of Mechanical Engineers Part H-Journal of Engineering in Medicine*, **220**(H1), 57–68
57. Friedel, T., Travitzky, N., Niebling, F., Scheffler, M. and Greil, P. (2005). Fabrication of polymer derived ceramic parts by selective laser curing. *Journal of the European Ceramic Society*, **25**(2–3), 193–197
58. Savalani, M. M., Hao, L. and Harris, R. A. (2006). Evaluation of CO2 and Nd: YAG lasers for the selective laser sintering of HAPEX (R). *Proceedings of the Institution of Mechanical Engineers Part B-Journal of Engineering Manufacture*, **220**(2), 171–182
59. Leukers, B., Gulkan, H., Irsen, S. H., Milz, S., Tille, C., Schieker, M. and Seitz, H. (2005). Hydroxyapatite scaffolds for bone tissue engineering made by 3D printing. *J Mater Sci-Mater Med*, **16**(12), 1121–1124
60. Seitz, H., Rieder, W., Irsen, S., Leukers, B. and Tille, C. (2005). Three-dimensional printing of porous ceramic scaffolds for bone tissue engineering. *J Biomed Mater Res B-Appl Biomater*, **74B**(2), 782–788

61. Cooper, K. G. (2001). *Rapid prototyping technology*. Marcel Dekker, New York.
62. Yang, S. F., Leong, K. F., Du, Z. H. and Chua, C. K. (2002). The design of scaffolds for use in tissue engineering. Part II. Rapid prototyping techniques. *Tissue Eng*, **8**(1), 1–11
63. Mapili, G., Lu, Y., Chen, S. C. and Roy, K. (2005). Laser-layered microfabrication of spatially patterned functionalized tissue-engineering scaffolds. *J Biomed Mater Res B-Appl Biomat*, **75B**(2), 414–424
64. Landers, R., Hubner, U., Schmelzeisen, R. and Mulhaupt, R. (2002). Rapid prototyping of scaffolds derived from thermoreversible hydrogels and tailored for applications in tissue engineering. *Biomaterials*, **23**(23), 4437–4447
65. Landers, R., Pfister, A., Hubner, U., John, H., Schmelzeisen, R. and Mulhaupt, R. (2002). Fabrication of soft tissue engineering scaffolds by means of rapid prototyping techniques. *J Mater Sci-Mater Med*, **37**(15), 3107–3116
66. Li, J. P., de Wijn, J. R., Van Blitterswijk, C. A. and de Groot, K. (2006). Porous Ti6Al4V scaffold directly fabricating by rapid prototyping: preparation and in vitro experiment. *Biomaterials*, **27**(8), 1223–1235
67. Pfister, A., Landers, R., Laib, A., Hubner, U., Schmelzeisen, R. and Mulhaupt, R. (2004). Biofunctional rapid prototyping for tissue-engineering applications: 3D bioplotting versus 3D printing. *Journal of Polymer Science Part A-Polymer Chemistry*, **42**(3), 624–638
68. Chu, T. M. G., Halloran, J. W., Hollister, S. J. and Feinberg, S. E. (2001). Hydroxyapatite implants with designed internal architecture. *J Mater Sci-Mater Med*, **12**(6), 471–478
69. Chu, T. M. G., Hollister, S. J., Halloran, J. W., Feinberg, S. E. and Orton, D. G. (2002). Manufacturing and characterization of 3-D hydroxyapatite bone tissue engineering scaffolds. *Reparative Medicine: Growing Tissues and Organs*, **961**, 114–117
70. Chu, T. M. G., Orton, D. G., Hollister, S. J., Feinberg, S. E. and Halloran, J. W. (2002). Mechanical and in vivo performance of hydroxyapatite implants with controlled architectures. *Biomaterials*, **23**(5), 1283–1293
71. Stampfl, J., Fouad, H., Seidler, S., Liska, R., Schwager, F., Woesz, A. and Fratzl, P. (2004). Fabrication and moulding of cellular materials by rapid prototyping. *International Journal of Materials & Product Technology*, **21**(4), 285–296
72. Woesz, A., Rumpler, A., Stampfl, J., Varga, F., Fratzl-Zelman, N., Roschger, P., Klaushofer, K. and Fratzl, P. (2005). Towards bone replacement materials from calcium phosphates via rapid prototyping and ceramic gelcasting. *Materials Science & Engineering C-Biomimetic and Supramolecular Systems*, **25**(2), 181–186
73. Chen, V. J., Smith, L. A. and Ma, P. X. (2006). Bone regeneration on computer-designed nano-fibrous scaffolds. *Biomaterials*, **27**(21), 3973–3979
74. Lee, M., Dunn, J. C. Y. and Wu, B. M. (2005). Scaffold fabrication by indirect three-dimensional printing. *Biomaterials*, **26**(20), 4281–4289
75. Limpanuphap, S. and Derby, B. (2002). Manufacture of biomaterials by a novel printing process. *J Mater Sci-Mater Med*, **13**(12), 1163–1166
76. Manjubala, I., Woesz, A., Pilz, C., Rumpler, M., Fratzl-Zelman, N., Roschger, P., Stampfl, J. and Fratzl, P. (2005). Biomimetic mineral-organic composite scaffolds with controlled internal architecture. *J Mater Sci-Mater Med*, **16**(12), 1111–1119
77. Wilson, C. E., de Bruijn, J. D., van Blitterswijk, C. A., Verbout, A. J. and Dhert, W. J. A. (2004). Design and fabrication of standardized hydroxyapatite scaffolds with a defined macro-architecture by rapid prototyping for bone-tissue-engineering research. *J Biomed Mater Res A*, **68A**(1), 123–132
78. Kalita, S. J., Bose, S., Bandyopadhyay, A. and Hosick, H. L. (2002). Porous calcium aluminate ceramics for bone-graft applications. *Journal of Materials Research*, **17**(12), 3042–3049
79. Sinha, V. R., Bansal, K., Kaushik, R., Kumria, R. and Trehan, A. (2004). Poly-epsilon-caprolactone microspheres and nanospheres: an overview. *International Journal of Pharmaceutics*, **278**(1), 1–23
80. Cao, T., Ho, K. H. and Teoh, S. H. (2003). Scaffold design and in vitro study of osteochondral coculture in a three-dimensional porous polycaprolactone scaffold fabricated by fused deposition modeling. *Tissue Eng*, **9**, S103–S112

81. Zein, I., Hutmacher, D. W., Tan, K. C. and Teoh, S. H. (2002). Fused deposition modeling of novel scaffold architectures for tissue engineering applications. *Biomaterials*, **23**(4), 1169–1185
82. Jones, A. C., Milthorpe, B., Averdunk, H., Limaye, A., Senden, T. J., Sakellariou, A., Sheppard, A. P., Sok, R. M., Knackstedt, M. A., Brandwood, A., Rohner, D. and Hutmacher, D. W. (2004). Analysis of 3D bone ingrowth into polymer scaffolds via microcomputed tomography imaging. *Biomaterials*, **25**(20), 4947–4954
83. Rai, B., Teoh, S. H., Ho, K. H., Hutmacher, D. W., Cao, T., Chen, F. and Yacob, K. (2004). The effect of rhBMP-2 on canine osteoblasts seeded onto 3D bioactive polycaprolactone scaffolds. *Biomaterials*, **25**(24), 5499–5506
84. Rai, B., Teoh, S. H., Hutmacher, D. W., Cao, T. and Ho, K. H. (2005). Novel PCL-based honeycomb scaffolds as drug delivery systems for rhBMP-2. *Biomaterials*, **26**(17), 3739–3748
85. Yang, S. F., Leong, K. F., Du, Z. H. and Chua, C. K. (2001). The design of scaffolds for use in tissue engineering. Part 1. Traditional factors. *Tissue Eng*, **7**(6), 679–689
86. Kricheldorf, H. R. and KreiserSaunders, I. (1996). Polylactides - Synthesis, characterization and medical application. *Macromolecular Symposia*, **103**, 85–102
87. Mikos, A. G., Lyman, M. D., Freed, L. E. and Langer, R. (1994). Wetting of Poly(L-Lactic Acid) and Poly(Dl-Lactic-Co-Glycolic Acid) foams for tissue-culture. *Biomaterials*, **15**(1), 55–58
88. Xiong, Z., Yan, Y. N., Zhang, R. J. and Sun, L. (2001). Fabrication of porous poly(L-lactic acid) scaffolds for bone tissue engineering via precise extrusion. *Scripta Materialia*, **45**(7), 773–779
89. Stampfl, J., Woss, A., Seidler, S., Fouad, H., Pisaipan, A., Schwager, F. and Liska, R. (2004). Water soluble, photocurable resins for rapid prototyping applications. *Macromolecular Symposia*, **217**, 99–107
90. Sachlos, E., Reis, N., Ainsley, C., Derby, B. and Czernuszka, J. T. (2003). Novel collagen scaffolds with predefined internal morphology made by solid freeform fabrication. *Biomaterials*, **24**(8), 1487–1497
91. Taylor, P. M., Sachlos, E., Dreger, S. A., Chester, A. H., Czernuszka, J. T. and Yacoub, M. H. (2006). Interaction of human valve interstitial cells with collagen matrices manufactured using rapid prototyping. *Biomaterials*, **27**(13), 2733–2737
92. Levy, R. A., Chu, T. M. G., Halloran, J. W., Feinberg, S. E. and Hollister, S. (1997). CT-generated porous hydroxyapatite orbital floor prosthesis as a prototype bioimplant. *American Journal of Neuroradiology*, **18**(8), 1522–1525
93. Keaveny, T. M., Morgan, E. F., Niebur, G. L. and Yeh, O. C. (2001). Biomechanics of trabecular bone. *Annual Review of Biomedical Engineering*, **3**, 307–333
94. Roschger, P., Grabner, B. M., Rinnerthaler, S., Tesch, W., Kneissel, M., Berzlanovich, A., Klaushofer, K. and Fratzl, P. (2001). Structural development of the mineralized tissue in the human L4 vertebral body. *Journal of Structural Biology*, **136**(2), 126–136
95. Roschger, P., Gupta, H. S., Berzanovich, A., Ittner, G., Dempster, D. W., Fratzl, P., Cosman, F., Parisien, M., Lindsay, R., Nieves, J. W. and Klaushofer, K. (2003). Constant mineralization density distribution in cancellous human bone. *Bone*, **32**(3), 316–323
96. Young, A. C., Omatete, O. O., Janney, M. A. and Menchhofer, P. A. (1991). Gelcasting of alumina. *Journal of the American Ceramic Society*, **74**(3), 612–618
97. Omatete, O. O., Janney, M. A. and Nunn, S. D. (1997). Gelcasting: From laboratory development toward industrial production. *Journal of the European Ceramic Society*, **17**(2–3), 407–413
98. Omatete, O. O., Janney, M. A. and Strehlow, R. A. (1991). Gelcasting - a new ceramic forming process. *American Ceramic Society Bulletin*, **70**(10), 1641–1649
99. Janney, M. A., Nunn, S. D., Walls, C. A., Omatete, O. O., Ogle, R. J., Kirby, G. H. and McMillan, A. D. (1998). Gelcasting. In: Rahaman, M. N. (ed.) *The handbook of ceramic engineering*. Marcel Dekker, New York.
100. Ortega, F. S., Valenzuela, F. A. O., Scuracchio, C. H. and Pandolfelli, V. C. (2003). Alternative gelling agents for the gelcasting of ceramic foams. *Journal of the European Ceramic Society*, **23**(1), 75–80

101. Ortega, F. S., Sepulveda, P. and Pandolfelli, V. C. (2002). Monomer systems for the gelcasting of foams. *Journal of the European Ceramic Society*, **22**(9–10), 1395–1401
102. Kurihara, N., Ikeda, K., Hakeda, Y., Tsunoi, M., Maeda, N. and Kumegawa, M. (1984). Effect of 1,25-Dihydroxyvitamin-D3 on Alkaline-Phosphatase activity and collagen-synthesis in osteoblastic cells, Clone Mc3t3-E1. *Biochemical and Biophysical Research Communications*, **119**(2), 767–771
103. Rumpler, M., Woesz, A., Varga, F., Manjubala, I., Klaushofer, K. and Fratzl, P. (2007). Three-dimensional growth behaviour of osteoblasts on biomimetic hydroxylapatite scaffolds. *Journal of Biomedical Materials Research Part A*, **81A**(1), 40–50
104. Woesz, A., Rumpler, M., Manjubala, I., Pilz, C., Varga, F., Stampfl, J. and Fratzl, P. (2005). The influence of the thermal treatment of hydroxyapatite scaffolds on the physical properties and the bone cell ingrowth behaviour. *Mater Res Soc Symp Proc*, **874**, L 7.9.1
105. Ashby, M. F., Gibson, L. J., Wegst, U. and Olive, R. (1995). The mechanical-properties of natural materials.1. Material property charts. *Proceedings of the Royal Society of London Series a-Mathematical and Physical Sciences*, **450**(1938), 123–140.
106. Tanner, K. E., Downes, R. N. and Bonfield, W. (1994). Clinical-applications of hydroxyapatite reinforced materials. *British Ceramic Transactions*, **93**(3), 104–107

Chapter 10
Laser Printing Cells

Bradley R. Ringeisen, Jason A. Barron, Daniel Young, Christina M. Othon, Doug Ladoucuer, Peter K. Wu and Barry J. Spargo

10.1 Introduction

Is there a laser-based tool that is used during surgery to regenerate tissue rather than cutting or cauterizing it? Not yet, but there are several innovative laser techniques under development that "print" cells and biomolecules layer-by-layer into organized three dimensional shapes that mimic the natural structure of living tissue. Akin to a biological mason that lays cells, extracellular matrix, and growth factors rather than bricks and mortar, these laser technologies are competing with ink jet, micropen, and electrospray methods to find a niche in tissue engineering applications. Some have gone as far to say that these printers could eventually be used to form pseudo-organs as an alternative to the current donor system, referring to this process as "organ printing" [1, 2].

The goal of tissue engineering research is to create in vitro 3D cell constructs, that once placed in an in vivo environment, differentiate to resemble natural systems as closely as possible (i.e., function, morphology, physical characteristics, etc.) [3–5]. Using cell printing to create heterogeneous 3D shapes with organized cell structure is an alternative to traditional tissue engineering approaches that involve homogeneous seeding of cells into porous three-dimensional scaffolding. Traditional scaffolds are made from highly sophisticated materials, usually biodegradable and/or biocompatible polymers, ceramics, or hydrogels [6–8]. In nearly all tissue engineering experiments, when cells meet scaffolding (seeding), the result is non-directed, homogenous coating of cells on the scaffold's surface [3–5]. Several cell types can be simultaneously seeded onto scaffolding, but there exists no way to place different cell types in different areas of the scaffolding, much less on a length scale comparable to the heterogeneity found in natural tissue (10–100 μm). Growth factors and extracellular matrices can also be incorporated into scaffolding, even released with different time profiles [9–16], but again, the spatial distribution of these molecules is nearly always uniform across the entire scaffold. Certain cell types used in tissue engineering experiments can utilize natural (from the body) or artificial (from the scaffold) signals to initiate differentiation (angiogenesis, neurogenesis, etc.), but a scaffold with built-in blood vessel-like structures or neural networks with the proper channel size, surrounding cell types, and biomolecules have not been realized. These are some of the challenges that could be addressed by laser printing cells into scaffolding rather than adding cells through random seeding.

There are two general approaches that use cell printing to create three dimensional scaffolds. The first can be referred to as "structural" cell printing and requires that the same tool print the scaffolding, cells, and biomolecules simultaneously or sequentially (Fig. 10.1a). This category of cell printing is most relevant to ink jet cell printers and has been popularized by several recent news outlets and research articles under the term "organ printing" [2, 17]. There are several stringent and sometimes unattainable requirements for inks that are to be used in structural cell printing, as they not only need to support living cells but also polymerize or gel into a structurally sound, contiguous 3D shape post-printing. The second approach is more relevant to laser-based cell printing and can be referred to as "conformal" cell printing. Conformal cell printing is a hybrid approach that prints high resolution patterns of cells and biomolecules on top of thin layers of prefabricated scaffolding (Fig. 10.1b). Both approaches are similar in that the finished printed cell construct would be fabricated from the bottom up (i.e., layer-by-layer or cell-by-cell) and has heterogeneous cell and biomolecular structure in three dimensions, as required to mimic natural tissue or organs.

Conformal cell printing is a hybrid cell printing approach that builds upon previously refined scaffold technology. Conformal cell printing would potentially be able to utilize many more types of materials than "structural" printing because the pre- and post-processing requirements of the scaffolding material itself would not need to be conducive to printing (prefabricated layers of scaffolding). It would therefore be possible to use highly sophisticated scaffolds with nano-scale structure or biological/chemical modifications [9–16]. Conformal cell printing also may have the advantage of higher resolution than structural printing, at least initially, based on published reports of laser-based cell printers that have achieved single cell resolution [18, 19]. Conversely, these approaches add complexity by requiring more than just a cell printer in the experimental design (i.e., automated scaffold slicer and handler) and also face possible problems with overall structural integrity. However,

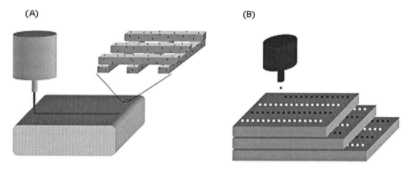

Fig. 10.1 Schematic representation of: (a) "structural" cell printing where both scaffold and cells are printed simultaneously or serially, and (b) "conformal" cell printing where cells alone are printed onto thin layers of pre-fabricated scaffolding. Laser-based cell printing was the first technique to demonstrate "conformal" printing

there is work describing how prefabricated scaffolds can be thinly sliced (similar to tissue sections) [20], while adhesives could be externally added or printed to help interconnect each layer, enhancing the structural integrity of the finished construct.

This chapter summarizes the investigations performed over the past 6 years into whether two laser-based rapid prototyping tools, matrix assisted pulsed laser evaporation direct write (MAPLE DW) and biological laser printing (BioLP), could be successfully used as high resolution cell printers. We will first describe the laser printing apparatus and the physical printing mechanism, specifically the laser-material interaction that enables orifice and capillary-free printing as well as potential damaging effects of the printing process (temperature rise, jetting, drying, shear forces, etc.). The cell printing experiments performed using both MAPLE DW and BioLP are then summarized, including evidence of substantial cell viability and retained phenotype and genotype post-printing. Further results have demonstrated that high resolution cell patterns can be printed into both 2D and 3D scaffolds. The chapter is concluded by discussing the future of cell printing and how laser-based printers compare with other techniques such as ink jet and electrospray.

10.2 Physical Mechanism of Printing

10.2.1 Laser Printing Apparatus

Both MAPLE DW and BioLP are techniques categorized as modified laser induced forward transfer (modified-LIFT) methods. All modified-LIFT experiments are comprised of three similar components: (1) a pulsed laser source, (2) a target from which a biomaterial is printed, and (3) a computer aided design/computer aided manufacturing (CAD/CAM) controlled receiving substrate that captures the printed material (Fig. 10.2). The target consists of two or three layers: a support that is transparent to the wavelength of light used in the experiment, a transfer layer that contains the biomaterial to be patterned or printed (i.e., cells), and an optional energy conversion layer between the support and transfer layers. A receiving substrate is positioned directly beneath the biological layer, spaced by 10–3000 μm. For cell printing experiments, the incident laser pulse is directed through the support layer and focused at the interface of the support and the transfer layer. Some portion of the target (matrix or absorption layer) absorbs the incident wavelength of light, resulting in energy being imparted to the transfer layer. The mechanism of this energy transfer depends of the wavelength and pulse length of the laser (photomechanical, photothermal, etc.), but the end result of the deposited energy is a jet of material being pushed "forward" ("forward transfer"), towards the receiving substrate and away from the support layer [21]. The amount of material printed to the substrate per laser pulse ranges between 500 fL to nL's, depending upon the size of the focused laser spot and the fluence (J cm^{-2}) used [21]. Modified-LIFT technologies are some of the only nozzle and orifice-free printing techniques and therefore may have advantages in some applications. Specific to cell printing, these

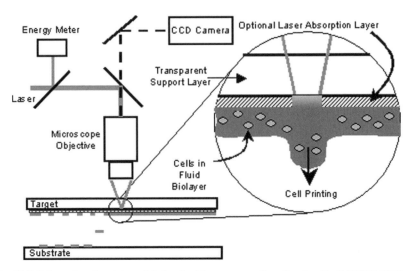

Fig. 10.2 Schematic of matrix assisted pulsed laser evaporation direct write (MAPLE DW) and biological laser printing (BioLP) as used to print living cells

techniques eliminate clogging and enable both viscous and non-viscous fluids to be printed while maintaining micron resolution.

There are two major categories of modified-LIFT techniques that have been used for cell printing. The first is referred to as MAPLE DW and relies upon a matrix material in the transfer layer to help promote light absorption and energy transfer [22–24]. This matrix material is homogeneously mixed with the biological material to be printed. For cell printing, the matrix is usually modified cell media (buffered aqueous solution, nutrients, glycerol, etc.) or a cell adherent surface such as a hydrogel or extracellular matrix. The second category of modified-LIFT techniques utilizes a three-layer target. By adding an energy conversion layer between the transparent support and the layer of cells, these approaches attempt to (a) protect the cells from the incident laser light, and (b) help make the energy conversion more reproducible, resulting in less spot-to-spot variation of the printed material. These three layer approaches have been referred to in the literature as absorbing film-assisted (AFA) LIFT and biological laser printing (BioLP). Some representative cell printing results for MAPLE DW and BioLP will be summarized in section 3 below [18, 25–31].

10.2.2 Laser-Material Interaction

The interaction between incident laser radiation and a solid/soft material can give rise to a large number of physical phenomena that occur simultaneously. To a first approximation, such an event can be simply described as the absorption of the laser light at the material and the resulting physical rearrangement of the solid/liquid/gel due to the energy provided by the laser. However, this process often occurs in a highly complex manner, and control over this event can be a crucial component of a "bio-friendly" printing method.

During the BioLP processes, the laser light often takes the form of a high-intensity optical pulse that impinges onto a thin metal film. Absorption will occur through physical mechanisms that have the potential to couple to the electromagnetic energy of the pulse. This can occur via the absorption of single photons, or by the simultaneous absorption of multiple photons. Generally, these absorption mechanisms produce either electronic or vibrational excited states in the material, or both. Under most conditions, these various excited states have the opportunity to relax to purely thermal vibrational modes through electron-phonon and/or phonon-phonon coupling, resulting in the conversion of the laser energy into localized heat. However, if the laser energy produces highly excited electronic or vibrational modes that persist through several lattice (or molecular) vibrational periods, then athermal processes such as desorption, ablation and radiative recombination can potentially occur [32]. It is expected that multiple relaxation pathways will occur simultaneously, and the exact nature of this process will vary widely depending on the material, laser wavelength and fluence, pulse width, and other process variables. In addition, simultaneously occurring phase changes such as melting and plasma generation further complicate the event. For example, the generation of a small plasma region by the leading edge of the laser pulse may enhance absorption of the trailing edge of the laser pulse, resulting in an exceptionally localized event [33].

Extensive modeling and experimental work have been done on the ablation of liquids due to laser absorption in materials, but there have been fewer studies into the nature of the controlled or focused material ejection in multi-phase systems, as is the case during the BioLP process [34]. There are two possible routes to controlled or focused ejection of solid/soft material by laser irradiance: (a) laser absorption can cause vaporization of the material which leads to partial or entire volatilization or (b) material expulsion can also occur via thermal-acoustical mechanisms initiated by photoabsorption of incident laser energy, either in the form of and ultrasonic atomization-type mechanism or as a shock-wave spallation effect. A combination of both mechanisms may also occur. However, ultra high-speed imaging has provided physical evidence of a prominent "bubble-and-bursting" effect during similar printing events [35]. Currently, the view of this process is that vaporization and vapor expansion are the dominant mechanisms of material acceleration in the BioLP process [33].

Modeling of the laser-material interactions that occur in the BioLP process can provide a physical picture of the printing mechanism. In order to study the BioLP process from first principles, the absorption of a typical laser pulse by the metal/oxide film should first be considered. As was stated in the description of the printing process, biomaterial transfer is achieved by focusing the laser at the interface of the laser absorption layer and the optically transparent support layer. The exponential absorption coefficient is based upon the calculated penetration depth (skin depth) for the metal. The calculated skin depths at 266 nm for the two metals used as targets are given in Table 10.1.

Table 10.1 Laser penetration depths into materials used as absorption layers in BioLP

Laser absorption material	Material thickness used in BioLP	Conductivity (σ) ($\Omega^{-1}m^{-1}$)	Skin depth(δ) (nm)	Laser absorption material
Au	35 nm	4.5e7	2.2	Au
Ti	75 nm	2.4e6	9.7	Ti
TiO$_2$	85 nm	–	10.5[a]	TiO$_2$

[a]"Skin Depth" for TiO$_2$ calculated from $\delta = 1/\alpha$, $\alpha = 4.\pi.k/\lambda$, $k = 2$ (average of $k_\perp$ and $k_\parallel$)

The metal layer thickness used on the targets is much greater than the skin depths. Thus, greater than 99.9% of the non-reflected incident laser energy is absorbed by the solid metal prior to arriving at the absorption layer-biomaterial interface. The TiO$_2$ layer also absorbs the radiation at 266 nm as indicated by the estimated "skin depth" ($\delta \equiv 1/\alpha$) calculated from the complex index of refraction (n+ik) and the absorption coefficient ($\alpha \equiv (4.\pi.k)/\lambda$) [36]. Using this calculation, we estimate that greater than 99.9% of the incident laser energy is also absorbed by the TiO$_2$ absorption layer prior to any interaction with the biomaterial.

In order to understand the role of heat transfer during the BioLP process, the energy transfer from the laser to the laser absorption layer was modeled by the time-dependent heat-transfer equation (parabolic formulation) with a source function given by the Lambert-Beer Law. The composite material heat-conduction equation is solved with a finite-element software package (FlexPDE). Calculations of laser penetration into the laser absorption material show that it is significantly less than the total layer thickness. Transient heat conduction models of laser irradiation of the laser absorption layer show temperature gains in the laser absorption layer of 1400 K for the typical laser fluence of 60 mJ cm^{-2} and an 85 nm thick Titanium absorption layer.

This calculated temperature change is greatly dependent upon the reflectivity and thermal conductivity of the absorption layer, and therefore can vary from a few hundred to over a thousand Kelvin. Heat transfer through the laser absorption layer showed little dependence on absorption material thickness, as long as the thickness of the absorption layer was greater than the penetration depth (skin depth) of the laser. The effects of thermal stress upon the solid materials, their spatial displacements and the formation of thermal shock waves are under investigation. It should be noted that thermal effects alone appear capable of generating a superheated vapor pocket capable of creating a printing event.

In contrast to the rapid energy conversion and transfer seen in the laser absorption layer, thermal penetration into the much thicker aqueous biolayer (10–100 μm thick) is observed to be negligible. Figure 10.3 shows the thermal penetration into a model aqueous biolayer as a function of distance at 50 and 500 ns after laser irradiation.

As can be seen in the graph, at 50 ns post-irradiation, thermal penetration is calculated to be less than 400 nm into the biolayer. Even 500 ns post-irradiation, the penetration is only 1.2 μm into the biolayer. This has been observed experimentally as well, when frozen crystals have been observed to be printed onto a room temperature substrate from a frozen target support. Further, this model is an upper

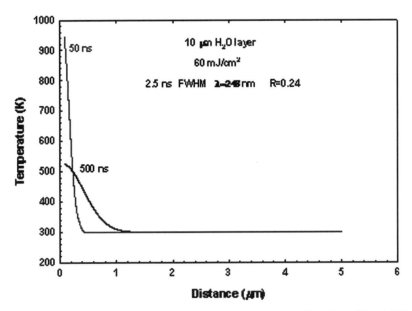

Fig. 10.3 Calculated thermal penetration into a liquid biolayer at specified times (50 and 500 ns) after incident exposure of the target to the laser pulse

limit for thermal penetration, as it uses an insulation model that does not allow for thermal cooling. In addition, the model does not account for the vaporization of the material that could potentially occur as soon as the biomaterial at the interface reaches slightly elevated temperatures. Vaporization at the interface will reduce the amount of thermal penetration by acting as an insulation layer and by reducing the ambient heat via heat of vaporization. While the current model predicts that the first 5% of the material nearest the interlayer is affected, it is likely that the amount of material actually affected is much less.

10.2.3 Fluid Jetting Effects

The material transfer in BioLP is initiated by the generation of the laser-induced superheated vapor pocket. This vapor pocket then expands and deforms the biomaterial layer. If the biomaterial takes the form of a sufficiently thin layer of fluid, then the vapor expansion will cause the detachment of a portion of the layer which is accelerated away from the fluid surface. The exact nature of this event is determined by the energetics of the vapor pocket combined with the rheology and surface tension of the fluid. Often, this even produces a coherent jet which then propagates onto the target substrate (Fig. 10.4). This jet may also coalesce into a single droplet, or stream of droplets, prior to impact with the substrate [21, 35].

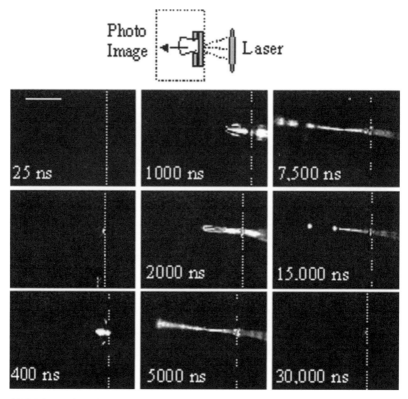

Fig. 10.4 BioLP of a BSA solution under optimal printing conditions at a mid-range fluence of 26 mJ cm^{-2}

Prior to significant material acceleration and jet/droplet generation, the vapor bubble may be considered to produce significant amounts of hydrostatic pressure inside the fluid layer. While these pressures may be quite high by macroscopic standards, the fluid is only subjected to these pressures for a relatively small amount of time before they are relieved by fluid flow away from the fluid layer. Given the short time span of this pressure and the fact the most biomaterials are relatively tolerant of hydrostatic pressures, it is expected that this portion of the jetting process is likely to be highly compatible with biomaterials.

As the high pressure forces fluid to flow into the form of a jet, there will also be significant velocity gradients in the system. The fluid situated directly above the center of the vapor bubble will be accelerated the most while there will be a steep drop-off in acceleration in the fluid closer to the vapor bubble edges. As a result of this, the fluid situated at the edges of the vapor bubble will experience significant shear flow. While this shear flow is unlikely to cause degradation of solvated biomolecules, it is possible that significant shear stress could be placed on suspensions of living cells. The membrane-based structures in mammalian cells are susceptible to rupturing via shear-strain, and their biological viability can be disrupted by low levels of mechanical shear stress, if shear strain is allowed to develop. While directional flow, acceleration and deceleration of the fluid and suspended biological structures are not problematic, any shear flow must be carefully considered when applying any jet-based method to biomaterials. Issues such as the vapor bubble size and temperature, fluid rheology and surface tension, volume and velocity of the resulting jet, and nature of the biomaterial must all be considered in order to determine the level of shear strain that will occur and if degradation will result. In practice, this determination is often theoretically complex and is often accomplished experimentally by quantitative measurement of the viability and phenotype of the printed cells.

Similar issues must be considered during the propagation of the jet through the atmosphere and any accompanying formation of discrete droplets. These processes may result in further shear flow, although the magnitude of shear stress during jet propagation is expected to be low compared to that experienced during the jet formation stage.

Other aspects of this process are generally considered benign. For example, acceleration and deceleration is minimal during this phase of the process, and the droplet propagation occurs over a sufficiently short period of time that desiccation is not expected, and has not been observed, to be a significant issue.

10.2.4 Substrate Impact Effects

The final interaction between the fluid jet/droplet and the substrate is often referred to as an "impact". It has been observed that significant deceleration and shear flow can occur as the momentum of the near-spherical droplet causes it to flatten against the surface of the substrate [37]. Again, the final profile of printed material will be determined by the fluid properties and the impact conditions. While this phenomenon has been studied using non-biological materials, extreme fluid flows have been observed during droplet and jet impact. This observation indicates that the final interaction between biomaterial and substrate in the BioLP process must be carefully considered as a possible source of material degradation. The authors consider the physical conditions during substrate-impact to be just as potentially energetic and damaging as the conditions during the initial biomaterial ejection.

10.3 Laser Cell Printing Results

10.3.1 Cell Viability and Damage Investigations

There are several elements of laser cell printing that could be harmful or fatal to the printed cells. A printed cell is removed from the target at accelerations of up to 10^8 m s^{-2} (or the equivalent of $10^7 \times g$, where $g = 9.8$ m s^{-2}), then passed through hundreds of microns of air at 5–50 m s^{-1}, and finally decelerated at the receiving substrate at 10^5–$10^6 \times g$, all within about 10 microseconds. The forces endured by a printed cell during this transfer process could be greater than 50 µN. Printed cells could also be exposed to deleterious heat and/or photon effects stemming from the incident laser energy. It is important to note that other cell printing techniques have similar hurdles to traverse as the exit velocity for most ink jet printers is also above 10 m s^{-1} [37]. Because of these potential damage pathways, a thorough course of study was performed to investigate cell viability as well as retained genotype and phenotype post-printing.

The first successful modified-LIFT cell printing experiment was performed by using MAPLE DW to print patterns of living *E. coli* [38]. These early experiments demonstrated viability and retained functionality of the printed bacteria through live/dead assays and detection of microbe-expressed green fluorescent protein. Later, the same research group demonstrated that mammalian cells could also be printed. An extensive study was published on pluripotent embryonal carcinoma cells (ATCC#P19) that demonstrated near-100% viability as well as retained genotype and phenotype post-printing [23].

Figure 10.5(a) is a micrograph of P19 cells printed by MAPLE DW to a dry, bare glass surface after exposure to a live/dead stain. The orange fluorescence in the micrograph results from exposure to a membrane-exclusion dye and indicates that all cell membranes were lysed under these conditions. The lysing was most likely from drying and/or shear stress during impact with this hard surface. When cell printing was performed with a glass receiving substrate coated with a thin layer (20–50 µm thick) of basement membrane hydrogel (Matrigel®), cells survived (green fluorescence = live cells) and proliferated post-printing (Fig. 10.5(b, c)). It appeared that by placing a hydrogel layer on the receiving substrate, both drying effects and shear forces could be counteracted. The percentage of viable cells post-printing is shown to be proportional to the thickness of the hydrogel layer, with 50% viability for a 20 µm thick hydrogel and near-100% viability for thicknesses greater than 40 µm. Similar results have since been demonstrated for many different cell types printed by BioLP as well [18, 27, 28].

Investigating any potential damage to the genotype of the printed cells is a prerequisite for a successful cell printer. To determine any potential damage to intracellular DNA during laser printing, the DNA of P19 cells printed by MAPLE DW was tested for strand breaks by a series of comet assays, an effective method to detect

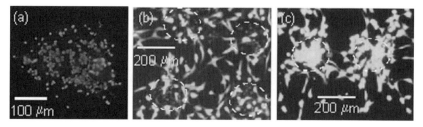

Fig. 10.5 (a) Cells printed to bare glass slide do not survive (green fluorescence = viable or alive; orange fluorescence = no viability or dead). (b) 50% viability when cells were printed to 20 µm-thick hydrogel. (c) Near-100% viability when cells were printed to 40 µm-thick hydrogel

DNA damage in mammalian cells induced by direct or indirect exposure to UV laser light [39, 40]. Both neutral and alkaline comet assays were performed in order to determine whether the MAPLE DW process induces double and single strand breaks, respectively, in the DNA of the transferred cells.

Both the neutral and alkaline comet tests performed on laser-transferred P19 cells showed no increase in DNA damage due to the laser printing process. Two cell sets were tested, those that were fixed immediately after printing and those that were incubated post-printing for 24 hrs [23]. The summary of the alkaline assay results is shown in Table 10.2. The parameters of tail moment (tail length integrated over intensity of the tail) and percent DNA (percent-migrated DNA) were used as indicators of the severity of DNA damage. Through these quantitative measures, both the immediately fixed cells and cells incubated 1 day then fixed show similar damage as non-printed control cells. The somewhat elevated data for control cells (7, 9.4) is a measure of the DNA that is constantly being damaged and repaired in normal cells. According to these comet assay parameters, the cells measured immediately after transfer show a slight increase in DNA damage when compared to the control cells, however, further statistical comparison via a one-tailed student's t test, data reveal no significant difference in the comet parameters for control and laser-transferred cells. This quantitative assessment of potential single strand DNA fragmentation indicates little, if any, damage induced during the laser printing process. Further, cells printed via BioLP are exposed to significantly fewer photons than those cells printed by MAPLE DW, as the absorptive interlayer used in the BioLP experiments blocks greater than 99% of the incident laser energy while the matrix used in this MAPLE DW experiment blocks less than 5% [27].

Another requirement for a successful cell printer is for cells to be deposited onto scaffold surfaces with minimal cell damage. The most likely source of cell damage during most types of cell printing experiments would occur via heat and/or shear stress at the point of ejection (heat and/or acceleration) or when contact is made with the receiving surface (deceleration) [17, 41, 42]. In order to determine whether primary cells (those cells more relevant to tissue engineering experiments) are exposed to heat and/or shear stress during BioLP, immunocytochemical staining experiments were performed to assay for heat shock protein expression (HSP60/70) from bovine aortic endothelial cells (BAECs) following printing into media [28]. Heat shock protein expression was assayed using anti-HSP60/70 following transfer

Table 10.2 Mean tail moment and percent DNA in the tail values for laser-transferred P19 cells as evaluated by the alkaline comet assay

Post-printing sample time	Exposure	Percent DNA ± SD	Tail moment ± SD
Immediate			
	Control	7.0 ± 0.7	0.79 ± 0.09
	Laser Printed	11 ± 2	1.3 ± 0.3
1 Day			
	Control	9.4 ± 0.8	1.1 ± 0.2
	Laser Printed	8.1 ± 0.9	0.9 ± 0.1

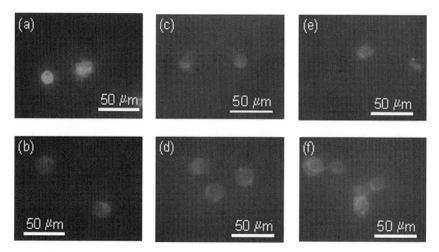

Fig. 10.6 Heat shock protein expression in control and BioLP printed BAECs (green fluorescence = white grayscale). (a) Fluorescent micrograph of positive control cells exposed to anti-HSP60/70 after incubation for 1 hr at 45 °C. (b) Negative control cells incubated for 1 hr at 37 °C. (c) BioLP printed cells at 0.15 J cm^{-2}, incubated for 1 hr at 37 °C, then immunoassayed. (d) 0.30 J cm^{-2}. (e) 0.75 J cm^{-2}. (f) 1.5 J cm^{-2}

at 1, 2, 5, and 10 times the threshold energy, that is, the energy below which there was no observable cells printed. Fig. 10.6 shows the results of this experiment, where panels (a) and (b) are positive and negative expression control cells and panels (c–f) show BAECs printed over a range of 10X in laser energy. After quantitative fluorescence image analysis, there was no statistical difference in HSP60/70 expression for printed and non-printed BAECs. Roughly 10% of cells printed at 10X in laser energy died, while the HSP expression was slightly higher than negative control cells but not to a statistically significant degree. These results indicate that regardless of the high levels of stress potentially present during the printing process, cells appear to remain unaffected, at least in terms of expressing damage from heat and/or shear stressors. Limited expression is most likely due to the short periods of time (estimated <10 μs) that printed cells may be exposed to elevated heat and/or shear stress [23–30].

Preliminary experiments investigating retained phenotype have also been performed on cells deposited by MAPLE DW [23]. Embryonic pluripotent P19 cells were printed and then allowed to grow and differentiate over several days down either a muscle or neural pathway. The cells were then immunocytochemically stained for the expression of either (a) myosin heavy chain (MHC) protein, a marker for muscle differentiation, or microtubular associated protein-2 (MAP-2), a marker for neural differentiation (Fig. 10.7). P19 cells were stimulated down a neural pathway by exposure to retinoic acid-doped media and a muscle pathway upon exposure to dimethyl sulfoxide (DMSO)-doped media. The fluorescence emanating from the printed P19 cells shown in Fig. 10.7 demonstrates that the cells retained the ability to express both MHC (panel a) and MAP-2 (panel b) post-printing under proper chemical stimulus. Therefore, the printing process did not prevent the cells from

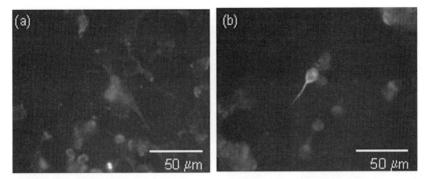

Fig. 10.7 Fluorescent microscopy images of MAPLE DW printed P19 pluripotent cells stained with (a) anti-MHC or (b) anti-MAP-2 monoclonal antibodies (green fluorescence = white in grayscale image) after growth with (a) DMSO or (b) retinoic acid for 5 days post-printing. Fluorescent images show positive MHC and MAP-2 expression indicating the ability of the printed P19 cells to undergo muscle and neural differentiation

undergoing subsequent differentiation. These experiments are excellent examples of the type of work that needs to be performed to demonstrate both cell viability and retained function prior to the onset of 2D and 3D cell printing studies.

10.3.2 Printing Two-Dimensional (2D) Cell Patterns

Large and reproducible arrays of both active proteins and cells have been printed by BioLP with average spot sizes ranging from 30 to 70 μm with single cell to tens of cells/spot resolution [18, 21, 27, 28]. One example of a viable cell array printed by BioLP is shown in Fig. 10.8(a). Twenty-seven spots of human osteosarcoma cells were printed to a 200 μm-thick hydrogel scaffolding layer from a "bioink" composed of 50% (v/v) DMEM, 45% (v/v) fetal bovine serum (FBS) and 5% glycerol (v/v) and 1.0×10^8 cells mL^{-1} [27]. A live/dead stain was exposed to the hydrogel-coated receiving substrate 24 hrs post-printing to demonstrate 100% viability of the osteosarcoma cells. Initial cell concentrations on the receiving substrate were approximately 3–10 cells per deposition spot. This relatively large distribution of # cells/spot corresponds to the statistical sampling that occurs during the printing process and does not reflect the spot-to-spot reproducibility of the printer, which was determined to be much lower (∼3%). Several other cell lines, both normal and carcinoma, have also been investigated and show near 100% viability when printed with several types of energy conversion layers [18, 25–31]. Adjacent patterns of different cell types have also been printed with resolution similar to the heterogeneous structure of natural tissue (10's μm) [27].

One unique aspect of cell printing experiments is that cells can be patterned on homogeneously adherent surfaces, eliminating the need to use lithography-based chemical surface modification. In lithographic cell patterning studies, cell growth and sometimes cell survival are controlled by the cell-surface interaction at the

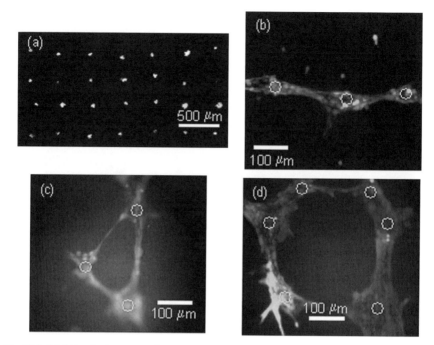

Fig. 10.8 (a) A 7 × 4 microarray of osteosarcoma cells printed by BioLP. Initial cell concentrations were 3–10 cells per printed pixel. The printed cells were exposed to a live/dead assay 24 hrs post-printing to show growth and cell viability (100%) at all deposition points (green fluorescence = live). (b–d) Fluorescent micrographs show patterns of bovine aortic endothelial cells (BAECs) printed by BioLP and exposed to a live/dead assay. Images show 100% cell viability 24 hrs post-deposition and retained cell-cell signaling as evidenced by the asymmetric growth (line, triangle, circle) demonstrated on the homogeneously adherent hydrogel surface

boundaries between the adhesive and non-adhesive surface structure [42–50]. In the study shown in Fig. 10.8(b–d), BioLP was used to form similar cell "islands" as are demonstrated by lithographic techniques, but the growth and migration of cells on laser printed islands are controlled by cell-cell interactions rather than chemically patterned surfaces (green fluorescence demonstrates 100% viability post-printing). Apparent cell-cell signaling and cell differentiation transformed the initially unconnected BAEC pattern (individual, separated spots highlighted by dashed circles) into an interconnected line (6b), triangle (6c), and circle (6d). This controlled growth was achieved without any surface modification and was a result of cell-cell interactions and cell differentiation only. This type of cell-cell signaling, surface migration, and cell differentiation is essential if cells printed into three-dimensional scaffolds are to grow into functional tissues.

Further studies have shown the ability to use BioLP to generate cell patterns with single cell resolution. Fig. 10.9 shows a micrograph of a 3×6 array of human osteosarcoma cells printed to a 50 μm thick hydrogel substrate. The micrograph shown in panel (a) was taken immediately after transfer, with 9 spots showing one cell transferred, 4 showing no cells transferred and 5 showing 2 cells transferred.

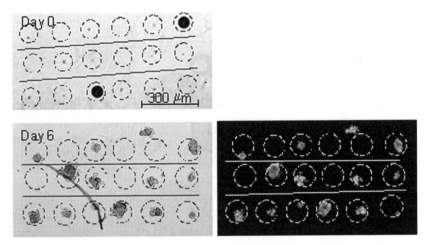

Fig. 10.9 6 × 3 microarray of printed osteosarcoma cells onto a layer of hydrogel. The printing process embeds the cells into the 100 mm thick gel layer. (a) Optical micrograph of the printed cells immediately after transfer. Single cell resolution is demonstrated. (b) Optical micrograph of the same microarray after incubation in culture media for 6 days. (c) Fluorescent micrograph of the cell microarray after 6 days incubation and exposure to a live/dead assay. The green fluorescence indicates near-100% viability

Bubbles in the matrigel substrate (black circles) were caused by the transfer and obscure two of the transfer spots in Fig. 10.9a. Based on subsequent pictures it was determined that 1 and 2 cells were transferred to those spots. The cells were then monitored while growing for 6 days (panels b, c). The ball-shaped growth observed is characteristic of healthy osteosarcoma cells that are growing embedded in a hydrogel surface. A live/dead assay was performed on day 6 to assess cell viability in the sample. As can be seen by the green fluorescence emanating from the osteosarcoma cells in panel (c), cell viability is near 100% after 1 week of growth with cellular activity at all transfer sites. Statistical analysis showed that the average number of cells transferred per laser pulse was 1.06 with a standard deviation of 0.725, with 1 cell per 121 pL deposited over an area of roughly 8×10^{-3} mm^2. It should be possible to further reduce the transfer volume, since in previous experiments we have shown transfer volumes in the hundreds of femtoliter range [21]. This would allow for much higher cell concentrations on the support and increased transfer efficiency.

10.3.3 Printing Three-Dimensional (3D) Patterns

One of the goals of cell printing is to create 3D scaffolds with cells patterned throughout the construct rather than homogeneously distributed, as is the case in most traditional tissue engineering experiments. Fig. 10.10 shows a fluorescent micrograph of three levels of a hydrogel scaffold with cells printed by BioLP into

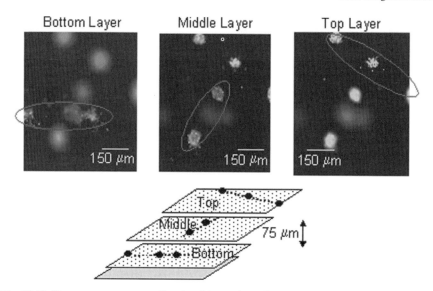

Fig. 10.10 Human osteosarcoma cells printed into a three-dimensional pattern by BioLP and exposed to a live/dead assay 24 hrs post-printing (green fluorescence = live). Viable cells were printed into different patterns in three distinct hydrogel layers

three distinct layers [26, 27, 31]. Three layers of basement membrane (Matrigel®) were manually spread prior to sequentially printing patterns of human osteosarcoma cells into each layer. Each layer of hydrogel was ~75 μm thick with a cell pattern resolution of ~100 μm. The cells were then incubated for 24 hrs and a live/dead assay was performed, resulting in the image shown in Fig. 10.10 (green fluorescence demonstrates 100% viability post-printing). This is the first demonstration of conformal cell printing, as BioLP was used to print viable cells throughout a prefabricated 3D hydrogel via a layer-by-layer approach.

10.4 Discussion

10.4.1 Laser Cell Printing Versus Other Cell Printers

Table 10.3 compares the demonstrated capabilities (resolution, print speed, cell throughput, load volume, and cell viability) of four jet-based cell printers as reported in the peer-reviewed literature. Perhaps the most important values are the percentage of viable cells achieved post-printing. Both thermal ink jet (70–95%) [51, 52] and laser printing (95–100%) techniques have reported a significant percentage of viable cells post-printing as assayed through fluorescent microscopy and live/dead cytochemical staining experiments. Observed cell viability in the thermal ink jet process is significantly reduced by the cytotoxicity of the printing medium, which is assumed to affect viability over time with shorter print times increasing viability (approaching 90%) and longer print times decreasing viability (approaching 70%).

Table 10.3 Comparison between demonstrated capabilities of different cell printers

	Spot size or resolution (μm)	Print speed	Maximum cell throughput (cells s^{-1})[a]	Load volume (mL)	Cell viability (%)[b]
Laser Printing	30–100	10^2 drops s^{-1}	10^4	0.5–20×10^{-3}	95–100
LGDW	1–30	Continuous (9×10^{-8} mL s^{-1})	0.04	Not Reported	–
Ink Jet	>300–	5×10^3 drops s^{-1}	850	0.3–0.5	75–90
Thermal Piezo-tip	–	1×10^4 drop s^{-1}	2	0.3–0.5	–
EHDJ	50–1000	Continuous (0.01 mL s^{-1})	2×10^4	2–5	–

[a] Demonstrated cell throughput per orifice or target. Often printheads can contain multiple orifices (see text below).
[b] Reported % viability. LG DW, Piezo-tip and EHDJ have reported cell viability without giving specific %

All other printers do not have reported data on the cytotoxicity of their print mediums over relevant periods of time. Laser guidance direct write (LG DW) [53, 54], piezo-tip ink jet [19, 55, 56], and electrohydrodynamic jetting (EHDJ) techniques [57–59] have demonstrated the ability to print some fraction of cells viably, but future experiments need to investigate the specific percentage of cells that remain viable post-printing in order to completely compare to other published methodologies.

Other data reported in Table 10.3 build towards the relevancy of cell printing to the tissue engineering community. Resolution, print speed, cell throughput and load volume will all have varying levels of relevancy depending upon the application (organ, tissue, injury site, etc.) being addressed. For example, if the application is to repair an injury to the retina or cornea, then a relatively small bridging scaffold may be needed that would require fewer cells to be deposited accurately (μm resolution) over small length scales (mm–cm). On the opposite end of the spectrum, if the objective is to print replacement tissues, faster print speeds (>10^5 cells s^{-1}) with less stringent resolution must be demonstrated, depending on the desired tissue to mimic. Ultimately, the best cell printers will be those that combine speed with high resolution (small pixel size), preferably fast enough to print large structures with pixel sizes on the order of a single cell (~10 μm). In all cases except EHDJ, spot-to-spot reproducibility is already at acceptable levels (~3%), and most printers have reached their fundamental limit in pixel size (i.e., matching size of orifice). The only exception to this statement are the laser printing techniques where spot sizes of 30 μm have been demonstrated (Table 10.3), while the fundamental limit in pixel size is controlled by the wavelength of light ($\lambda/2$) and is on the nm-scale.

Load volume may become an important factor in cell printing for two reasons. First, if relevant cell types are difficult to harvest or are slow or even impossible to grow in vitro, laser cell printing (down to 500 nL load volume) would be preferable, enabling limited numbers of cells to be concentrated into ultra-small volumes [21, 27, 28]. Laser cell printing has the added benefit of being an orifice/capillary-free technique (no potential for clogging) that enables orders of magnitude more concentrated cell solutions to be used for printing (10^8 cells mL^{-1}) compared to other jet-based technologies. Contrary to this, if a large cell construct is to be built, techniques that utilize mL-scale cell reservoirs such as ink jet and EHDJ may be advantageous.

Another important consideration in cell printing is whether the technique is suitable for scale-up, most notably the ability to perform experiments in parallel. Ink jet printers are renowned for the ability to print multiple types of "inks" in parallel by either using multiple printheads and/or printheads with multiple chambers. A recent report presented the ability of a thermal ink jet cell printer to deposit cells from 50 independent chambers, increasing print speeds up to 250,000 drops s^{-1} ($\sim 4 \times 10^4$ cells s^{-1}) [51]. By using multiple printheads, it may be possible to increase the print speed even further. EHDJ has also been used in configurations with multiple needles to enable parallel or simultaneous deposition [60, 61]. These types of approaches may lead to not only higher cell throughput but also unique 3D element printing such as multi-element cell sheets or tubes. Laser cell printing techniques have also reported depositing different cells from multiple adjacent wells on a single target [27]. In addition, because cell laser printing techniques are based on directed laser energy, enhancements can be made by utilizing different laser sources (beam splitting for parallel deposition, faster pulse rates, etc.). Optically based cell printers also can enable specific cell types to be directly selected from the target [29]. Overall, we believe next generation cell printers hold promise to achieve even faster deposition rates with the ability to deposit many different types of materials (cells, growth factors, etc.) in parallel.

10.4.2 Future of Laser Printing Cells

Successful biological printing experiments using laser-based approaches have been reported by several independent groups around the world [22, 23, 25, 27, 29, 30, 41, 62]. Deposition speed and resolution are adequate for the tissue engineering community (2000–100,000 cells sec^{-1} possible, 10–100 micron resolution), and as discussed in this chapter, recent laser cell printing experiments have gone beyond testing cell viability by demonstrating retained cell phenotype and genotype post-printing. Therefore, the next step will be to begin depositing relevant three-dimensional cell constructs through structural or conformal printing methods. "Relevant" in this sense means printing proper scaffolding materials, biomolecular cues, and cell types to form heterogeneous scaffolds that would address current problems or inadequacies in traditional tissue engineering experiments. Cell growth and differentiation in heterogeneous 3D environments would be studied first in vitro with subsequent in vivo experimentation.

If laser cell printers are to be useful to tissue engineers and ultimately in the medical community, there is a simple outline to follow for success. First, demonstrate a high percentage yield of viable cells that have been printed in 2D or 3D patterns. All laser printing experiments have passed this first hurdle. The next step is to demonstrate that cells are minimally damaged during printing and that they maintain their specific genotype and phenotype. Recent studies, as described above, have begun to study the affect of stress incurred by the cells during the laser printing process, with minimal effects on the cells observed. The next step is to design cell printing experiments that address a specific need in tissue engineering such as angiogenesis, neurogenesis, ocular repair or tissue grafts (skin, cartilage, etc.). Two-dimensional experiments may be investigated first, as there is much to learn about cell-cell interactions on homogeneous surfaces. Most experiments that have probed cell-cell and cell-surface interactions were performed on heterogeneous (chemically modified and patterned) 2D surfaces that often used physical (wells, channels, etc.) or chemical guides to create cell patterns. These guides are dissimilar to the natural cell environment, not only because they are true 2D surfaces but also because they force the cells to respond to external cues rather than relying on their own signals. Cell printers can be used to form patterns onto more diverse surfaces, thereby enabling novel approaches to study cell-cell and cell-surface interactions [28].

Once a particular system is investigated in a 2D environment, cell printers could then be used to form heterogeneous 3D cell constructs. Again, it will be important to study relevant problems in the tissue engineering community. Cell printing needs to focus on its strengths: the ability to construct complicated structures cell-by-cell or layer-by-layer with the possibility of interweaving cell and biomolecular diversity on the micron-scale. This concept could take shape by improving upon a current product, such as a skin graft, by printing more realistic tissue with structural or biological heterogeneity (nerves, cell layers, pores, etc.). Another possibility would be to investigate highly complex systems (e.g., nervous or vascular system) or even organs, attempting to create cell constructs that mimic natural tissue. Just as cell printers may enable new experiments to be performed in two-dimensions, the formation of heterogeneous 3D cell constructs should enable tissue engineers to uniquely study how cells interact with each other and materials/scaffolds in this more realistic 3D environment. Simultaneously, 3D cell constructs could be used for in vivo studies. No matter the experiment, particular importance should be placed on qualitatively and quantitatively demonstrating differences between traditionally seeded scaffolds and cell printed scaffolds, both in vitro and in vivo. Until improvements over traditional tissue engineering approaches are shown, cell printing will be considered a fringe technology with potential rather than a mainstream scientific tool for mass use and study.

References

1. Boland T, Mironov V, Gutowska A, Roth Elisabeth A, Markwald Roger R (2003) Cell and organ printing 2: fusion of cell aggregates in three-dimensional gels. Anat Rec A Discov Mol Cell Evol Biol 272:497
2. Mironov V, Boland T, Trusk T, Forgacs G, Markwald RR (2003) Organ printing: computer-aided jet-based 3D tissue engineering. Trends Biotechnol 21:157
3. Cortesini R (2005) Stem cells, tissue engineering and organogenesis in transplantation. Transpl Immunol 15:81

4. Saltzman WM (2004) Tissue engineering: engineering principles for the design of replacement organs and tissues. Oxford University Press (USA, New York 1st ed.), New York, NY.
5. Williams D, Sebastine I (2005) Tissue engineering and regenerative medicine: manufacturing challenges. IEE Proc Nanobiotechnol 152:207
6. Murugan R, Ramakrishna S (2006) Review article: nano-featured scaffolds for tissue engineering: a review of spinning methodologies. Tissue Eng. 12:435
7. Norman J, Desai T (2006) Methods for fabrication of nanoscale topography for tissue engineering scaffolds. Annals of Biomedical Eng. 34:89
8. Yang SF, Leong KF, Du ZH, Chua CK (2002) The design of scaffolds for use in tissue engineering. Traditional factors. Part 1. Tissue Eng. 7:679
9. Chen VJ, Ma PX (2004) Nano-fibrous poly(l-lactic acid) scaffolds with interconnected spherical macropores. Biomaterials 25:2065
10. Lee KY, Mooney DJ (2001) Hydrogels for tissue engineering. Chem. Rev. 101:1869
11. Lee KY, Peters MC, Mooney DJ (2001) Controlled drug delivery from polymers by mechanical signals. Advanced Mater 13:837
12. Ma PX (2004) Scaffolds for tissue fabrication. Materials Today 7:30
13. Ma PX, Choi J-W (2001) Biodegradable polymer scaffolds with well-defined interconnected spherical pore network. Tissue Eng. 7:23
14. Ma PX, Zhang R (1999) Synthetic nanoscale fibrous extracellular matrix. J Biomed Mater Res 46:60
15. Shea L, Smiley E, Bonadio J, Mooney DJ (1999) DNA delivery from polymer matrices for tissue engineering. Nat Biotechnol 17:551
16. Vasita R, Katti D (2006) Growth factor-delivery systems for tissue engineering: a materials perspective. Expert Rev Med Devices 3:29
17. Jakab K, Neagu A, Mironov V, Forgacs G (2004) Organ printing: fiction or science. Biorheology 41:371
18. Barron JA, Krizman David B, Ringeisen Bradley R (2005) Laser printing of single cells: statistical analysis, cell viability, and stress. Ann Biomed Eng 33:121
19. Saunders R, Bosworth L, Gough J, Derby B, Reis N (2004) Selective cell delivery for 3D tissue culture and engineering. Eur Cell Mater 7(Suppl. 1):84
20. Othman S, Xu H, Royston T, Magin R (2005) Microscopic magnetic resonance elastography (microMRE). Magn Reson Med 54:605
21. Barron J, Young H, Dlott D, Darfler M, Krizman D, Ringeisen B (2005) Printing of protein microarrays via a capillary-free fluid jetting mechanism. Proteomics 5:4138
22. Chrisey D, Pique A, McGill R, Horwitz J, Ringeisen B, Bubb D, Wu P (2003) Laser deposition of polymer and biomaterial films. Chem Rev 103:553
23. Ringeisen BR, Kim H, Barron JA, Krizman DB, Chrisey DB, Jackman S, Auyeung RYC, Spargo BJ (2004) Laser printing of pluripotent embryonal carcinoma cells. Tissue Eng 10:483
24. Young HD, Auyeung RCY, Ringeisen BR, Chrisey DB, Dlott DD. Jetting behavior in the laser forward transfer of rheological systems. In *U.S. Pat. Appl. Publ.*; (The United States of America as represented by the Secretary of the Navy, USA). Us, 2003; p 16
25. Barron JA, Rosen R, Jones-Meehan J, Spargo BJ, Belkin S, Ringeisen BR (2004) Biological laser printing of genetically modified Escherichia coli for biosensor applications. Biosens Bioelectron 20:246
26. Barron JA, Spargo BJ, Ringeisen BR (2004) Biological laser printing of three dimensional cellular structures. Appl Phys Mater Sci. Process 79:1027
27. Barron JA, Wu P, Ladouceur HD, Ringeisen BR (2004) Biological laser printing: a novel technique for creating heterogeneous 3-dimensional cell Patterns. Biomedical Microdevices 6:139
28. Chen CY, Barron JA, Ringeisen BR (2006) Cell patterning without chemical surface modification: Cell-cell interacftions between bovine aortic endothelial cells (BAEC) on a homogeneous cell-adherent hydrogel. Applied Surface Science 252(24):8641–8645

29. Hood BL, Darfler MM, Guiel TG, Furusato B, Lucas DA, Ringeisen BR, Sesterhenn IA, Conrads TP, Veenstra TD, Krizman DB (2005) Proteomic analysis of formalin-fixed prostate cancer tissue. Mol Cell Proteomics 4:1741
30. Hopp B, Smausz T, Kresz N, Barna N, Bor Z, Kolozsvari L, Chrisey D, Szabo A, Nogradi A (2005) Survival and proliferative ability of various living cell types after laser-induced forward transfer. Tissue Eng 11:1817
31. Ringeisen BR, Barron JA, Spargo BJ (2004) Novel seeding mechanisms to form multilayer heterogeneous cell constructs. Materials Research Society Symposium Proceedings EXS-1:105
32. Hagland R (1998) Mechanisms of Laser0Induced Desorption and Ablation. In: Miller J, Haglund R (ed) Laser-Induced Desorption. Academic Press, Chestnut Hill, MA, p 15
33. Young D, Auyeung R, Piqué A, Chrisey D, Dlott D (2002) Plume and jetting regimes in a laser based forward transfer process as observed by Time-Resolved optical microscopy. Appl Surf Sci 197:181
34. Dou Y, Zhigilei L, Postawa A, Winograd N, Garrison B (2001) Nuclear Instruments and Methods in Physics Research B 180:105
35. Young D, Auyeung R, Piqué A, Chrisey D, Dlott D (2001) Time resolved optical microscopy of a Laser-Based forward transfer process. Appl Phys Lett 78:3139
36. Palik E (1985) Handbook of optical constants of solids. Academic Press, Chestnut Hill, MA,
37. van Dam D, Clerc C (2004) Experimental study of the impact of an impact of an Ink-Jet droplet of a solid substrate. Phys Fluids 16:3403
38. Ringeisen BR, Chrisey DB, Pique A, Young HD, Modi R, Bucaro M, Jones-Meehan J, Spargo BJ (2001) Generation of mesoscopic patterns of viable Escherichia coli by ambient laser transfer. Biomaterials 23:161
39. DeWith A, Leit G, Greulich KO (1994) UV-B-Laser-induced DNA-damage in lymphocytes observed by Single-cell Gel-electophoresis. J Photochem Photobiol B 24:47
40. Mohanty SK, Rapp A, Monajembashi S, Gupta PK, Greulich KO (2002) Comet assay measurements of DNA damage in cells by laser microbeams and trapping beams with wavelengths spanning a range of 308 nm to 1064 nm. Radiat Res 157:378
41. Fernandez-Pradas JM, Colina M, Serra P, Dominguez J, Morenza JL (2004) Laser-induced forward transfer of biomolecules. Thin Solid Films 27:453–454
42. Folch A, Jo BH, Hurtado O, Beebe DJ, Toner M (2000) Microfabricated elastomeric stencils for micropatterning cell cultures. Biomed Mater Res 52:346
43. Falconnet D, Csucs G, Michelle Grandin H, Textor M (2006) Surface engineering approaches to micropattern surfaces for cell-based assays. Biomaterials 27:3044
44. Folch A, Toner M (2000) Microengineering of cellular interactions. Annu Rev Biomed Eng 2:227
45. Kane RS, Takayama S, Ostuni E, Ingber DE, Whitesides GM (1999) Patterning proteins and cells using soft lithography. Biomaterials 20:2363
46. Lahiri J, Ostuni E, Whitesides GM (1999) Patterning ligands on reactive SAMs by microcontact printing. Langmuir 15:2055
47. Spargo BJ, Testoff MA, Nielsen TB, Stenger DA, Hickman JJ, Rudolph AS (1994) Spatially controlled adhesion, spreading, and differentiation of endothelial cells on self-assembled molecular monolayers. Proc Natl Acad Sci US A 91:11070
48. Tien J, Nelson C, Chen C (2002) Fabrication of aligned microstructures with a single elastomeric stamp. Proc Natl Acad Sci USA 99:1758
49. Wilbur JL, Kumar A, Biebuyck HA, Kim E, Whitesides GM (1996) Microcontact printing of self-assembled monolayers: applications in microfabrication. Nanotechnology 7:452
50. Xia Y, Whitesides GM (1998) Soft lithography. Annu Rev Materials Sci 28:153
51. Xu T, Gregory CA, Molnar P, Cui X, Jalota S, Bhaduri SB, Boland T (2006) Viability and electrophysiology of neural cell structures generated by the inkjet printing method. Biomaterials 27:3580
52. Xu T, Jin J, Gregory C, Hickman JJ, Boland T (2005) Inkjet printing of viable mammalian cells. Biomaterials 26:93

53. Odde DJ, Renn MJ (1999) Laser-guided direct writing for applications in biotechnology. Trends Biotechnol 17:385
54. Odde DJ, Renn MJ (2000) Laser-guided direct writing of living cells. Biotechnology and Bioengineering 67:312
55. Saunders R, Derby B, Gough J, Reis N (2004) Ink-jet printing of human cells. Materials Research Society Symposium Proceedings EXS-1:95
56. Saunders R, Gough J, Derby B (2005) Ink jet printing of mammalian primary cells for tissue engineering applications. Materials Research Society Symposium Proceedings 845:57
57. Eagles PAM, Qureshi AN, Jayasinghe SN (2006) Electrohydrodynamic jetting of mouse neuronal cells. Biochem J 394:375
58. Jayasinghe SN, Eagles PAM, Qureshi AN (2006) Electric field driven jetting: an emerging approach for processing living cells. Biotechnol J 1:86
59. Jayasinghe SN, Qureshi AN, Eagles PAM (2006) Electrohydrodynamic jet processing: An advanced electric-field-driven jetting phenomenon for processing living cells. Small 2:216
60. Bocanegra R, Galan D, Marquez M, Loscertales I, Barrero A (2005) Multiple electrosprays emitted from an array of holes. J Aerosol Sci 36:1387
61. Loscertales G (2002) Micro/nano encapsutation via electrified coaxial liquid jets. Science 295:1695
62. Colina M, Serra P, Fernandez-Pradas JM, Sevilla L, Morenza JL (2005) DNA deposition through laser induced forward transfer. Biosensors & Bioelectronics 20:1638

Chapter 11
Selective Laser Sintering of Polymers and Polymer-Ceramic Composites

Suman Das

11.1 Introduction

It is widely accepted that rapid prototyping methods based on solid freeform fabrication (SFF) [1–5] have the potential to address the challenges associated with manufacturing scaffolds for tissue engineering. SFF refers to a group of manufacturing technologies that are capable of fabricating geometrically complex objects directly from computer models (or three-dimensional digital representations) without part-specific tooling or knowledge [6]. Many SFF techniques are based on an additive manufacturing process in which objects are constructed by the sequential and controlled deposition of material (liquid, powder, solid sheets) in a layer-by-layer fashion. Recently, there has been an increasing interest in using SFF methods to manufacture tissue engineering scaffolds [5, 7–23].

Selective Laser Sintering (SLS) is a laser-based SFF technique in which an object is built layer-by-layer using powdered materials, radiant heaters, and a computer controlled laser [24]. The process is characterized by particle coalescence in the scanned regions of a powder bed resulting from the application of laser and thermal energy. The reduction in surface free energy associated with particle coalescence is the main driving force for the sintering process [25]. In SLS a digital representation of an object is first mathematically sliced into a number of thin layers. The object is then created by selectively fusing (sintering or melting, depending on powder material composition) patterns into sequentially deposited layers of powder with a scanning laser beam. Each layer of scanned powder is fused to its underlying layer and corresponds to a cross-section of the object as determined from the mathematical slicing operation. Figure 11.1 provides a general schematic of the SLS process.

11.2 Fundamentals of Selective Laser Sintering

Particle coalescence in selective laser sintering can be described by mechanisms similar to those in conventional sintering. A schematic of two particles in contact, undergoing sintering is shown in Fig. 11.2.

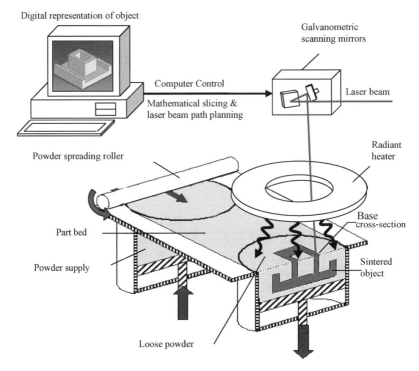

Fig. 11.1 Selective Laser Sintering process schematic

One of the earliest models for describing viscous sintering was developed by Frenkel [26]. Frenkel derived an equation to describe the time evolution of the dimensionless neck radius (neck radius x normalized by the particle radius a) as follows:

$$\left(\frac{x}{a}\right)^2 = \frac{3}{2\pi}\frac{\sigma}{a\eta}t \tag{11.1}$$

In Frenkel's model, the sintering rate (or the degree of neck formation as a function of time) is dependent on σ, the surface tension, μ, the viscosity, and a, the particle radius. It states that sintering proceeds faster for polymers with a higher surface tension, lower viscosity, and smaller particle radius. However, this model assumes that particle radius remains nearly constant, and is thus valid only for the

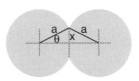

Fig. 11.2 Neck formation by sintering of two particles in contact

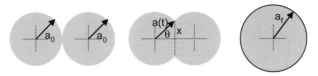

Fig. 11.3 Sintering of two particles to eventually form a single larger particle

initial stage of sintering, when the neck radius is much smaller than the particle radius.

To fully appreciate sintering phenomena in which particle coalescence occurs to a greater extent, including the extreme situation of complete coalescence of two particles to form a single larger particle, it is necessary to resort to a more sophisticated model. Pokluda's model [27] describes this phenomenon. A schematic for describing Pokluda's model is shown in Fig. 11.3.

According to Pokluda's improved sintering model, the rate of change of the angle θ is given by:

$$\frac{d\theta}{dt} = \frac{\sigma}{a_0 \eta} \frac{2^{-\frac{5}{3}} \cos(\theta) \sin(\theta) [2 - \cos(\theta)]^{\frac{1}{3}}}{[1 - \cos(\theta)][1 + \cos(\theta)]^{\frac{1}{3}}} \quad (11.2)$$

As a result, the particle radius as a function of time is given by:

$$a(t) = a_0 \left(\frac{4}{\{1 + \cos[\theta(t)]\}^2 \{2 - \cos[\theta(t)]\}} \right)^{\frac{1}{3}} \quad (11.3)$$

Consequently, the dimensionless neck radius (ratio of neck radius x normalized by the instantaneous particle radius a) is given by:

$$\frac{x(t)}{a_0} = \sin(\theta) \left(\frac{4}{\{1 + \cos[\theta(t)]\}^2 \{2 - \cos[\theta(t)]\}} \right)^{\frac{1}{3}} \quad (11.4)$$

A plot of the dimensionless neck radius as a function of dimensionless time is given in Fig. 11.4.

Figure 11.4 reveals that at complete particle coalescence, the new particle radius is 1.2599 times the original particle radius, which is consistent with volume conservation considerations, thus confirming that Pokluda's model satisfies complete densification. Figure 11.4 also reveals that time for complete coalescence is directly proportional to surface tension, and is inversely proportional to viscosity and initial particle radius.

There are however, additional considerations pertinent to polymer processing by SLS. The viscosity of most polymers is known to have an Arhennius rate dependence on temperature. Thus, temperature dependence of polymer viscosity is given by [28]:

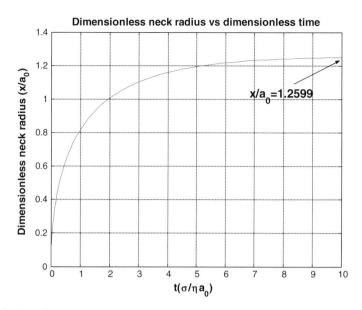

Fig. 11.4 Dimensionless neck radius as a function of dimensionless time

$$\eta = \eta_0 e^{\left(-\frac{\Delta E}{\Re T}\right)} \quad (11.5)$$

As is evident from Eq. (11.5), the polymer melt viscosity is a very sensitive function of temperature. Thus, semi-crystalline polymers superheated to temperatures beyond their melting range but below their decomposition temperatures will undergo a sharp drop in viscosity, aiding better polymer flow during the short timescales associated with the scanning laser beam interacting with the powder during SLS.

A further consideration must be taken into account with respect to polymer viscosity. Viscosity of polymers is known to depend on the molecular weight as follows [28]:

$$\eta = K_H Z_W^{3.4} \quad (11.6)$$

Equation 11.6 reveals that the polymer melt viscosity is a very sensitive function of molecular weight. In the context of SLS where a low melt viscosity if desired in order to enable the polymer melted by the laser to flow and spread quickly, clearly, lower molecular weight polymers will have a greater ability to form smooth, flat, and dense layers than polymers with high molecular weights. Furthermore, the dependence of the polymer melt viscosity on molecular weight also has implications on batch to batch control of part quality. A small change in the molecular weight of a polymer from one batch of material to another will significantly impact the ability to consistently make high quality parts using the same set of optimal processing

11 Selective Laser Sintering of Polymers and Polymer-Ceramic Composites

parameters which have presumably been arrived at by extensive testing on a previous batch of material.

The SLS process is a thermal manufacturing process. Heat transfer considerations play an important role in process control for SLS. One of the simplest models for the SLS process involves the laser as a heat source incident on a semi-infinite solid medium delivering a heat flux q_0 for a time duration τ. The closed form solution for the one-dimensional heat conduction equation describing this situation is [29]:

$$T - T_i = \left(\frac{2q_0\sqrt{\alpha}}{k}\right)\left\{\sqrt{t}\,ierfc\left[\frac{z}{\sqrt{\alpha t}}\right]\right. \\ \left. - K\sqrt{t-\tau}\,ierfc\left[\frac{z}{\sqrt{4\alpha(t-\tau)}}\right]\right\} \quad (11.7)$$

where $K = \begin{cases} 0 \text{ for } t \leq \tau \\ 1 \text{ for } t > \tau \end{cases}$

where T is the temperature at depth z, and ierfc is the integral of the complementary error function.

Define a dimensionless temperature θ^*, dimensionless time t^* and dimensionless depth z^* as follows:

$$\theta^* = \frac{k(T - T_i)}{q_0\sqrt{4\alpha t}}, \quad t^* = \frac{t}{\tau}, \quad z^* = \frac{z}{\sqrt{4\alpha t}} \quad (11.8)$$

Then, the temperature as a function of time and depth is as follows:

$$\theta^* = \begin{cases} \sqrt{\frac{t^*}{\pi}}\exp(-\frac{z^{*2}}{t^*}) - z^*\left[1 - erf\left(\frac{z^*}{\sqrt{t^*}}\right)\right] & \text{for } t^* \leq 1 \\ \sqrt{\frac{t^*}{\pi}}\exp(-\frac{z^{*2}}{t^*}) + z^*erf\left(\frac{z^*}{\sqrt{t^*}}\right) - \\ \sqrt{\frac{t^*-1}{\pi}}\exp(-\frac{z^{*2}}{t^*-1}) - z^*erf\left(\frac{z^*}{\sqrt{t^*-1}}\right) & \text{for } t^* > 1 \end{cases} \quad (11.9)$$

Plots of θ^* versus t^* for different values of z^* are shown in Fig. 11.5.
By setting the following variables:

$$\tau = \frac{d}{v}, \quad q_0 = \frac{4(1-R)P}{\pi d^2}, \quad T = T_m, \quad \alpha = \frac{k}{\rho c_p}, \quad St = \frac{c_p(T_m - T_i)}{\lambda},$$

$$\theta^*_{max} = \frac{k(T_m - T_i)}{q_0\sqrt{4\alpha t}} \quad (11.10)$$

and using a third order curve fit to express z^* as a function of θ^*_{max}, the following expression is obtained for the dimensionless melt depth, i.e. the depth at which the temperature is equal to the melting point T_m.

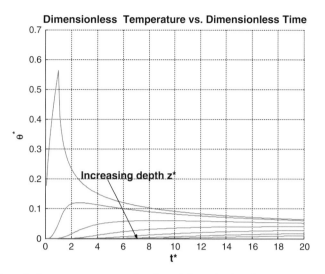

Fig. 11.5 Dimensionless temperature as a function of dimensionless time

$$z^* = -0.31 + 0.17 \left(\frac{1}{\theta^*_{max}}\right) - 0.0018 \left(\frac{1}{\theta^*_{max}}\right)^2 + 6 \times 10^{-6} \left(\frac{1}{\theta^*_{max}}\right)^3 \quad (11.11)$$

Figure 11.6 shows the third order curve fit compared with the close form solution data obtained by simply inverting Fig. 11.5 for $T = T_m$.

By substituting the variable definitions of Eq. 11.10 and into Eq. 11.11, the following expression is obtained for melt depth.

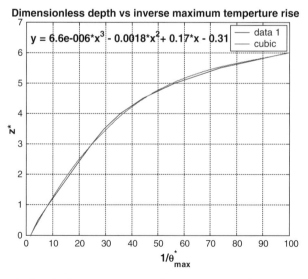

Fig. 11.6 Dimensionless depth versus inverse maximum temperature rise

$$z = -0.31\sqrt{\frac{4\alpha d}{v}} + \frac{0.68}{\pi}\frac{4(1-R)}{\rho\lambda}\frac{\lambda}{c_p(T_m - T_i)}\frac{P}{dv} + \text{higher order terms} \quad (11.12)$$

Equation 11.12 reveals that the melt depth is a function of the following controllable process parameters: beam diameter d, the laser scan speed v, the laser power P, and the initial temperature T_i. Clearly, using a larger beam diameter and slower scanning speed yields a larger melt depth. Similarly, using a higher laser power also results in greater melt depth. These predictions are consistent with experimental observations. Furthermore, the initial substrate temperature T_i plays a significant role as well. A higher initial substrate temperature, or equivalently, a small Stefan number yields a greater melt depth. This is also consistent with experimental observations.

The selective laser sintering process builds parts by fusing layers of a powder using a scanning laser beam. The situation of a polymer melt created on top of a previously formed layer of the same polymer constitutes homologous wetting. Homologous wetting occurs when a molten material spreads on a solid substrate of its own kind, and has been extensively studied by Schiaffino and Sonin [30, 31]. It involves simultaneous heat transfer, fluid flow and solidification. Figure 11.7 shows a schematic of homologous wetting.

In homologous wetting, the molten material superheated above its melting temperature wets and spreads over a solid substrate of its own kind while simultaneously losing latent heat to the substrate. During this process, a solidification front grows into the spreading material, and eventually, spreading is arrested when the contact angle of the solidification front θ_s is equal to the instantaneous angle of contact θ_a of the spreading material. Schiaffino and Sonin termed this angle θ^* and showed that for molten microcrystalline wax, θ^* is function of the Stefan number St as follows:

$$\theta^* = 85\sqrt{St} \quad (11.13)$$

Thus, a small angle of arrest is achieved when the Stefan number is small, or equivalently, when the initial temperature of the substrate is close to the melting

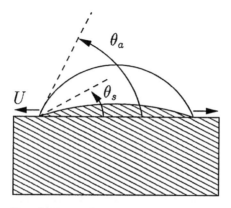

Fig. 11.7 A schematic of homologous wetting

temperature. If similar behavior is assumed for other semi-crystalline polymers, then in the context of SLS, we can infer that maintaining a substrate temperature close to the melting temperature of the polymer will ensure that the molten polymer wets and spreads well over a previously formed layer and achieves a low angle of contact at the time of solidification so that a flat, smooth new layer is formed. In practice, the substrate preheat temperature during SLS of semi-crystalline polymers is maintained within approximately 15 °C of the melting temperature of the polymer.

From the above considerations relating to polymer sintering, heat transfer and fluid flow, it can be inferred that both polymer physical properties and SLS process parameters have a strong impact on the ability to process a given polymer into three-dimensional shapes by SLS. Polymers with high surface tension, low melt viscosity, and small powder particle size have higher sintering rates, resulting in faster processes for full densification. The laser power, beam diameter, scan speed, and the substrate preheat temperature all control the melt depth. The substrate preheat temperature also affects the wetting and spreading of the molten polymer on a previously formed layer of the same material.

11.3 Selective Laser Sintering of Biomaterials

SLS of calcium phosphate, a resorbable material suitable for bone replacement was investigated by Lee [32] and Vail [33]. SLS of non-resorbable materials has been investigated by several groups worldwide for biomedical applications [34–38]. More recently, the fabrication of drug delivery devices [39] and tissue engineering scaffolds [40–42] has been attempted via SLS of polymeric biomaterials and their composites. While the efforts in references [40–42] have shown promise by documenting the feasibility of fusing powder particles together by laser sintering of such materials, they have not visually demonstrated the capability of fabricating complex three-dimensional structures in these materials.

In a previous article [43], our group at the University of Michigan demonstrated a bone tissue engineering approach using polycaprolactone scaffolds fabricated by SLS. Mechanical properties of such scaffolds were shown to be within the reported lower range of properties for trabecular bone. However, in that work, scaffolds were shown to be incompletely dense in the design solid regions, i.e. struts of the scaffolds resulting in approximately 20% porosity in the design solid regions of the scaffolds. Furthermore, the designed porosity, i.e. the orthogonal porous channels were not faithfully reproduced according to design due to excess powder being sintered and bonded to the pore channel interior surfaces, resulting in actual porosity of the scaffold being less than the design porosity. These effects were attributed to sub-optimal SLS processing parameters, including the laser power, the scan speed and the powder bed preheat temperature. Subsequently, we conducted a thorough study to develop optimal processing parameters based on a design of experiments approach. This study resulted in the development of SLS processing parameters that enabled us to achieve density in excess of 94% relative density in the solid regions of the scaffold while faithfully constructing the overall scaffold and the pore geometry to within 3–8% of designed dimensions. Two-level factorial design of experiments procedures allowed us to achieve optimal SLS processing parameters for PCL. Scaffolds built using these parameters were subsequently characterized by compressive mechanical testing. The following sections report on the recent work done by the Solid Freeform Fabrication laboratory in selective laser sintering of PCL and PCL-TCP composites.

11.4 Materials and Methods

11.4.1 Polycaprolactone and Tricalcium Phospahte

Polycaprolactone (PCL) powder marketed under the brand name CAPA® 6501 (Solvay Caprolactones, Warrington, UK) was used for this study. CAPA® 6501 is a high molecular weight ($\overline{M}_n \approx 50,000$) linear, biocompatible, bioresorbable thermoplastic belonging to the aliphatic polyester family. It is a white, odorless powder with a rated particle size distribution in which 99% of all particles are less than 100 μm in size (no particles greater than 125 μm). It is a semi-crystalline (56%) polymer having a melting point of 58–60 °C, a glass transition temperature of approximately −60 °C, and a decomposition temperature of around 350 °C [44]. Homopolymers of PCL have degradation times on the order of two years but rates can be tailored via copolymerization with other aliphatic polyester monomers (i.e. lactide and glycolide) [45, 46]. The toxicology of PCL has been extensively studied [47] and it is currently regarded as a nontoxic hard and soft tissue compatible material [45].

The Tricalcium Phosphate powder, chemical formula $Ca_5(PO_4)_3OH$, used in our studies was obtained from Astaris LLC (St. Louis, MO). This particular powder (CAS NO. 1306-06-5) is a FCC grade TCP conditioner with an average particle size of 45 μm and a broad distribution (no particles > 150 μm).

11.4.2 Preparation and Processing

The PCL powder and PCL-TCP composite powder blends used in our studies were processed using a Sinterstation® 2000 commercial SLS machine (3D Systems Inc., Valencia, CA). PCL-TCP powder blends were created by physically blending pure PCL and TCP powders in a rotary tumbling-mixer prior to SLS processing. Our investigations focused on processing pure PCL powder and PCL-TCP composites with 10–30% volume fraction of TCP. Individual layers of the powders were sequentially spread across the part build area and preheated prior to laser scanning. A low power CO_2 laser ($\lambda = 10.6$ μm, continuous wave, power <10W) focused to a 450 μm spot was then scanned across the part bed. The resulting application of laser and thermal energy induced sintering to take place, followed by rapid melting and resolidification of the polymer within and across adjacent powder layers to produce solid, three-dimensional scaffolds.

11.5 Selective Laser Sintering Parameter Development

Two-level factorial Design of Experiments (DOE) techniques were employed to determine optimal SLS parameters for processing PCL. Optimal processing parameters are defined as those that result in parts as close to ideal SLS processed parts as possible. An ideal SLS processed part is defined as one that is (a) fully dense in the regions where material is present in the part's design, (b) dimensionally accurate, and (c) easily removed from its support powder.

11.5.1 Test Geometry Design to Replicate Typical Scaffold Features

A test geometry was chosen that represents the typical features and feature sizes found in bone tissue engineering scaffolds intended to replace or repair an anatomic defect [48–50]. A schematic of the test geometry consisting of a half-porous, half-solid 11.5 mm × 14.2 mm × 10.8 mm cube is shown in Fig. 11.8. The porous section consists of an orthogonal network of rectangular channels and is representative in

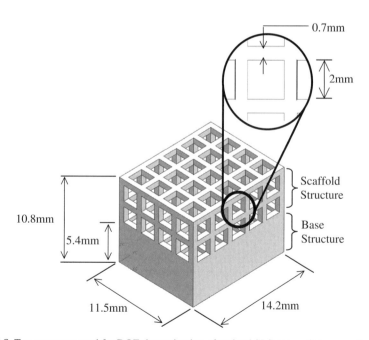

Fig. 11.8 Test geometry used for DOE determination of optimal SLS processing parameters

design and shape to scaffold designs incorporating periodic, interconnected pore architectures. The solid section of the test geometry is representative of surgical fixation base structures found in anatomic scaffold designs. This test geometry was chosen so that optimal processing parameters could be determined for part slice cross-sections that are large or small in area. The digital solid model for this test geometry was constructed using Unigraphics® CAD/CAM/CAE software and was exported as an STL file for upload to the Sinterstation® 2000.

11.5.2 SLS Process Parameters and Ranges Investigated

The SLS process parameters investigated in this study were laser power, scan speed, scan spacing, part bed temperature, and powder layer delay time. Table 11.1 lists

Table 11.1 SLS parameters and ranges investigated

Parameter		Description	Unit	Low Level	High Level
X_1	Laser power	The amount of laser power delivered to the part bed when the laser is filling in the cross-sectional slice of a part	Watts	4.1	5.4
X_2	Scan speed	The speed at which the laser beam moves over the part bed during the scanning operation	mm/sec (in /sec)	1079.5 (42.5)	1231.9 (48.5)
X_3	Scan spacing	The distance between parallel scans during a laser scanning operation	μm (in)	76.2 (0.003)	152.4 (0.006)
X_4	Part bed temp	The surface temperature of the powder at the part bed during the laser scanning operation. Thermal energy is supplied from a radiant heater located above the part bed and the temperature is monitored by an IR sensor.	°C	46	48
X_5	Powder layer delay	The amount of time a scanned layer is exposed to the radiant heater before a new layer of powder is added.	sec	0	8

these parameters along with a brief description and the corresponding ranges that were investigated.

The upper and lower ranges for the process parameters of Table 11.1 were chosen as follows: Combinations of laser power and scan speed were chosen so that at the PCL polymer powder was visibly strongly melted by the laser beam and was on the verge of decomposition with the emission of small amounts of smoke at the combination of the high level of laser power and low level of scan speed. At the low level of laser power and high level of scan speed, a coupon scanned by the laser beam was weakly sintered and barely held its shape together. The upper and lower ranges of scan spacing were chosen so that the beam overlap factor (beam diameter divided by scan spacing) ranged between approximately 3 and 6. Further, the lower level of scan spacing is the smallest scan spacing allowable by the Sinterstation 2000 machine control software. It is well known that in selective laser sintering of semi-crystalline polymers, the part bed temperature must be kept as close to the melting point as possible while avoiding part bed caking or incipient melting of the polymer powder before laser sintering. Maintaining the part bed preheat temperature close to the melting point of the polymer ensures that the thermal gradients in the solidifying molten polymer are minimized, thermomechanical distortion, i.e. curl is minimized and the polymer has sufficiently low viscosity so as to flow and spread properly and that the polymer has sufficient time to relax [6]. In practice, the powder bed preheat temperature is usually maintained between 3 and 15 degrees Celsius below the melting temperature of the semi-crystalline polymer. The melting point of CAPA® 6501 is 56 °C. With this in mind, we kept the low level of part bed temperature at 46 °C while the high level was a mere 2 °C higher at 48 °C. We found the solid state sintering characteristics of CAPA® 6501 to be extremely sensitive to temperature in this temperature range and at temperatures even slightly above 48 °C, the powder in the part build area began to cake severely. At part bed temperatures below 46 °C, thermomechanical distortion and out-of-plane curl was so severe that distortion of the first scanned layer prevented the deposition of additional powder and the fabrication of additional layers on top of the first layer. Hence the low and high levels of powder bed preheat temperature were selected as 46° and 48 °C respectively. The default value of the powder layer delay in the Sinterstation 2000 control software is set to zero. However, we decided to vary this parameter and make the powder layer delay longer to test whether additional time exposure of a scanned layer would allow further polymer relaxation and densification, particularly for sections of the scaffold structure with designed porous channels wherein thin sections, i.e. the struts are separated by regions of relatively insulating powder. Incorporating a powder layer delay causes the overall time for part fabrication to increase, as there is a delay included for each layer sintered by the laser. Therefore, if a powder delay is incorporated, it should be as short as possible, so that the overall build time of a part is not significantly increased. In preliminary experiments, we found that a powder layer delay beyond 8s did not provide any additional polymer densification, as determined by optical microscopy and void fraction analysis.

11.5.3 Design Matrix and Experimental Procedures

In order to analyze the effects of five processing parameters in a two-level DOE, a complete factorial design requires that $2^5 = 32$ unique test conditions must be conducted. Process parameters were coded such that the high level for each parameter setting was denoted by +1 and the low level by −1. Table 11.2 outlines the 2^5 *design matrix* that was followed in conducting the DOE for this work. The design matrix, given in both coded and uncoded form, defines the 32 distinct test recipes used in the design of this experiment. The particular arrangement of this matrix guarantees that all columns are orthogonal to each other so that independent estimates of all main and parameter interaction effects may be easily obtained. One experimental

Table 11.2 Design matrix used to conduct the DOE

	Coded Test Conditions					Actual Test Conditions				
Test	X_1	X_2	X_3	X_4	X_5	Laser Power (W)	Laser Scan Speed (mm/sec)	Laser Scan Spacing (μm)	Part Bed Temp (°C)	Powder Layer Delay (sec)
1	−1	−1	−1	−1	−1	4.1	1079.5	76.2	46	0
2	+1	−1	−1	−1	−1	5.4	1079.5	76.2	46	0
3	−1	+1	−1	−1	−1	4.1	1231.9	76.2	46	0
4	+1	+1	−1	−1	−1	5.4	1231.9	76.2	46	0
5	−1	−1	+1	−1	−1	4.1	1079.5	152.4	46	0
6	+1	−1	+1	−1	−1	5.4	1079.5	152.4	46	0
7	−1	+1	+1	−1	−1	4.1	1231.9	152.4	46	0
8	+1	+1	+1	−1	−1	5.4	1231.9	152.4	46	0
9	−1	−1	−1	+1	−1	4.1	1079.5	76.2	48	0
10	+1	−1	−1	+1	−1	5.4	1079.5	76.2	48	0
11	−1	+1	−1	+1	−1	4.1	1231.9	76.2	48	0
12	+1	+1	−1	+1	−1	5.4	1231.9	76.2	48	0
13	−1	−1	+1	+1	−1	4.1	1079.5	152.4	48	0
14	+1	−1	+1	+1	−1	5.4	1079.5	152.4	48	0
15	−1	+1	+1	+1	−1	4.1	1231.9	152.4	48	0
16	+1	+1	+1	+1	−1	5.4	1231.9	152.4	48	0
17	−1	−1	−1	−1	+1	4.1	1079.5	76.2	46	8
18	+1	−1	−1	−1	+1	5.4	1079.5	76.2	46	8
19	−1	+1	−1	−1	+1	4.1	1231.9	76.2	46	8
20	+1	+1	−1	−1	+1	5.4	1231.9	76.2	46	8
21	−1	−1	+1	−1	+1	4.1	1079.5	152.4	46	8
22	+1	−1	+1	−1	+1	5.4	1079.5	152.4	46	8
23	−1	+1	+1	−1	+1	4.1	1231.9	152.4	46	8
24	+1	+1	+1	−1	+1	5.4	1231.9	152.4	46	8
25	−1	−1	−1	+1	+1	4.1	1079.5	76.2	48	8
26	+1	−1	−1	+1	+1	5.4	1079.5	76.2	48	8
27	−1	+1	−1	+1	+1	4.1	1231.9	76.2	48	8
28	+1	+1	−1	+1	+1	5.4	1231.9	76.2	48	8
29	−1	−1	+1	+1	+1	4.1	1079.5	152.4	48	8
30	+1	−1	+1	+1	+1	5.4	1079.5	152.4	48	8
31	−1	+1	+1	+1	+1	4.1	1231.9	152.4	48	8
32	+1	+1	+1	+1	+1	5.4	1231.9	152.4	48	8

run was conducted for each of the 32 tests resulting in the manufacture of one test part for each of the 32 test recipes.

A powder layer thickness of 101.6 μm (0.004") was used for each of the test builds. This is the smallest layer thickness that can be used for processing the powders used in this study and it is physically limited by the largest average powder particle size of 100 μm for the CAPA® 6501 PCL. The process chamber for all builds was kept at ambient temperature ($\sim$ 20°C). The average build time for fabricating each of the 32 test parts was approximately 2 hours. Parts were allowed to cool-down for 45 minutes after each build was complete. Parts were subsequently removed from the machine and cleaned to remove all support powder. All runs were conducted on the same machine using similar powder sources (i.e. same type, from same manufacturer/shipment/lot, having similar levels of previous use, thermal history and storage time).

11.5.4 SLS Part Quality Characterization

SLS part quality was defined in terms of three critical characteristics: (1) ease of part break-out, (2) degree of dimensional accuracy, and (3) the resulting void fraction content of the part. These three characteristics were used to assess the part quality for each base and scaffold structure comprising each DOE test part. Their importance and method of measurement are described in greater detail below.

11.5.4.1 Break-out Quality Metric

Due to the nature of the SLS process, parts are encased in a mass of powder upon completion of a build. As a result, ease of break-out is a qualitative measure of the effort involved in removing all of the powder surrounding a completed part. If processing conditions are not optimal, excess support powder can partially sinter and adhere to exterior part surfaces. This thermal growth can lead to dimensional inaccuracy, clogging of pores, and even parts that are impossible to recover.

For the 32 test parts produced we attempted to quantify this quality characteristic by a numerical value termed the break-out quality metric (BQM). A BQM value between 0 and 1 was assessed and assigned to each base and each scaffold section for every test part using the scale shown in Table 11.3 as a guideline. This number represents an attempt to quantify the relative ease and/or difficulty with which a built part can be post-processed (i.e. removed and cleaned).

11.5.4.2 Dimensional Accuracy Quality Metric

Dimensions of the scaffold and base structures for each of the 32 test parts were measured. The length and width of each base structure was measured in several locations using a pair of Mitutoyo digital calipers with a resolution of 0.01 mm.

11 Selective Laser Sintering of Polymers and Polymer-Ceramic Composites   243

Table 11.3 BQM assessment guidelines

BQM	Base Structure	Scaffold Structure
0.0	All powder is able to be brushed off base exterior	All powder is able to be removed from porous channels using a compressed air gun
0.1	Most powder is able to be brushed off base exterior, very light scraping required to remove partially sintered powder "skin"	All powder is able to be removed from horizontally aligned porous channels (side pores) using a compressed air gun, unable to completely blow powder out of vertically aligned pores (top pores)
0.3	Slight scraping required to remove partially sintered layer of powder from base exterior	Most support powder is able to be removed from side pores using a compressed air gun, unable to blow powder out of top pores
0.5	Moderate scraping required to remove partially sintered layer of powder from base exterior	Able to blow roughly half of powder from side pores using a compressed air gun, unable to blow powder out of top pores
0.7	Heavy scraping required to remove partially sintered layer of powder from base exterior	Unable to remove most powder from pores using a compressed air gun, unable to blow powder out of top pores
1.0	Unable to remove completed part from powder bed and/or unable to completely remove partially sintered powder from base exterior without excessive force or part damage	Unable to remove any amount of powder from either side or top pores using a compressed air gun

Measurements for each feature were then recorded, averaged, and compared with their target dimension by calculating the deviation as a percent difference. A dimensional quality metric (DQM) was then calculated for the base structure of each test part by averaging the percent difference in deviation from both the length and width. A DQM was calculated for the scaffold structure of each test part in a similar manner. In addition, measurements were also made of the individual pore geometry for each scaffold structure. Each test part was placed on its side under a Nikon SMZ-2T stereo microscope (Nikon Instruments Inc., Melville, NY) equipped with a COHU CCD camera (Cohu Inc., Electronics Division, San Diego, CA). The scaffold structure for each test part was examined under 10X magnification and a digital photomicrograph was taken for each sample. The width and height of several pores for each scaffold structure were then measured using ImageJ (http://rsb.info.nih.gov/ij/) image analysis software. Measurements obtained for both the scaffold exterior and pore geometry were recorded, averaged, and then compared with their target dimensions by calculating their percent difference of deviation. The average percent deviation of these measurements was assigned as the scaffold DQM. The DQM's calculated for each base and scaffold structure fell between 0 and 1 for all test parts produced (the highest DQM recorded for either structure was 0.36). The dimensional accuracy quality metric provides an average measure and quantitative indication of the dimensional accuracy of a SLS fabricated test part. A DQM value of zero indicates a structure that is dimensionally perfect, while a value of 1 would indicate on average that a structure is approximately twice its specified size.

11.5.4.3 Void Fraction Quality Metric

One of our goals is to produce parts that are fully dense or as close to fully dense as possible in those regions where material is present in the part's design. The SLS process is characterized by particle coalescence in scanned regions of a powder bed resulting from the application of laser and thermal energy. Inadequate energy levels, time scales, and unfavorable material sintering characteristics can result in incomplete particle densification and poor layer-to-layer bonding resulting in the formation of unintentional, manufacturing-induced porosity. Optical microscopy and image analysis techniques were used to characterize the resulting microstructure and manufacturing induced porosity of SLS fabricated test parts. Fabricated specimens were cleaved parallel and perpendicular to the build direction using a single-edged razor blade. Each cleaved surface was stained using a black colored dry-erase marker. The ink from the marker remained in the void spaces when the excess was wiped off from each of the cleaved surfaces. Each cleaved surface for each base and scaffold structure was then examined under a Nikon Optiphot optical microscope equipped with a DS-5M-L1 digital sight camera system. Several photomicrographs were taken of each cleaved surface (one in the center and one near the periphery). Figure 11.9 shows the stained cross-section of a DOE test part cleaved perpendicular to the build direction along with the resulting photomicrographs taken of the base and scaffold structure.

The photomicrographs for each base and scaffold structure were then analyzed using ImageJ software to determine their void fraction content. Each RGB image

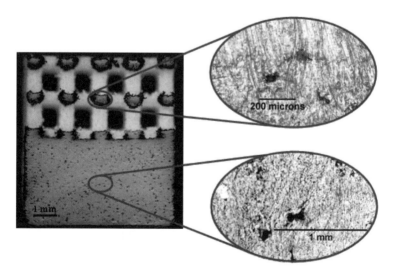

Fig. 11.9 Cleaved cross-section and resulting microstructure of optimally processed PCL test part

was converted to a binary image and subsequent thresholding operations were performed to isolate voids from the surrounding fully dense material matrix. The planar area-based void fraction for each specimen was determined using the in-built 'Analyze Particles' capabilities of ImageJ. A void fraction quality metric (VQM) was then assigned to each base and each scaffold structure by averaging the void fraction estimates for respective structures that were obtained using ImageJ. This method provided a representative estimate of the manufacturing induced porosity in both the base and the scaffold structures for each test part.

11.5.5 Total Quality Metric

We defined a total quality metric (TQM) for each base and each scaffold structure. The TQM weighs together the three previously defined critical quality characteristics into one numerical value between 0 and 1 that can be viewed as a representative measure of overall part quality. A TQM value of 0 indicates a perfect part, while a value of 1 indicates a poorly produced part. Equal weighting was given to each of the three critical quality metrics with the reasoning that all three are equally important in evaluating the overall quality of an SLS produced part. The following relationship was chosen to represent the measured TQM quality response of an SLS produced part:

$$TQM(BQM, DQM, VQM) = \frac{1}{3}BQM + \frac{1}{3}DQM + \frac{1}{3}VQM \qquad (11.14)$$

11.5.6 Mathematical Process Model Development

11.5.6.1 Calculation of Main and Interaction Parameter Effects

The first step toward developing a statistically significant mathematical model capable of predicting the SLS part quality response defined by the TQM involves calculating all main and interaction effects for the five process parameters under investigation. The average or main effect of a parameter is the amount of change witnessed in the quality response, on the average, when only that parameter is changed from its low to high level. Similarly, parameter interaction effects measure the dependency of one or more parameter effects on the level of another over the investigated parameter ranges. Using the data from a 2^5 factorial experiment, the following parameter effects shown below in Table 11.4 can be estimated.

The 32 variable effect estimates were calculated according to the well established and standardized methods for analyzing two-level factorial design of experiments. A brief description of the process used to calculate the variable effects is outlined below; for a more detailed description, see [51]. In order to obtain the necessary effect estimates, the design matrix listed in Table 11.2 was augmented by adding

Table 11.4 2^5 Factorial DOE effect estimates

Description	Number of effects
Mean response	1
Main effects	5
Two-factor interaction effects	10
Three-factor interaction effects	10
Four-factor interaction effects	5
Five-factor interaction effects	1
TOTAL	32

columns for the two-, three-, four-, and five-factor interactions. These new columns were formed from the design matrix by taking all possible products of the main parameter columns two, three, four, and five at a time. The next step involved adding a column of all "+1" values under the column heading I (identity column). This column was used to estimate the mean system response. The completed matrix is referred to as the *calculation matrix* [51]. Using the calculation matrix, all main and interaction effects were tabulated by multiplying a column containing the measured quality response for each test (TQM) with the corresponding ±1 values of each effect column, algebraically summing the result, and then dividing that total by 16 (total number of tests conducted divided by 2).

11.5.6.2 Statistically Significant Effect Estimates

Normal probability plots of the calculated effect estimates were then used to isolate those effects that are truly important (statistically significant) from those that appear to be driven solely by the forces of random variation. Normal probability plots for the base and scaffold structure effect estimates were constructed using Matlab and are shown in Figs. 11.10 and 11.11.

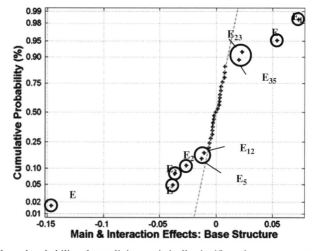

Fig. 11.10 Normal probability plot outlining statistically significant base structure effect estimates

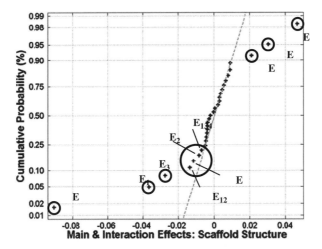

Fig. 11.11 Normal probability plot outlining statistically significant scaffold structure effect estimates

Several key observations can be noted from these figures. First, the majority of effect estimates in each plot tend to fall along or about a straight line centered near zero (indicated on plots by dashed lines). This indicates that these parameter effect estimates can be considered to arise from a normal distribution, and as such, have a statistically insignificant affect on the resulting quality response of a SLS produced PCL part. Second, it can be seen that there are several effect estimates in each plot that fall well off the dashed line. These effect estimates have been circled and marked on each respective figure (E denotes the effect estimate and the subscript identifies the particular parameter or parameters responsible). We can conclude that these effects are statistically significant and do not arise from a normal distribution with a mean of zero. As such, they can be considered to have the largest influence on the SLS processing of PCL.

11.5.6.3 Building the Process Model for Quality Improvement

The important effect estimates previously identified were used to build a statistically significant mathematical model capable of characterizing and predicting the quality response for base and scaffold structures processed out of PCL using SLS. The model relates the influence of the five process parameters under investigation to resulting part quality through a relationship taking on the form of Fig. 11.12, and described by Eqs. (11.15) and (11.16), and Table 11.5 as follows:

$$TQM_{base} = b_0 + b_1 x_1 + b_2 x_2 + b_3 x_3 + b_4 x_4 + b_5 x_5 + b_{12} x_1 x_2$$
$$+ b_{13} x_1 x_3 + b_{23} x_2 x_3 + b_{34} x_3 x_4 + b_{35} x_3 x_5 \quad (11.15)$$
$$TQM_{scaffold} = b_0 + b_1 x_1 + b_3 x_3 + b_4 x_4 + b_{12} x_1 x_2 + b_{13} x_1 x_3 + b_{14} x_1 x_4$$
$$+ b_{24} x_2 x_4 + b_{34} x_3 x_4 + b_{134} x_1 x_3 x_4 + b_{1345} x_1 x_3 x_4 x_5 \quad (11.16)$$

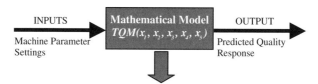

Fig. 11.12 Schematic of mathematical model

The x_i's in Eqs. (11.15) and (11.16) represent the settings of the five parameters studied in this investigation. The b_0 coefficient in Eqs. (11.15) and (11.16) is the mean system response and is the mean of all 32 measured TQM's for each respective structure (base or scaffold). The remaining b coefficients in Eqs. (11.14) and (11.15) are model coefficients derived from the statistically significant effect estimates previously identified and are calculated according to Eq. (11.16) as

$$b_i = \frac{E_i}{2} \qquad (11.16)$$

where b_i is a particular model coefficient and E_i the corresponding effect estimate. The calculated b_i coefficients are exactly one-half the value of the corresponding E_i. effect estimates due to the fact that each b_i coefficient measures the incremental change in the TQM per a one-unit change in the corresponding x_i (i.e. from 0 to 1), while the effect E_i measures the incremental change for a two-unit change in x_i (i.e. from -1 to $+1$).

Upon completion of the DOE and after analysis of the results, a set of confirmatory tests were conducted by repeating the set of experiments in the design matrix and experimentally measuring TQM values. The predicted quality response of the mathematical models generated in Eqs. 11.2 and 11.3 were compared with the experimentally measured base and scaffold TQM values. The average model error between the measured and predicted quality response for the DOE test runs was found to be zero for both models, confirming the legitimacy of the model. In

Table 11.5 Model coefficients

Model coefficients			
Base		Scaffold	
b_0	0.139	b_0	0.132
b_1	0.037	b_1	0.023
b_2	-0.013	b_3	-0.045
b_3	-0.073	b_4	0.015
b_4	0.027	b_{12}	-0.007
b_5	-0.006	b_{13}	-0.018
b_{12}	-0.005	b_{14}	0.011
b_{13}	-0.019	b_{24}	-0.004
b_{23}	0.011	b_{34}	-0.014
b_{34}	-0.018	b_{134}	-0.004
b_{35}	0.010	b_{1345}	-0.006

addition, the range of model error was also found to be quite low (0.039 for base structure model and 0.053 for scaffold structure model). As a result, we concluded that the empirically determined models provided a good fit to the experimentally obtained results, and can therefore be used to adequately predict the quality response of an SLS produced PCL part within the range of process parameters that were investigated.

11.5.6.4 Optimal SLS Parameters for Processing PCL

Nonlinear programming optimization techniques were used to determine the set of optimal SLS parameters for processing CAPA® 6501 PCL. Matlab® software (Mathworks, Natick, MA) was used to optimize the objective functions defined by Eqs. 11.2 and 11.3 to determine the optimal SLS process parameters for base and scaffold structures. A Matlab® script utilizing the *fmincon* function contained within Matlab's optimization toolbox was used to minimize the predicted quality response (defined by the mathematical models generated in Eqs. 11.2 and 11.3) over the tested variable range. The function *fmincon* finds the constrained minimum of a scalar function of several variables and its output, in our case, consisted of the minimum predicted TQM and the corresponding parameter values (i.e. optimal process parameter settings).

Table 11.6 shows the optimal process parameter settings determined by Matlab® along with a comparison of the corresponding predicted TQM with that of the actual measured TQM for the part produced at these settings. These settings represent the best possible set of process parameters we were able to develop for fabricating the highest quality base and scaffold structures in CAPA® 6501 PCL using SLS.

Figure 11.13 shows three views of the PCL test part fabricated at these settings. It can be noted that both the exterior geometry of the part and the internal geometry of the channels have been faithfully reproduced, as shown in Figs. 11.13a–c. Only on the topmost surface of the scaffold shown in the top view of Fig. 11.13c, a small amount of powder remains bonded to the inner surface of the channels giving the appearance of circular holes. This is due to the fact that after completion of the part, an additional ten layers of powder are swept on top of the part to cover the part and enable more uniform cooling and to avoid post-build thermal distortion. In this process, a small amount of powder gets bonded only at the openings of the rectangular channels at the top surface of the part, causing the channels in the top view to appear circular.

The process models developed in Eqs. 11.2 and 11.3 were also extrapolated to predict the quality response for an enlarged process parameter window. The previously defined Matlab® optimization algorithm was extended to include parameter settings within a range up to 50% larger than the actual test conditions. The minimum TQM's predicted for these extrapolated settings were found to be slightly lower than those falling within the tested range, suggesting that potentially higher quality parts could be produced at parameter settings both higher and lower than those values tested. Several test parts then were produced at "optimally" identified process parameter settings falling outside the original tested range. However, the

Table 11.6 Optimal SLS parameters and associated quality response for processing PCL

	Parameter	Setting Coded	Setting Uncoded	Predicted Quality Response	Measured Quality Response	
Base Structure	Laser power	−1	4.1 W			
	Laser scan speed	−1	1079.5 mm/sec	TQM	BQM = 0	TQM
	Laser scan spacing	+1	152.4 μm	0.032	DQM = 0.033	0.033
	Part bed temperature	−1	46 °C		VQM = 0.065	
	Powder layer delay	−1	0 sec			
Scaffold Structure	Laser power	−1	4.1 W			
	Laser scan speed	−1	1079.5 mm/sec	TQM	BQM = 0	TQM
	Laser scan spacing	+1	152.4 μm	0.071	DQM = 0.082	0.049
	Part bed temperature	−1	46 °C		VQM = 0.066	
	Powder layer delay	+1	8 sec			

experimentally measured TQM's for these parts were observed to be higher than both the predicted and measured TQM values reported in Table 11.6. This result offers further evidence that the parameter settings listed in Table 11.6 are indeed optimal, and indicates that the predictive capability of the processing model developed in Eqs. 11.2 and 11.3 is valid only over the studied parameter ranges.

As expected, the optimal process parameters developed for processing base and scaffold geometries out of PCL are very similar. The process settings are identical

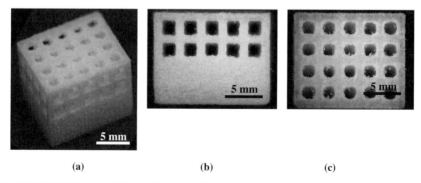

Fig. 11.13 DOE Test part fabricated at optimal process parameter settings

with the exception of an 8 second powder layer delay used when fabricating scaffold type structures. This difference may be explained by the fact that scaffold structures consist of small isolated cross-sections that do not have a large volume of underlying solidified material to act as a significant "heat sink" (as opposed to base structures). As a result, the additional time spent under the radiant heater may provide additional thermal energy needed for the material to properly flow and stress relax during the solidification process [6].

The measured quality responses for both the base structure and the scaffold structure indicate that the CAPA® 6501 PCL parts produced by SLS using the optimal parameter settings are very easily recovered from the part bed and easy to clean, are dimensionally accurate to within 3% for base and within 8% for scaffold, and are highly dense in the design solid regions of the scaffold, with density being 94% relative to full density. Figure 11.13a–c show isometric, side and top views of a DOE test part fabricated at these settings and a cross-sectional photomicrograph depicting its resulting microstructure respectively.

11.6 Fabrication of Anatomic PCL Scaffolds

As a proof-of-concept, we successfully built several PCL scaffolds using the optimal SLS process parameters. Shown in Fig. 11.14 a is a 3-D rendition of a minipig mandibular condyle scaffold design with bounding box dimensions of 22.9 mm × 24.4 mm × 14.75 mm. Figure 11.14b and 11.14c show different views of the corresponding PCL scaffold produced by SLS. Figure 11.15 a shows the 3-D rendition of a human condyle scaffold design with bounding box dimensions of 29.9 mm × 18.3 mm × 44.5 mm. Figures 11.15b–e show different views of the corresponding PCL scaffold built by SLS. Both the minipig and human condyle scaffolds were designed using image-based design methods developed by Hollister et al. [10, 17, 52]. The exterior geometry of these scaffolds was derived from volumetric 3D reconstructions of actual bone computed tomography (CT) scan images.

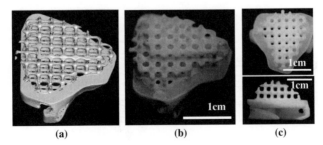

Fig. 11.14 (a) 3-D rendition of a minipig mandibular condyle scaffold design. (b) Actual PCL scaffold produced by SLS. (c) Top and side views of scaffold

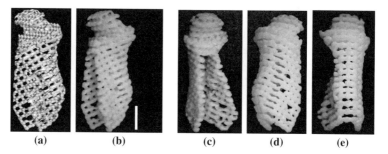

Fig. 11.15 (a) 3-D rendition of a human condyle scaffold design. (b)-(e) Actual PCL scaffold produced by SLS

A porous architecture was then digitally superimposed in selected regions to form the scaffold design with exterior anatomic shape.

11.7 Characterization of Scaffolds

11.7.1 Mechanical Properties of SLS Processed PCL Scaffolds

Tensile and compressive testing on SLS processed PCL was conducted in accordance with ASTM standards D638-03 and D695-02a [4]. Mechanical test specimens designed with solid and porous gage sections were fabricated by SLS using CAPA® 6501 PCL. Testing was conducted both parallel and perpendicular to the SLS build direction (i.e. across and along direction of layer stacking). The porous test specimens incorporated a network of 2 mm × 2 mm cubical channels in one, two, and three-dimensions. Schematics outlining the test geometries, dimensions, build orientation, and loading directions are shown in Figs. 11.16–11.19. Tensile specimens were tested using a displacement controlled Instron 4206 tensile testing machine (Instron Corp., Canton, MA) at a displacement rate of 50 mm/min. Each

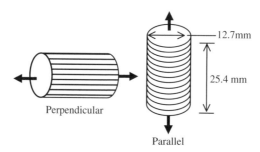

Fig. 11.16 Geometry, dimensions, build direction, and testing direction for compression mechanical test specimen

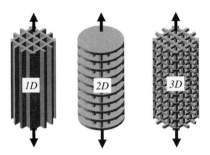

Fig. 11.17 Designs of compression test specimens with 1-D, 2-D, and 3-D porous architectures in the gauge section

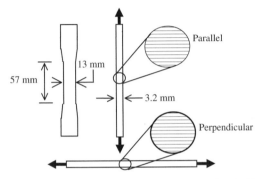

Fig. 11.18 Geometry, dimensions, build direction, and testing direction for Type 1 tensile mechanical test specimen

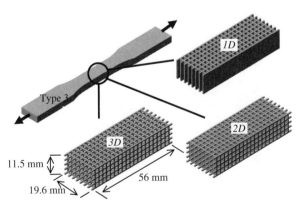

Fig. 11.19 Designs of Type 3 tensile test specimens with 1-D, 2-D, and 3-D porous architectures in the gauge section

Table 11.7 Mechanical properties of bulk and porous SLS processed PCL

	Property	Unit	SLS				
			Solid gage section		Porous Gage Section[1]		
			Parallel	Perpendicular	1D	2D	3D
Tension	Elastic Modulus, E	Mpa	363.4 ± 71.6	343.9 ± 33.2	140.5 ± 19.6	42.0 ± 6.9	35.5 ± 5.8
	0.2% Offset Yield Strength, σ_Y	Mpa	8.2 ± 1.0	10.1 ± 1.5	3.2 ± 0.6	0.67 ± 0.08	0.67 ± 0.06
	Strain at Yield, ε_Y	mm/mm	0.024 ± 0.006	0.031 ± 0.002	0.024 ± 0.001	0.017 ± 0.002	0.020 ± 0.002
	Ultimate Strength, σ_{UT}	Mpa	–	16.1 ± 0.3	4.5 ± 0.4	1.2 ± 0.2	1.1 ± 0.1
	Strain at Break, ε_B	mm/mm	0.043 ± 0.007	> 7.90	0.095 ± 0.022	0.092 ± 0.022	0.096 ± 0.025
Compression	Elastic Modulus, E	Mpa	299.0 ± 9.2	317.1 ± 3.9	133.4 ± 2.6	12.1 ± 0.5	14.9 ± 0.6
	0.2% Offset Yield Strength, σ_Y	Mpa	12.5 ± 0.4	10.3 ± 0.2	4.25 ± 0.05	0.45 ± 0.01	0.42 ± 0.03
	Strain at Yield, ε_Y	mm/mm	0.052 ± 0.003	0.037 ± 0.002	0.0370 ± 0.000	0.0376 ± 0.001	0.0268 ± 0.003

[1]Porous compression specimens tested parallel to SLS build direction, porous tensile specimens tested perpendicular to SLS build direction

specimen was loaded to failure or until the maximum allowable crosshead travel (433 mm) was reached. Compression specimens were mechanically tested using a MTS Alliance RT30 test frame (MTS Systems Corp., Eden Prairie, MN). Specimens were compressed to 50% strain between two steel platens at a rate of 1 mm/min after an initial preload of 6.7 N (1.5 lb) was applied. Tensile and compressive properties reported for specimens incorporating porous gage sections correspond to effective values of stress and strain.

The mechanical property measurements for SLS processed PCL loaded both parallel (across layers) and perpendicular (along layers) to the build direction are listed in Table 11.7. These measurements provide baseline mechanical properties of bulk PCL and important information on the mechanical performance of porous PCL scaffold structures. Even though SLS processed PCL exhibits signs of anisotropy, it was found to be quite strong/stiff, having bulk mechanical properties comparable to those measured for PCL processed via injection molding and falling within the range measured for trabecular bone [53].

11.8 Microstructure Characterization by Micro-computed Tomography

Figure 11.20 illustrates a representative volume-rendered micro-computed tomography (μ-CT) image of a 3-D porous compression test specimen [4]. Table 11.8 lists the results of μ-CT structural analysis of PCL scaffolds processed by SLS and confirms that both solid and designed porous architecture sections are near fully dense.

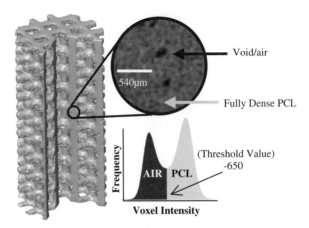

Fig. 11.20 Micro-computed tomography (?-CT) volume-rendered porous test specimen along with close-up of resulting microstructure and thresholding

Table 11.8 Density analysis by micro-computed tomography

	ASTM Specimen	Designed porosity (%)			Process Induced Porosity (%)	% Fully Dense
		Measured[1]	Target[2]	Diff		
Solid	D638 ∥				0.3	99.7
	D638 ⊥		Not Applicable		0.9	99.1
	D695 ∥				0.4	99.6
	D695 ⊥				0.4	99.6
Porous	D638 1D	44.1	51.1	7.0	4.8	95.2
	D638 2D	57.5	68.5	11.0	2.6	97.4
	D638 3D	77.3	80.9	3.6	1.4	98.6
	D695 1D	44.8	56.9	12.1	2.9	97.1
	D695 2D	61.9	67.4	5.5	7.3	92.7
	D695 3D	76.5	83.3	6.8	3.4	96.6

[1] Calculated using GEMS Microview,
[2] Calculated using Unigraphics

11.9 Conclusions

SLS is a very promising technique for manufacturing tissue engineering scaffolds. Significant progress has been made in SLS of biocompatible and bioresorbable polymers and polymer composites over the past decade. It is now possible to fabricate anatomically shaped resorbable scaffolds with designed porous architectures in PCL using an optimized SLS process. Such scaffolds exhibit near-full density in all SLS processed regions and faithfully reproduce the scaffold design to with 3–8% of design dimensions. Furthermore, the compressive mechanical properties of the PCL scaffolds fall within the lower range of mechanical properties for human trabecular bone, suggesting their potential use for reconstruction in load-bearing anatomical structures.

However, much work remains to be done. Tissue engineering scaffolds produced by SLS must be and implanted and tested in large animals in order to evaluate their regenerative potential and their long-term load-bearing potential under normal living conditions. SLS processed scaffolds must also be tested for their effectiveness in conveying cells and growth factors into an implant site for regeneration of multiple tissues and tissue interfaces. SLS processing of composites of bioresorbable polymers with inorganic osteoinductive materials such as tricalcium phosphate must be investigated to achieve improved mechanical properties and bone regeneration potential. It is anticipated that these steps will allow development of SLS techniques capable of creating functionally tailored 3D scaffolds incorporating gradients in both material composition and porosity. Such scaffolds are projected to play an extremely important role for simultaneously reconstructing and regenerating multiple different tissues in the body.

Acknowledgment The work reported here was funded in part by the National Institute of Dental and Craniofacial Research (NIDCR) of the National Institutes of Health (NIH) under grant R21 DE014736.

Nomenclature

K_H	Material specific constant
R	Reflectivity
$\Re$	Universal gas constant
T	Absolute temperature
T_i	Initial temperature
T_m	Melting temperature
Z_w	Aa measure of the number of atoms along a polymer's backbone.
a	Particle radius
a_0	Initial particle radius in Pokluda's model
c_p	Specific heat
d	Laser beam diameter
k	Thermal conductivity
q_0	Heat flux
t	Time
v	Laser scan speed
x	Neck radius
z	Depth into substrate
ΔE	Activation energy for viscous flow
α	Thermal diffusivity
η	Viscosity
η_0	Reference Viscosity
θ	Angle between particle radius and horizontal
λ	Latent heat of fusion
ρ	Density
σ	Surface tension
τ	Beam-material interaction time, duration of heat incident flux

References

1. D. W. Hutmacher, "Scaffold design and fabrication technologies for engineering tissues-state of the art future perspectives," J. Biomater Sci Polymer Edn **12,** 107–124 (2001).
2. D. W. Hutmacher, M. Sittinger, and M. V. Risbud, "Scaffold-based tissue engineering: rationale for computer-aided design and solid free-form fabrication systems," Trends Biotechnol **22,** 354–62 (2004).
3. E. Sachlos and J. T. Czernuska, "Making tissue engineering scaffolds work. Review on the application of solid freeform fabrication technology to the production of tissue engineering scaffolds," European Cells and Materials **5,** 29–40 (2003).
4. B. Partee, S. J. Hollister, S. Das. Selective Laser Sintering of Polycaprolactone Bone Tissue Engineering Scaffolds. Materials Research Society Symposium Proceedings, p 845 (2005).
5. T.-M. G. Chu, J. W. Halloran, S. J. Hollister, and S. E. Feinberg, "Hydroxyapatite implants with designed internal architecture," Journal of Materials Science: Materials in Medicine **12,** 471–478 (2001).

6. J. J Beaman, J. W. Bourell D. L. Barlow, R. H. Crawford, H. L. Marcus, and K. P. McAlea, *Solid Freeform Fabrication: A New Direction in Manufacturing* (Kluwer Academic Publishers, Boston 1997).
7. M. N. Cooke, J. P. Fisher, D. Dean, C. Rimnac, and A. G. Mikos, "Use of stereolithography to manufacture critical-sized 3D biodegradable scaffolds for bone ingrowth," J Biomed Mater Res **64B,** 65–9 (2003).
8. S. Das and S. J. Hollister, "Tissue engineering scaffolds" in *Encyclopedia of Materials: Science and Technology*, K. H. J. Buschow, R. W. Cahn, M. C. Flemings, B. Ilschner, E. J. Kramer, and S. Mahajan, eds. (Elsevier, 2001).
9. R. A. Giordano, B. M. Wu, S. W. Borland, L. G. Cima, E. M. Sachs, and M. J. Cima, "Mechanical properties of dense polylactic acid structures fabricated by three dimensional printing," J Biomater Sci Polym Ed **8,** 63–75 (1996).
10. S. J. Hollister, R. D. Maddox, and J. M. Taboas, "Optimal design and fabrication of scaffolds to mimic tissue properties and satisfy biological constraints," Biomaterials **23,** 4095–4103 (2002).
11. D. W. Hutmacher, "Scaffolds in tissue engineering bone and cartilage," Biomaterials **21,** 2529–2543 (2000).
12. S. Limpanuphap and B. Derby, "Manufacture of biomaterials by a novel printing process," Journal of Materials Science: Materials in Medicine **13,** 1163–1166 (2002).
13. Jill K. Sherwood, Susan L. Riley, Robert Palazzolo, Scott C. Brown, Donald C. Monkhouse, Matt Coates, Linda G. Griffith, Lee K. Landeen, and Anthony Ratcliffe, "A three-dimensional osteochondral composite scaffold for articular cartilage repair," Biomaterials **23,** 4739–4751 (2002).
14. R. Sodian, M. Loebe, A. Hein, T. Lueth, D. P. Martin, E. V. Potapov, F. Knollmann, and R. Hetzer, "Application of stereolithography for scaffold fabrication for tissue engineering of heart valves," ASAIO Journal: 46th Annual Conference and Exposition of ASAIO, Jun 28–Jul 1 2000 **46,** 238 (2000).
15. Ralf Sodian, Matthias Loebe, Andreas Hein, David P. Martin, Simon P. Hoerstrup, Evgenij V. Potapov, Harald Hausmann, Tim Lueth, and Roland Hetzer, "Application of stereolithography for scaffold fabrication for tissue engineered heart valves," ASAIO Journal **48,** 12–16 (2002).
16. Steidle, Cheri, Klosterman, Don, Graves, George, Osborne, Nora, and Chartoff, Richard. Automated fabrication of nonresorbable bone implants using laminated object manufacturing (LOM). Solid Freeform Fabrication Symposium Proceedings. 1998. Austin, University of Texas.
17. J. M. Taboas, R. D. Maddox, P. H. Krebsbach, and S. J. Hollister, "Indirect solid free form fabrication of local and global porous, biomimetic and composite 3D polymer-ceramic scaffolds," Biomaterials **24,** 181–194 (2003).
18. Wang, F., Shor, L., Darling, A., Sun, W., Guceri, S., and Lau, A. Precision extruding deposition and characterization of cellular poly-e-caprolactone tissue scaffolds. Solid Freeform Fabrication Symposium. 2003. Austin, Texas, University of Texas.
19. C. E. Wilson, W. J. A. Dhert, C. A. Van Blitterswijk, A. J. Verbout, and J. D. De Bruijn, "Evaluating 3D bone tissue engineered constructs with different seeding densities using the alamarBlue [trademark] assay and the effect on in vivo bone formation," Journal of Materials Science: Materials in Medicine **13,** 1265–1269 (2002).
20. B. M. Wu, S. W. Borland, R. A. Giordano, L. G. Cima, E. M. Sachs, and M. J. Cima, "Solid free-form fabrication of drug delivery devices," J Controlled Release **40,** 77–87 (1996).
21. Zhuo Xiong, Yongnian Yan, Shenguo Wang, Renji Zhang, and Chao Zhang, "Fabrication of porous scaffolds for bone tissue engineering via low-temperature deposition," Scripta Materialia **46,** 771–776 (2002).
22. Zhuo Xiong, Yongnian Yan, Renji Zhang, and Lei Sun, "Fabrication of porous poly(-lactic acid) scaffolds for bone tissue engineering via precise extrusion," Scripta Materialia **45,** 773–779 (2001).
23. Yongnian Yan, Rendong Wu, Renji Zhang, Zhuo Xiong, and Feng Lin, "Biomaterial forming research using RP technology," Rapid Prototyping Journal **9,** 142–149 (2003).

24. C. R. Deckard, Selective Laser Sintering. University of Texas, Austin (1988).
25. M. M. Sun, Physical Modeling of the Selective Laser Sintering Process. University of Texas, Austin, Texas (1991).
26. J. Frenkel, "Viscous flow of crystalline bodies under the action of surface tension," Journal of Physics **IX**, 385–391 (1945).
27. C. T. Bellehumeur J. Vlachopoulos O. Pokluda, "Modification of Frenkel's Model for Sintering," AIChe Journal **43**, 3253–3256 (1997).
28. L. H. Sperling, in *Introduction to Physical Polymer Science*, L. H. Sperling, ed. (John Wiley and Sons, 2005).
29. H.S. Carslaw and J.C. Jaeger, *Conduction of Heat in Solids* (Oxford University Press, 1986).
30. Schiaffino, Stefano and Sonin, Ain A. "Motion and arrest of a molten contact line on a cold surface: An experimental study. Physics of Fluids," 9(8), 2217–2226. 1997/08/00/. AIP.
31. Schiaffino, Stefano and Sonin, Ain A. "On the theory for the arrest of an advancing molten contact line on a cold solid of the same material," Physics of Fluids 9(8), 2227–2233. 1997/08/00/. AIP.
32. G. Lee, Selective Laser Sintering of calcium phosphate materials for orthopedic implants. University of Texas, Austin, Texas (1997).
33. N. K. Vail, L. D. Swain, W. C. Fox, T. B. Aufdlemorte, G. Lee, and J. W. Barlow, "Materials for biomedical applications," Materials & Design **20**, 123–132 (1999).
34. E. Berry, J. M. Brown, M. Connell, C. M. Craven, N. D. Efford, A. Radjenovic, and M. A. Smith, "Preliminary experience with medical applications of rapid prototyping by selective laser sintering," Med Eng Phys **19**, 90–96 (1997).
35. S. Das, S. J. Hollister, C. Flanagan, A. Adewunmi, K. Bark, C. Chen, K. Ramaswamy, D. Rose, and E. Widjaja, "Freeform Fabrication of Nylon-6 Tissue Engineering Scaffolds," Rapid Prototyping Journal **9**, 43–49 (2003).
36. Das, Suman, Hollister, Scott J., Flanagan, Colleen, Adewunmi, Adebisi, Bark, Karlin, Chen, Cindy, Ramaswamy, Krishnan, Rose, Daniel, and Widjaja, Erwin. "Computational design, freeform fabrication and testing of Nylon-6 tissue engineering scaffolds". Rapid Prototyping Technologies, Materials Research Society Symposium Proceedings, **758**, 205–210, (2003).
37. J. T. Rimell and P. M. Marquis, "Selective laser sintering of ultra high molecular weight polyethylene for clinical applications," J. Biomater Res Part B **52**, 414–420 (2000).
38. I. V. Shishkovsky, E. Yu. Tarasova, L. V. L. V. Zhuravel, and A. L. Petrov, "The synthesis of a biocomposite based on nickel titanium and hydroxyapatite under selective laser sintering conditions," Technical Phys Lett **27**, 211–213 (2001).
39. K. H. Low, K. F. Leong, C. K. Chua, Z. H. Du, and C. M. Cheah, "Characterization of SLS parts for drug delivery devices," Rapid Prototyping Journal **7**, 262–268 (2001).
40. C. K. Chua, K. F. Leong, K. H. Tan, F. E. Wiria, and C. M. Cheah, "Development of tissue scaffolds using selective laser sintering of polyvinyl alcohol/hydroxyapatite biocomposite for craniofacial and joint defects," Journal of Materials Science: Materials in Medicine **15**, 1113–1121 (2004).
41. K. H. Tan, C. K. Chua, K. F. Leong, C. M. Cheah, P. Cheang, M. S. Abu Bakar, and S. W. Cha, "Scaffold development using selective laser sintering of polyetheretherketone-hydroxyapatite biocomposite blends," Biomaterials **24**, 3115–3123 (2003).
42. K. H. Tan, C. K. Chua, K. F. Leong, C. M. Cheah, W. S. Gui, W. S. Tan, and F. E. Wiria, "Selective laser sintering of biocompatible polymers for applications in tissue engineering," Biomed Mater Eng **15**, 113–24 (2005).
43. J. M. Williams, A. Adewunmi, R. M. Schek, C. L. Flanagan, P. H. Krebsbach, S. E. Feinberg, S. J. Hollister, and S. Das, "Bone tissue engineering using polycaprolactone scaffolds fabricated via selective laser sintering," Biomaterials **26**, 4817–27 (2005).
44. Solvay Caprolactones. Properties & Processing of CAPA® Thermoplastics. [1]. 2001.
45. D. W. Hutmacher, "Polymers for Medical Applications," Encyclopedia of Materials: Science and Technology 7664–7673 (2001).
46. B. Saad and U.W. Suter, "Biodegradable polymeric materials," Encyclopedia of Materials: Science and Technology 551–555 (2001).

47. P. D. Darney, S. E. Monroe, C. M. Klaisle, and A. Alvarado, "Clinical evaluation of the Capronor contraceptive implant: preliminary report," Am J Obstet Gynecol **160,** 1292–1295 (1989).
48. S. J. Hollister, C. Y. Lin, C. Y. Lin, R. D. Schek, J. M. Taboas, C. L. Flanagan, E. Saito, J. M. Williams, S. Das, T. Wirtz, and P. H. Krebsbach, "Design and fabrication of scaffolds for anatomic bone reconstruction," Med J Malaysia **59**(Suppl B), 131–132 (2004).
49. S. J. Hollister, C. Y. Lin, E. Saito, C. Y. Lin, R. M. Schek, J. M. Taboas, J. M. Williams, B. Partee, C. L. Flanagan, A. Diggs, E. N. Wilke, Van Lenthe G.H. , R. Muller, T. Wirtz, S. Das, S. E. Feinberg, and P. H. Krebsbach, "Engineering craniofacial scaffolds," Orthodontics and Craniofacial Research **8,** 162–73 (2005).
50. R. C. Thomson, M. C. Wake, M. J. Yaszemski, and A. G. Mikos, "Biodegradable polymer scaffolds to regenerate organs" in *Biopolymers II*, Nicholas A. Peppas and R.S. Langer (eds.), Springer-Verlag GmbH & Company KG, Berlin, Germany, (1995).
51. DeVor, R. E., Chang, T., and Sutherland, J. W. *Statistical Quality Design and Control: Contemporary Concepts and Methods* (New Jersey, Prentice-Hall, pp. 543–605, 1992).
52. S. J. Hollister, R. A. Levy, T. M. Chu, J. W. Halloran, and S. E. Feinberg, "An image-based approach for designing and manufacturing craniofacial scaffolds," Int J Oral Maxillofac Surg **29,** 67–71 (2000).
53. R. W. Goulet, S. A. Goldstein, M. J. Ciarelli, J. L. Kuhn, M. B. Brown, and L. A. Feldkamp, "The relationship between the structural and orthogonal compressive properties of trabecular bone," J Biomechanics **27,** 375–389 (1994).

Chapter 12
Design, Fabrication and Physical Characterization of Scaffolds Made from Biodegradable Synthetic Polymers in combination with RP Systems based on Melt Extrusion

D. W. Hutmacher, M. E. Hoque and Y. S. Wong

12.1 Introduction

The more recently coined term "Regenerative Medicine" represents a shift in emphasis from current methods to replace tissues with medical devices and artificial organs to more biological approaches which focus on regeneration rather than replacement or repair. Regenerative medicine envelops several biomedical fields; however it can be argued that cell therapy and tissue engineering are on the forefront in the twenty-first century.

It can be argued that the beginning of the "scaffold-based tissue engineering concept"—as we know it today—was in the mid-1980s when the pediatric surgeon J. Vacanti of the Children's Hospital approached the chemical engineer R. Langer of MIT with an idea to design scaffolds for cell delivery as opposed to seeding cells onto or mixing cells into available naturally occurring matrices having physical and chemical properties that was difficult to be manipulated, thus resulting in wide variations of the results produced in vitro and in vivo.

Today's scaffold-based tissue engineering concepts involve the combination of viable cells, biomolecules and a scaffold to promote the repair and/or regeneration of tissues as depicted schematically in Fig. 12.1. The science behind engineering tissue engineered constructs (TEC) is still under intense investigation and various approaches and strategies are currently developed by a plethora of tissue engineering laboratories and institutes around the globe. However, after studying the scaffold literature it must be concluded is by no means clear what defines an ideal scaffold/cell or scaffold/neo-tissue construct, even for a specific tissue type (Hutmacher et al., 2007). The considerations are complex and scaffolds in tissue engineered constructs will have certain minimum requirements for biochemical as well as chemical and physical properties. These include biocompatibility, angiogenesis, vascularisation and chemotaxis issues, the scaffold must not be an agent for allergic reaction and disease transmission, it must posse's suitable gross architectural qualities able to be

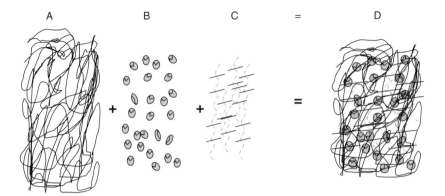

Fig. 12.1 A paradigm shift is taking place in regenerative medicine from using synthetic implants and tissue grafts to engineering a tissue engineered construct (D) that uses biodegradable scaffolds (A) combined with cells (B) or biological molecules (C) to repair and/or regenerate defect sites

produced via a reproducible processing platform. There are also sterilization and administrative issues to contend with.

12.2 Basic Principles of Scaffold Design and Characterization

12.2.1 Background

It is important to emphasize, at the outset, that the field of scaffold-based tissue engineering is still in its infancy, and many different approaches are under experimental investigation. Thus, it is by no means clear what defines an ideal scaffold/cell or scaffold/neo-tissue construct, even for a specific tissue type. Indeed, since some tissues perform multiple functional roles, it is unlikely that a single scaffold would serve as a universal foundation for the regeneration of even a single tissue. Hence, the considerations for scaffold design are complex. They include; material composition, porous architecture, structural mechanics, surface properties, degradation properties and their products (degradation rate strongly depends on polymer type, impurities, manufacturing process, sterilization, device size, and the local environment), together with the composition of any biological component which may have been added to the scaffold to improve function. Furthermore one must also consider the behaviours and the consequences of how all of the aforementioned factors may change with time.

Hollister (2006) stated that approaches in scaffold design must be able to create hierarchical porous structures to attain desired mechanical function and mass transport (permeability and diffusion) properties, and to produce these structures within arbitrary and complex three-dimensional (3D) anatomical shapes. Hierarchical refers to the fact that features at scales from the nanometre to millimetre level will determine how well the scaffold meets conflicting mechanical function and mass-transport needs. Material chemistry together with processing determines the maximum functional properties that a scaffold can achieve, as well as how cells interact with the scaffold. However, mass-transport requirements for cell nutrition pore interconnections for cell migration, and surface features for cell attachment necessitate minimal requirements for scaffold surface and morphology. The porous structure dictates that achievable scaffold properties will fall between the theoretical maximum set by the material and the theoretical minimum of zero predicted by composite theories. The critical issue for design is then to compute the precise value of mechanical as well as mass-transport properties at a given scale based on more microscopic properties and structure. (Fig. 12.2 from Nature paper)

On the other hand, the challenge for tissue engineers is that cell and tissue remodeling is important for achieving stable biomechanical conditions and vascularization at the host site. Hence, the 3-D scaffold/tissue construct should maintain sufficient structural integrity during the in vitro and/ or in vivo growth and remodelling process. The degree of remodelling depends on the tissue itself, and its host anatomy and physiology. Hence, the degradation and resorption kinetics of the scaffold material need to be chosen based on the relationships of mechanical properties, molecular weight (Mw/Mn), mass loss, and tissue development. A number of studies (Hutmacher et al., 2007) have demonstrated the dependence of mechanical properties on scaffold porosity when designing and fabricating scaffolds via rapid prototyping. Practically, as porosity increases at the design stage, its mechanical properties would decrease correspondingly. This decline has been found to follow a power law relationship.

In addition to these essentials of mechanics and geometry, a suitable construct will possess surface properties which are optimized for the attachment and migration of cell types of interest (depending on the targeted tissue). The external size and shape of the construct must also be considered; especially if the TEC needs to be customised for an specific defect site.

12.2.2 Scaffold Design & Characterization

The ASTM terminology for porous materials is classified into three groups: interconnecting (open pores), non-connecting (closed pores), or a combination of both. When the pores are open, the foam material is usually drawn into struts forming the pore edges. A network of struts produced a low-density solid with pores connecting to each other through open faces. When the pores were closed, a network of interconnected plates produced a higher-density solid. The closed pores were sealed off from adjacent neighbours.

In their widely studied textbook Gibson and Ashby (1997) classified porous solids into two general groups: foams and honeycombs (Fig. 12.3). A honeycomb consisted of a regular two-dimensional array of polygonal pores each defined by a wall shared between adjacent pores. The pores were packed to in planar arrays like

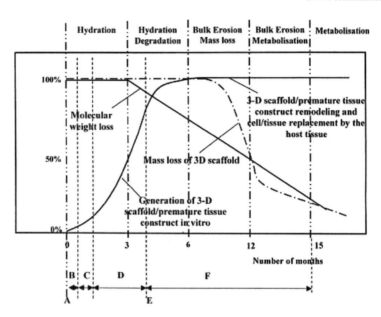

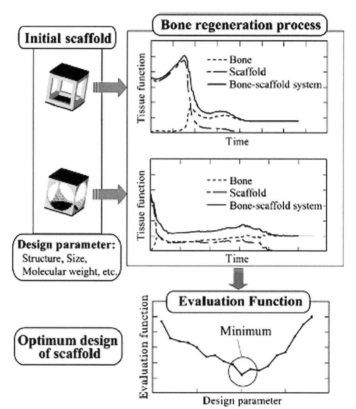

Fig. 12.3 SEM picture of a foam-like scaffold produced by salt-leaching combined with solvent casting (left). Scaffold morphology shows large pores but small pore interconnections. SEM graph of a scaffold with a honeycomb architecture produced by Fused Deposition Modeling (FDM). The channel like design provides large pores and large pore interconnections

the hexagonal cells of a honeycomb. Ashby and Gibson describe the mechanical properties of a porous solid to depend mainly on its relative density, the properties of the material that make up the pore edges or walls and the anisotropic nature, if any, of the solid. In general, the stiffness (E^*) and yield strength (σ^*) in compression, of porous solids were each related to the relative density by a power law relationship. Given that most constructs will require a high degree of porosity to accommodate mass transfer and tissue development, the volume fraction of the scaffold will necessarily be low. In all but the most biomechanically challenging applications, it is likely that the test for the scaffold engineer will be to achieve sufficient stiffness and strength in a highly porous structure to provide adequate mechanical integrity. One of the most demanding applications will be the repair and generation of musculoskeletal tissues, particularly bone, where scaffolds need to have a high elastic modulus in order to provide temporary mechanical support without showing symptoms of fatigue or failure, to be retained in the space they were designated for and to provide the tissue with adequate space for growth. One

◂─────────────────────────────────

Fig. 12.2 a and b Cell and tissue remodelling is important for achieving stable biomechanical conditions and vascularization at the host site. Hence, the 3-D scaffold/tissue construct should maintain sufficient structural integrity during the in vitro and/ or in vivo growth and remodelling process. The degree of remodelling depends on the tissue itself (e.g. skin 4–6 weeks, bone 4–6 months), and its host anatomy and physiology. Scaffold architecture has to allow for initial cell attachment and subsequent migration into and through the matrix, mass transfer of nutrients and metabolites and provision of sufficient space for development and later remodelling of organised tissue. Hutmacher (2000) was first which introduced that the degradation and resorption kinetics of the scaffold need to be designed based on the relationships of mechanical properties, molecular weight (Mw/Mn), mass loss, and tissue development described in Fig. 12.2a (Image reproduced with permission from Elsevier [Hutmacher, 2000]). Hollister's group (2006) was then the first which introduced a framework for optimal design of porous scaffold microstructure by computational simulation of bone regeneration. (Image reproduced with permission from Elsevier [Hollister et al., 2006])

of the fundamental challenges of scaffold design and material selection concerns the achievement of high enough initial strength and stiffness; the scaffold material must have both a sufficiently high interatomic and intermolecular bonding and/or a physical and chemical structure which must in turn allow for hydrolytic attack and breakdown, *in vivo,* as the scaffold degrades over time.

Porosity: A pore can be defined as a void space within a scaffold, whereas porosity can be considered as a collection of pores. Pore size and porosity are important scaffold parameters. Macro-pores (i.e. >50 µm) are of an appropriate scale to influence tissue function, for example, pores greater than 300 µm in size are typically recommended as optimal for bone in-growth in relation to vascularisation of the construct. Micro-pores (i.e. <50 µm) are of a scale to influence cell function (e.g. cell attachment) given that mammalian cells typically are 10–20 µm in size. Nano-porosity refers to pore architectures or surface textures on a nano-scale (i.e. 1–1000 nm).

There is often a compromise between porosity and scaffold mechanical properties. Increasing porosity may provide more accessible pore volume for cell infiltration as well as ECM formation and subsequently in vivo tissue formation, except for there is a concomitant decrease in mechanical properties of the scaffold itself in accordance to a power-law relationship (Gibson & Ashby, 1997).

In addition, pore interconnectivity is a critical factor and is often overlooked in scaffold design and characterisation. A scaffold may be porous, but unless the pores are interconnecting (i.e. voids linking one pore to another), they serve no purpose and become superfluous within a scaffold intended for tissue engineering. The "interconnecting pore size" is more critical than "pore size", and should be suitably large to support cell migration and proliferation in the initial stages and subsequent ECM infiltration of desired tissue. It is preferable that scaffolds for tissue engineering have 100% interconnecting pore volume, thereby also maximizing the diffusion and exchange of nutrients (e.g. oxygen) and the eliminations of waste throughout the entire scaffold pore volume.

As a measure of pore interconnectivity, the accessible pore volume, or permeability, of a scaffold can be measured. Accessible pore volume can be defined as the total volume of pores which can be infiltrated from all peripheral borders to the interior of the scaffold. Scaffold permeability can be measured by determining the flow rate of fluid flow through interconnecting pores. However, this technique is not suitable in scaffolds with large, 100% interconnected pore volumes as the scaffold provides no resistance to fluid flow. Alternatively, accessible pore volume, as well as porosity, pore size distribution and scaffold surface area to volume ratio (i.e. volume fraction) can be characterised using techniques such as mercury intrusion porosimetry, micro-computed tomography (µCT) or image analysis (Ho & Hutmacher, 2005). Mercury porosimetry is a popular technique which is based on the principle that the pressure required to force a non-wetting liquid such as mercury into pores, against the resistance of liquid surface tension, is indicative of the pore size, assuming the pores are cylindrical in shape. However, the resolution of the technique is severely limited in scaffolds with large pore sizes (>500 µm) where low mercury intrusion pressures are necessary, and it has limitations when applied to materials that have irregular pore geometries.

Alternative techniques such as micro computed tomography (μCT), have been developed for analysing bone architecture and more recently scaffold architecture, through utlisation of 3D CT imaging to generate computer models of porous materials. Using 3D μCT techniques, a much greater amount of information can be obtained to characterise pore architectures containing features ranging from 6 μm to >1500 μm, without the physical limitations associated with mercury porosimetry (Ho & Hutmacher, 2005).

12.2.3 Scaffold Materials

One of the fundamental issues with regard to tissue engineering is the choice of suitable material. Currently, polymeric materials have drawn great attention from the scientific and medical communities for tissue engineering applications (Maquet & Jerome, 1997). Natural polymers, such as collagens, glycosaminoglycan, starch, chitin, and chitosan have been used to repair nerves, skin, cartilage, and bone (Mano et al., 1999). These naturally occurring biomaterials might most closely simulate the native cellular milieu. However, large batch-to-batch variation upon isolation from biological tissues and availability are the main limitations tissue engineers have to take into consideration. Poor mechanical performance is also a drawback for transplanted scaffolds made from natural polymers, too. On top of that, natural polymers such as collagen and glycosaminoglycans carry the potential to provoke adverse tissue reactions and immune responses.

Synthetic polymers have been developed to overcome the aforementioned shortcomings associated with natural polymers. Synthetic polymers are well known for their enormous availability, high processability, and controllable mechanical and biochemical properties. Most synthetic polymers degrade via chemical hydrolysis and insensitive to enzymatic processes so that their degradation behaviour does not vary from patient to patient. Many synthetic bioresorbable polymers, such as poly (α-hydroxy ester)s, polyanhydrides, polyorthoesters, and polyphosphazens, have been studied for temporary surgical and pharmacological applications (Vert et al., 1992). These polymers have been found to be suitable to built scaffolds for tissue engineering applications. Properties of these different polymers are summarized in Table 12.1.

12.2.3.1 PCL Copolymers

Presently, poly(ε-caprolactone) (PCL) is regarded as a soft- and hard-tissue compatible biodegradable material and often selected as a suitable material for thermoplastic processing of scaffolds for tissue engineering (Perrin & English, 1998). The first generation of bioresorbable scaffolds for tissue-engineering applications has been fabricated from synthetic polymers of the aliphatic polyester family (Hutmacher, 2000a; Vats et al., 2003). However, the number of such bioresorbable polymers is limited when polymers with different properties are needed for the

Table 12.1 Properties of Biodegradable Polymers (Shalaby et al., 1994; Maquet et al., 1997; Perrin and English, 1997; Middleton et al., 1998; Ali and Hamid, 1998; http://physics.iisc.ernet.in)

Polymer type	Melting point(°C)	Glass transition temp (°C)	Degradation time (Months) [a]	Density (G/CC)	Tensile strength (MPA)	Elongation %	Modulus (GPA)
PLGA	Amorphous	45–55	6–12	1.27–1.34	41.4-55.2	3–10	1.4–2.8
DL-PLA	Amorphous	55–60	12–16	1.25	27.6–41.4	3–10	1.4–2.8
L-PLA	173–178	60–65	>24	1.24	55.2–82.7	5–10	2.8–4.2
PGA	185–225	25–65	6–12	1.53	>68.9	15–20	>6.9
PCL	59–64	−65	>24	1.11	20.7–34.5	300–500	0.21–0.34
PEG	67–69	−72	–	1.05	–	–	–
PCL-PLA	65	−55	>36	–	–	–	–
PCL-PEG	67	−69	>24	–	–	–	–
PCL-PEG-PCL	67	−69	>24	–	–	–	–
PLA-PCL-PLA	67	−67	>24	–	–	–	–
PEG-PCL-PLA	65	−68	>6–12	–	–	–	–

[a] Time also depends on size & geometry of specimen.

design and fabrication of devices and scaffolds adapted to specific applications (Saltzman, 1999; Hutmacher, 2001a). Polymers such as poly(ethylene glycol) (PEG), poly(ε-caprolactone) (PCL) and poly(DL-lactide) (P(DL)LA) have been used to make in vivo degradable medical and drug-delivery devices with Food and Drug Administration approval (Pitt, 1990). Polyester–polyether block co-polymers composed of PCL or PLA and PEG have been considered as suitable because they offer possibilities to vary the ratio of hydrophobic/hydrophilic constituents by copolymerization and to modulate degradability and hydrophilicity of corresponding matrices and surfaces (Rashkov et al., 1996; Li et al., 2002). Despite of favorable rheological properties and thermal stability in molten state, PCL degrades very slowly due to its high hydrophobicity and crystallinity (Pitt, 1990). Introduction of hydrophilic blocks and/or fast degrading blocks into PCL main chains can be a means to prepare novel degradable and bioresorbable polymers. Hydrophilic polyether blocks such as poly(ethylene glycol) (PEG) are introduced into PCL chains to enhance the hydrophilicity of the parent PCL homo-polymer (Lee et al., 2001; Li et al., 2002). Likewise, block co-polymerization of PCL with faster degrading polyesters, such as poly(lactide) (PLA), allowed to modify the degradability of the parent PCL homo-polymer (Feng et al., 1983; Deng et al., 1997). However, both types of co-polymers present specific disadvantages. PLA is a hydrophobic polymer, whereas PEG is hydrophilic but not degradable in vivo. Therefore, Hoque et al. (2005) synthesized PCL-based co-polymer (PEG-PCL-PLA) by combining both PEG and PLA blocks with PCL chains to produce novel hydrophilic and bioresorbable co-polymer, with the aim of enhancing hydrophilicity and degradability. The feasibility of fabricating PEG-PCL-PLA scaffold using a desktop robot based melt extrusion rapid prototyping system and in vitro characterization for tissue engineering was studied. The results demonstrated the suitability of the copolymer to be processed into 3D scaffolds with honeycomb-like architectures having completely interconnected and controlled pore channels. Preliminary study on cell response to the as-fabricated scaffolds using primary human fibroblasts also showed the cell biocompatibility of the copolymer (Hoque et al., 2005).

Similarly, if ε-caprolactone is copolymerised with ethylene oxide (EO) or poly(ethylene glycol) (PEG) to prepare PCL/PEG(PEO) block copolymers, their hydrophilicity and biodegradability can also be improved, and thus they may find much wider applications. Recently, several research groups (Li et al., 1996; Dobrzynski et al., 1999; Yuan et al., 2000; Longhai et al., 2003; Huang et al., 2004) prepared bioresorbable polyester–PEG diblock or triblock copolymers by using a monohydroxy or α,ω-dihydroxy PEG as initiator for the polymerization of lactone monomers employing various techniques.

Longhai et al. (2003) synthesized and characterized the PCL-PEG-PCL triblock copolymers by ring-opening polymerization of ε-caprolactone (CL) in the presence of poly(ethylene glycol) using calcium catalyst. The differential scanning calorimetry and wide-angle X-ray diffraction analyses revealed the micro-domain structure in the copolymer. The melting temperature, T_m and crystallization temperature, T_c of the PEG domain were observed to be influenced by the relative length of the PCL blocks. They mentioned that it was because of the strong covalent interconnection between the two domains. Huang et al. (2004) performed degradation and cell culture studies on PCL homopolymer and PCL/PEG diblock and triblock copolymers prepared by ring-opening polymerization of ε-caprolactone in the presence of ethylene glycol or PEG using zinc metal as catalyst. They cultured primary human and rat bone marrow derived stromal cells (hMSC, rMSC) on the scaffolds manufactured with PCL homopolymer and PCL/PEG diblock and triblock copolymers via solid free form fabrication. Light, scanning electron and confocal laser microscopy as well as immunocytochemistry, showed cell attachment, proliferation and extracellular matrix production on the surfaces along with inside the scaffold architectures of all polymers. However, the copolymers showed better performance in the cell culture studies than the PCL homopolymer.

12.3 Techniques for Scaffold Fabrication

12.3.1 Background

In the early days of tissue engineering, FDA approved devices and implants made of polymers of synthetic origin, such as sutures, meshes etc. were used (Hutmacher, 2001b). Later, techniques were developed to fabricate polymeric scaffolds based on either heating macromolecules or dissolving them in a suitable organic solvent. In these techniques, the viscous behavior of the polymers above their melting temperatures, and their solubility in various organic solvents are the important characteristics, which determine the type of process that will be used. Based on the use of organic solvents, a number of techniques have been developed to design and fabricate porous 3D bioresorbable scaffolds for tissue engineering applications (Lu & Mikos, 1996). These polymer-processing techniques, called conventional techniques include fiber bonding (Wang et al., 1993; Brauker et al., 1995), phase separation (Lo et al., 1995; Ma and Zhang, 1999b), solvent casting/particulate leaching (Mikos et al., 1993b; Holy et al., 2000), membrane lamination (Mikos et al., 1996), melt molding (Thomson et al., 1995a), gas foaming/high pressure processing (Baldwin et al., 1995; Mooney et al., 1996), hydrocarbon templating (Shastri et al., 2000), freeze drying (Whang et al., 1995; Healy et al., 1998) and combinations of these techniques (e.g., gas foaming/particulate leaching (Harris et al., 1998), etc.). The working principles, procedures, applications, potentials and shortcomings of so called conventional techniques can be found in several other works (Widmer & Mikos, 1998; Hutmacher, 2000b; Thomson et al., 2000; Yang et al., 2001; Sachlos & Czernuszka, 2003).

However, these techniques often remain impractical to manufacture scaffolds with complex & reproducible architectures, and compositional variation across the entire structure. To overcome this hurdle, rapid prototyping (RP), called solid free form (SFF) fabrication techniques have been introduced with high encouragement for scaffold fabrication.

RP is a fast growing popular technology that enables quick and easy fabrication of customized forms directly from computer aided design (CAD) model to solid model. The needs for quick product development, decreased time to market, and highly customized and low quantity parts are driving the demand for RP technology (Anna, 1997). The flexible manufacturing capabilities of RP techniques have been applied to biomedical engineering applications ranging from the production of scale replicas of human bones (D'Urso et al., 2000) and body organs (Sanghera et al., 2001) to advanced customized drug delivery devices (Leong et al., 2001) and other areas of medical sciences including anthropology (Recheis et al., 1999), palaeontology (Zollikofer & De Leon, 1995) and medical forensics (Vanezi et al., 2000). Today, RP technique is regarded as an efficient tool to reproducibly generate scaffolds with suitable properties on a large scale (Hollister et al., 2000; Sachlos & Czernuszka, 2003; Hutmacher & Cool 2007).

Solid free-form fabrication (SFF) and rapid prototyping (RP) (Hutmacher et al., 2006) have been applied to fabricate complex shaped TECs. Unlike conventional machining which involves constant removal of materials, SFF is able to build scaffolds by selectively adding materials, layer-by-layer, as specified by a computer program. Each layer represents the shape of the cross-section of the computer-aided design (CAD) model at a specific level. Today, SFF is viewed as a highly potential fabrication technology for the generation of scaffold technology platforms (Hutmacher et al., 2006). In addition, one of the potential benefits offered by SFF technology is the ability to create parts with highly reproducible architecture and compositional variation across the entire matrix due to its computer controlled fabrication. It is beyond the scope of this chapter to discuss all SFF technologies and therefore the authors decided to focus on melt extrusion based systems.

12.3.2 Rapid Prototyping Systems Based on Extrusion Technology

The speed, customizability, complex geometric capability and range of material variation lead to the utilization RP techniques for scaffold fabrication. An additional advantage is that basic structures can be achieved without applying high temperature or pressure in many RP techniques. This allows fabrication of scaffolds with materials or material additives that otherwise decompose under the harsh conditions of normal fabrication procedure. Currently, a number of RP techniques have been utilized for fabricating tissue engineering scaffolds. These techniques can be classified based on their processing technologies such as systems based on laser technology (Stereolithography Apparatus, SLA & Selective Laser Sintering, SLS), systems based on print technology (Three-dimensional Printing, 3DP), system based on assembly technology (Shape Deposition Manufacturing, SDM) and system based on extrusion technology (Hutmacher et al., 2004; Leong et al., 2003; Sachlos & Czernuszka, 2003). Of the aforementioned, the extrusion-based technologies are focused in this chapter.

This technology involves the extrusion of a melt to build a 3D scaffold through a jet or nozzle in a layered fashion guided by the computer model. This technique includes multiphase jet solidification (MJS), 3D plotting, fused deposition modeling (FDM), precise extrusion manufacturing (PEM) and desktop robot based rapid prototyping (DRBRP) technique. A variety of biomaterials can be utilized to fabricate scaffolds based on the type of machine used.

12.3.2.1 Multiphase Jet Solidification (MJS)

The basic principle of the multiphase jet solidification (MJS) method involves the extrusion of a melted material through a jet. The MJS process is usually used to produce high density metallic and ceramic parts. A feedstock consists of powder-binder-mixture is heated in a process chamber above the melting point of the binder, and thus only the binder is liquefied during the process. A piston squeezes out the low viscous mixture through an x-y-z controlled jet. The feed rate of the piston controls the material flow. The material is deposited layer-by-layer by moving the jet and solidifies when touches the base platform or the previous layer due to temperature decrease. Following the aim to build up metallic and ceramic parts, most investigations were performed using powder-binder-mixtures of stainless steel. However, for medical applications the constraints and aims of using MJS are quite different. In this case, the modeling material used is a polymeric material instead of powder-binder-mixture and supplied as powders, pellets or bars.

Koch and co-workers (1998) studied the use of the MJS process to build 3D hollow scaffolds made of poly (D, L)-lactide for bone and cartilage tissue engineering. The scaffold pore size was found to be in the range of 300–400 µm and the structure supported ingrowth of human bone tissues. However, there was no report on the detailed scaffold morphology using microscope analyses or on any mechanical study of the scaffold properties. Calvert and Crockett (1997) developed an in-house extrusion-based solid free-form fabrication technique to manufacture scaffolds. They built scaffolds with typical layer heights of 0.2–1.0 mm and a resolution of ~0.5 mm. Xiong et al. (2001) built an RP technique called precision extrusion manufacturing (PEM) to fabricate 3D scaffolds. They manufactured and tested porous scaffolds with poly L-lactic acid (PLLA) and tricalcium phosphate (TCP) for bone tissue engineering.

12.3.2.2 Three-Dimensional Plotting

This system was developed by the researchers at the University of Freiburg (Landers & Muellhaupt, 2000) and was termed as "bioplotter". This technique involves a moving extruder head (x-y-z control) and uses compressed air to force out a liquid or paste like plotting medium. The process generates an object by building micro strands or dots in a layered fashion. Depending on the type of dispenser head, a wide variety of polymer hotmelts as well as pastes, solutions and dispersions of polymers and reactive oligomers can be processed on the machine. The versatility of this technique has been elaborated in other studies, too (Landers et al., 2002a, 2002b; Huang et al., 2004).

Further modification of bioplotter enabled the researchers to extrude highly viscous polyethyleneoxide-terephtalate and polybutylene-terephtelate (PEOT/PBT) copolymers into 3D scaffolds by varying fiber diameter, fiber spacing, layer thickness, fiber orientation etc (Moroni et al., 2006). Pfister et al. (2004) compared 3DP and 3D bioplotting in manufacturing biodegradable polyurethane scaffolds using aliphatic polyurethanes based on lysine ethyl ester diisocyanate and isophorone diisocyanate. In 3DP, scaffolds are built up layer-by-layer by bonding starch particles together followed by infiltration and partial cross linking of starch with lysine ethyl ester diisocyanate. On the other hand, the 3D bioplotting allows 3D dispensing and reactive processing of oligoetherurethanes derived from isophorone diisocyanate, oligoethylene oxide, and glycerol. The 3D plotting method generally involves low cost with minimal specialized equipment. However, resolution is the primary limiting factor that is determined by the size of dispensing tip.

12.3.2.3 Fused Deposition Modeling (FDM)

A traditional FDM machine consists of a head-heated liquefier attached to a carriage moving in the horizontal XY plane (Comb et al., 1994b). The function of the liquefier assembly is to heat the filament material to a semi-liquid state and pump through a nozzle directly onto the build platform following a programmed model. The extrusion head moves in the X- and Y- axes, while the platform lowers in the Z-axis for each new layer to form. The extruded material solidifies, laminating to the preceding layer and the process repeats until the object is completed. In the past, users only made use of FDM technique to process a few non-resorbable polymeric materials, such as polyamide, ABS, and some other resins.

An interdisciplinary group at NUS (Hutmacher et al., 2000b, 2001b; Zein et al., 2002) has studied and patented the parameters to process PCL and several composites (PCL/HA, PCL/TCP etc.) by the FDM technique. For more than 6 years this first generation of scaffolds (PCL) has been studied with, and without cells in a clinical setting (Fig. 12.4). To the authors knowledge this is the first type of biodegradable polymer scaffold which were fabricated by a RP technique which received FDA approval for oral and maxillofacial applications (Osteopoe Int.).

The second generation of FDM scaffolds for bone tissue engineering was fabricated from different polymers and CaP by FDM (Hutmacher & Cool 2007). These composite scaffolds ascertained favorable mechanical, degradation and resorption kinetics and biochemical properties. The strength is conferred via ceramic phase whereas, toughness and plasticity are attained via polymer phase. In addition, these scaffolds offer improved cell seeding, and enhanced incorporation and immobilization of growth factors. Among several investigators Endres et al. (2003), Rai et al. (2004) and Zhou et al. (2007) have evaluated these PCL/CaP composite scaffolds in vitro and in vivo and reported encouraging results.

Woodfield et al. (2004) used a FDM-like fiber deposition technique for manufacturing 3-D poly(ethylene glycol)-terephthalate-poly(butylene terephthalate) (PEGT/PBT) block co-polymer scaffolds through to engineer articular cartilage. It was reported that this technique allowed them to "design-in" scaffolds with desired

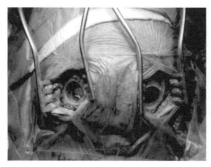 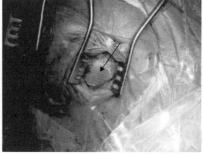

Fig. 12.4 Reconstruction of a bone defect with a medical grade PCL scaffold (arrow) of a patient which was treated for the evacuation of a hemorrhage

characteristics by controlling the deposition of molten co-polymer fibers from a pressure-driven syringe onto a computer controlled table. By varying PEGT/PBT composition, porosity and pore geometry, 3D-deposited scaffolds were produced with a range of mechanical properties. A range of mechanical properties with 100% interconnected pore network was obtained by varying PEGT/PBT composition, porosity and pore geometry of the scaffolds. The equilibrium modulus and dynamic stiffness were reported to be in the range of 0.05–2.5 MPa and 0.16–4.33 MPa, respectively, which were closer to the native articular cartilage explants (0.27 and 4.10 MPa, respectively). A small animal model test was performed with this (PEGT/PBT) scaffold, which showed encouraging results. In this model, scaffolds were seeded with bovine articular chondrocytes that were homogeneously distributed into the scaffolds, and subsequently formed cartilage-like tissue. This tissue construct was cultured in vitro and further subcutaneously implanted in nude mice.

Wang et al. (2004) developed a variation of FDM process called precision extruding deposition (PED) system to fabricate interconnected 3D scaffolds at Drexel University. PED system fused pellet-formed PCL by two electrical band heaters connected with two respective thermo-couples. Then the PCL melt was extruded by the pressure created by a turning precision screw to build the scaffold layer by layer. The scaffold morphology, internal micro-architecture, mechanical properties and biocompatibility of the as-fabricated scaffolds were evaluated. The test results reported the structural integrity, controlled pore size (250 μm), pore interconnectivity, a favorable mechanical property and basic biocompatibility of the PED-fabricated PCL scaffolds.

To date, FDM technique has shown promising success in the fabrication of 3D scaffolds in terms of pore reproducibility and interconnectivity. However, FDM technique requires pre-formed filaments with specific size and material properties to build scaffold. The materials should be ductile and flexible enough to be made into filament and still will have to provide sufficient mechanical strength to be compressed into the heater or liquefier. As a result, FDM technique has a narrow material processing window (Landers et al., 2002b), and thus remains tedious and costly. That's why to exclude the complexity and cost related to fabrication of filament precursor, the techniques has been developed (Wang et al., 2004; Woodfield et al., 2004; Hoque et al., 2005) which apply the same modeling principles as FDM but employ different driving forces like, the external gas pressure or the pressure created by turning screw to extrude the polymer melt. The major contribution of the further developed pressure driven systems is that the scaffolding material can be directly melted and extruded without filament preparation, which greatly widens the material processing window.

However, different pressure driven systems have inherent advantages and shortcomings. For example, the system that extrudes the polymer melt by external gas pressure is easy to operate and maintain cleanliness. But there is a possibility that if there are air bubbles in the polymer melt they may remain entrapped because of lack of appropriate mixing, which may cause imperfect filament deposition. In that case, maintaining a bit higher super heat in the polymer melt may facilitate easy escape of the air bubbles. On the other hand, the system that extrudes the polymer melt using the pressure created by turning screw is tedious to maintain the cleanliness, and without appropriate cleaning of the system the scaffold might get contaminated. However, the advantage is that the polymer melt is mixed thoroughly and thus avoid the possibility of entrapment of air bubbles.

12.3.3 Conclusion

Scaffolds are of great importance for tissue engineering since they enable the fabrication of functional living implants out of cells obtained from cell culture. As the scaffolds for tissue engineering will be eventually be implanted in the human body, the scaffold materials should be non-antigenic, non-carcinogenic, non-toxic, and possess high cell/tissue biocompatibility so that they will not trigger pathological reactions after implantation. Besides biomaterial issues, the macro- and micro-structural properties of the scaffold are also very important. In general, the scaffolds require individual external shape and well-defined internal structure with interconnected porosity to host most cell types. From a biological point of view, the designed matrix should serve functions, including (1) as an immobilization site for transplanted cells, (2) formation of a protective space to prevent unwanted tissue growth into the wound bed and allow healing with differentiated tissue, (3) directing migration or growth of cells via surface properties of the scaffold, and (4) directing migration or growth of cells via release of soluble molecules such as growth factors, hormones, and/or cytokines. Future work has to provide further evidence that some of these techniques offer the right balance of capability and practicality to be suitable for fabrication of materials in sufficient quantity and quality to move holistic tissue engineering technology platforms into the clinical application.

In conclusion, melt extrusion based RP technologies add further versatility and detail in scaffold design and fabrication. Today these RP systems allow the usage of a wide range of biomaterials or a combination of materials (composites or co-polymers). At the same time as the bioplotter was developed at the University of Freiburg a similar machine concept was developed at the National University of Singapore (Hoque et al., 2005) by the interdisciplinary groups of the two senior authors of this chapter. The following section will describe in detail the design, fabrication and characterization of scaffolds by utilizing this type of RP machine.

12.4 Scaffold Processing using DRBRP Technique

12.4.1 Introduction

The requirements for tissue engineering scaffolds like, morphological, mechanical and biochemical properties are becoming more specific. One of the major challenges in the design of tissue engineering scaffold is to mimic the structure and biological functions of the natural ECM. Essentially, the scaffolds should be degradable, three-dimensional and highly porous with an interconnected pore network to favor flow transport of nutrients and metabolic waste as well as tissue integration and vascularization. As described above these requirements can be fulfilled by the combination of synthetic biopolymers and rapid prototyping technology. Based on this background, the authors of this chapter employed PCL and PCL-based block copolymers to develop scaffolds using the in house built DRBRP system (Fig. 12.5). We studied how the structural features (lay-down-pattern & filament distance, FD) and process conditions (pressure, temperature and deposition speed) influenced the morphological and mechanical characteristics of the scaffolds (Fig. 12.6). The viscoelastic property (i.e. the effect of change of cross head speed during compression) and in vitro degradation behavior of the scaffolds were also studied. The test results

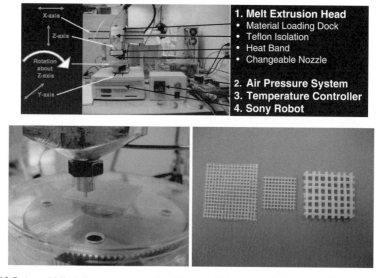

Fig. 12.5 A multidisciplinary group at the National University of Singapore did develop a RP system which can extrude natural and synthetic polymers (Hoque et al., 2005). The melt extrusion sep up (bottom left) allows the dispensing of various biodegradable and thermoplastic polymers

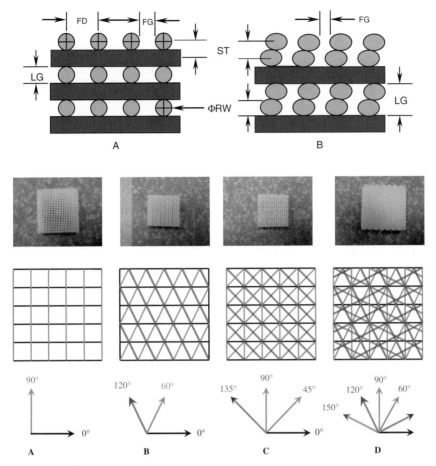

Fig. 12.6 (a) Models of lay-down patterns viewed in cross-section. (A): 0/90°, (B): 0/60/120° Symbols are RW: road width, FG: fill gap, LG: layer gap, ST: slice thickness. (b) Graphical illustration of 4 different lay down patterns and the pictures of actual scaffold produced by the RPBOD system. (A): 0/90°, (B): 0/60/120°, (C): 0/45/90/135° and (D): 0/30/60/90/120/150°

were compared between two polymers. All the polymers utilized were synthesized at the Centre de Recherche sur les Biopolymères Artificiels (UMR CNRS 5473, Montpellier, France). The details of the synthesis and characterization of the polymers were reported elsewhere (Huang et al., 2004; Hoque et al., 2005).

12.4.2 Scaffold Processing

The polymers were melted by electrical heating and directly extruded out through a minute nozzle of 500 μm internal diameter by means of compressed purified air to build 3D porous scaffolds layer by layer. The scaffolds were fabricated with three lay-down patterns (0/90°, 0/60/120° and 0/30/60/90/120/150°) and three FDs (1.00, 1.25 and 1.50 mm) by using appropriate positioning of the robotic control system. This system provides four-axes (Fig. 12.5) movement consisting of three simultaneous translational movements along the x-, y- and z-axes with an additional rotary motion about the z-axis. The three translational movements had positioning accuracy of up to 0.05 mm and a minimum step resolution of 0.014 mm. For all studies except to investigate the effect of process conditions, the liquefier temperature, extrusion pressure and deposition speed were set at 90 °C, 4.0 bars and 300 mm/min, respectively while the ambient temperature was maintained at 25 ± 2 °C. To investigate the influence of process parameters, a range of liquefier temperatures (80, 90 & 100 °C), extrusion pressures (3, 4 & 5 bars) and deposition speeds (240, 300 & 360 mm/min) was employed while the ambient temperature was also maintained at 25 ± 2 °C. The bulk scaffolds (e.g. 50.0 × 50.0 × 5.0 mm) were built on a flat plastic platform and upon fabrication were removed, and cut into smaller blocks (e.g. 6.0 × 6.0, × 5.0 mm) with an ultra sharp blade for further analyses. It is to note that as both polymers have very close melting temperatures (~65 °C) for convenience, the same process conditions were applied to them.

12.4.3 Characterization of Scaffolds

12.4.3.1 Scaffold Gross Morphology

The various scaffold architectures (lay-down patterns) are schematically shown in Fig. 12.6a/b. Square pores were resulted from the 0/90° lay-down pattern. Both 0/60° and 0/45° lay-down patterns produced triangular pores, whereas the 0/30° pattern formed complex polygonal pores. The micro-structural formability and internal morphology of as-fabricated scaffolds were evaluated by the scanning electron microscope (SEM) JSM-5800LV (JEOL USA, Peabody, MA). The gold-coated scaffold samples were observed both from top and cross-sectional views under SEM operated under high vacuum at an accelerating voltage of 15kV and a current of 60–90 mA.

The pores created were regular and coherent to the designs that indicated the feasibility of processing these polymers into 3-D porous scaffolds with fully interconnected, integrated and reproducible pore networks. It was also observed that filaments fused evenly at the junctions (Fig. 12.7 circles) that prevented interlayer delamination. This phenomenon evidenced the suitability of PCL and PCL-PEG rheological properties for melt extrusion and the capability of DRBRP technique to process these polymers into 3D porous scaffolds.

The pore openings or pore sizes formed in different directions of fabrication process varied in dimensions. In the x or y direction, pore openings were formed in-between the intercrossing of filaments, and are determined mainly by user-defined parameter settings like, FD. Whereas, the pore openings in the z-direction, were formed from voids produced by the stacking of filament layers, and hence their sizes were mainly dependent on the layer gap (Fig. 12.7). In case of scaffolds fabricated

12 Design, Fabrication and Physical Characterization of Scaffolds

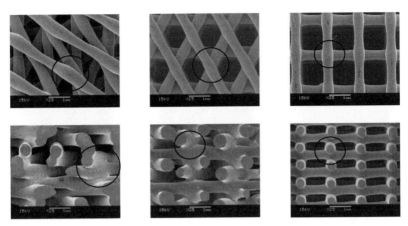

Fig. 12.7 SEM images of scaffold made of PCL and processed with a: FD = 1.5 mm, nozzle = 0.5 mm Circles show that at the bars and struts of the 6 angle lay down pattern show wider variation in diameter compared to the 3 and 2 angle pattern. The processing condition have to be optimized therefore for each pattern. E.g. processing a 6 angle pattern means that the contact surface for the extruded bar on the strut below is much higher compared to the 2 angle pattern

with various lay-down patterns but fixed FD, the pore openings (conventionally, FG) in x or y direction did not vary much due to change of filament angle orientation rather than only a little which was also due to the variation of filament diameter. On the contrary, the pore openings in z-direction (conventionally, LG) altered significantly due to the change of filament angle orientation; smallest (500±20 µm) in 2-angle pattern and largest (2500±100 µm) in 6-angle pattern (Fig. 12.7). Because, with the increase of number of angle-orientation the edge-to-edge vertical distance between layers of same filament alignment increases.

In case of scaffolds fabricated with various FDs but same lay-down pattern, the pore width in x or y direction varied significantly from 500±30 to 1000±50 µm, depending on the FD (Fig. 12.8a/b), whereas the pore openings in z-direction remained almost same, 250±25 µm irrespective of FD. The pore openings in z-direction irrespective of lay-down pattern and FD, can be increased by increasing the slice thickness (ST) and/or laying down more than one layer of filaments at the same angle orientation (Moroni et al., 2006). The ST can be increased only by a small amount which is ultimately regulated by nozzle diameter but laying down more than one layer of filaments at the same angle orientation can increase significantly.

In principle, the filament should be uniform in diameter throughout its length. However, at the junctions they became thicker compared to in-between the junctions. Most likely, it is due to the phenomenon that during fabrication, the solidified filaments of previous layer provide support at the junctions for the molten filaments of next layer. As a result, the molten polymer spreads at the junctions to some extent prior to solidification. Likewise, the road width (i.e. filament diameter) should correspond to the nozzle size. However, the filaments extruded through nozzle of 500 µm

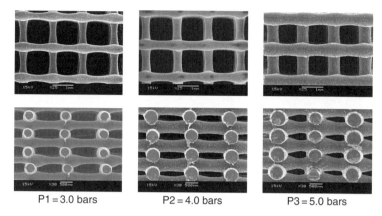

Fig. 12.8a SEM images and analysis of PCL scaffolds demonstrating the influence of extrusion pressure on their morphologies (nozzle size: 500 μm & FD: 1.5 mm); top row: plan view & bottom row: x-sectional view. (b) SEM images of PCL scaffolds demonstrating the influence of FD on pore morphologies (nozzle size: 500 μm & Pattern 0/90); top row: plan view & bottom row: x-sectional view

varied slightly as reported in. This may be because of the influence of process parameters such as liquefier temperature, extruding pressure and deposition speed.

The process parameters were varied primarily to study their influences on the filament diameter, in turn on scaffolds' porous characteristic that can result in a range of scaffold properties, and ultimately to find their best suited combination for targeted properties. The best selection of process parameters to fabricate 3D scaffold by a solid freeform fabrication technique involves complex interactions among hardware, software and material properties (Comb et al., 1994a). The experimental investigations revealed that the liquefier temperature, extruding pressure and deposition speed had direct influence on the material flow, in turn on road width and

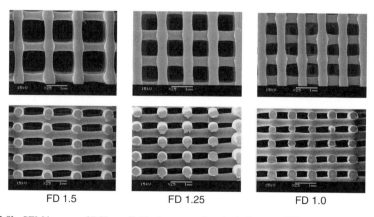

Fig. 12.8b SEM images of PCL scaffolds demonstrating the influence of FD on pore morphologies (nozzle size: 500 μm & Pattern 0/90); top row: plan view & bottom row: x-sectional view

consequently on the morphological and mechanical characteristics of the scaffolds. It was hypothesized that both polymers (PCL & PCL-PEG) have same rheological properties as they have almost same melting temperatures (~65 °C). Hence, to find best combination of process parameters, three values of each parameter were used (Temperatures 80, 90 & 100 °C; Pressures 3, 4 & 5 bars; Speed 240, 300 & 360 mm/min). To determine a best-suited value of any parameter for a given nozzle diameter of 500 µm, initially it was focused to achieve a target raster RW equivalent to nozzle diameter (i.e. 500 µm) with minimum fabrication time while structural integrity and reproducibility of the scaffold were maintained.

Increase of liquefier temperature resulted in thickening of extruded filaments and consequently, decreased the pore size and porosity at a specific extruding pressure and deposition speed. It is because of the fact that the fluidity of the polymer melt increases with the increase of temperature that makes easy and excessive dispensing of the polymer. Likewise, increase of extruding pressure caused the filaments to be thicker and thus decreased the pore size and porosity at a given condition of temperature and speed. Because at increased pressure excessive polymer melt extrudes out whereas, the nozzle moves at the same speed which makes the filament thicker. In contrast, at increased deposition speed the filaments became thinner and accordingly, the pore size and porosity were decreased. In this case the polymer melt comes out at a specific flow rate under certain conditions whereas, the nozzle drugs the melt at a faster rate which makes the filament thinner. It can also be explained by means of mathematical relationship as given below:

$$V_1 S_1 = V_2 S_2 \qquad (12.1)$$

where, V is the volume of melt that is in tern equivalent to the filament diameter and S is the deposition speed. If S increases the volume, i.e. diameter of the filament must decrease to maintain specific flow rate.

In case of selecting the best combination of these parameters, some factors need to be considered. For example, if the temperature is too low the polymer melt becomes too viscous to be dispensed and thus requires increased pressure which might impair the system. In addition, at low temperature there might be inadequate fusion (i.e. bonding) at the filament junctions that may lead to delamination of the layers and irregularity (Fig. 12.7) in structure because of insufficient superheat in the polymer melt. On the other hand, if the temperature raises too high polymer property might degrade even the hardware accessories (e.g. plastic tubing etc.) may be fused/burnt off. Likewise, the insufficient pressure and/or inappropriate dispensing speed might cause irregularity in filament deposition and ultimately the structural integrity will be lost. Entrapped air-bubbles (if any) might be another problem in generating consistent filaments because they interrupt the continuity of polymer flow. To overcome this, the polymer melt should be held at a constant liquefier temperature for 20 minutes to provide the opportunity for the air-bubbles to break out.

12.4.3.2 Scaffold Porosity

The porosities of the different scaffolds were measured as:

$$P = [(V_a - V_t)/V_a] \times 100\% \qquad (12.2)$$

where, V_a (mm^3) is the apparent volume of each scaffold block calculated from geometry, and V_t (mm^3) is the true volume of scaffold specimen measured by a gas pycnometer (Quantachrome Ultrapycnometer 1000, Quantachrome Corporation, Boynton Beach, FL) at 25 °C in pure argon.

The microtomography (μ-CT) analysis of as fabricated scaffold was also performed by Skyscan 1072 micro-CT desktop scanner (Skyscan, Belgium). The micro-CT was set at 19 μm resolution. 2D analyses and 3D reconstructions of core regions of the samples were performed which enabled to calculate the porosity and interconnectivity of the scaffolds. Strut and pore widths were measured in 2D images from each sample. Mimics software (Materialise, Belgium) by pre-processing with ImageJ was used for the 3D reconstructions. Sixty-two sequential 200 × 200 pixel images were cropped from the serial images taken from the center of each sample. Imported into Mimics, these serial core images were reconstructed into 3D volumetric models. Volume fraction and surface per unit volume were determined in 3D analysis, and relative area was measured in randomly selected slice images in 2D analysis. Thresholds of –872 to –185 Houndsfiled (HU) were inverted to allow measurement of the volume of all pore spaces within the model. Subsequently, a region-growing operation was performed, creating a mask consisting only of interconnected pore spaces. The volume for this region-grown mask was determined and the ratio of region-grown volume to the total volume was calculated. The percentage of this ratio is defined as the degree of interconnectivity. There was only a little difference in the porosity values measured by pycnometer method and micro-CT method.

The architectural influence (Table 12.2) on the morphology (pore shape, size and porosity) was studied by the scaffolds fabricated with various lay-down patterns (0/90°, 0/60/120° and 0/30/60/90/120/150°) but fixed FD (1.5 mm). The lay-down patterns of 0/90°, 0/60/120° and 0/30/60/90/120/150° are also called 2-angle, 3-angle and 6-angle patterns, respectively. The measured porosities were found to be in the range of 62–72%. Scaffold with large-angle or 2-angle pattern (0/90°) was composed of large square pores and thus produced high porosity (72%) i.e. more porous structure. On the other hand, the small-angle or 6-angle pattern (0/30/60/90/120/150°) produced small pores and consequently resulted in low porosity (62%) i.e. dense structure. Therefore, the lay-down pattern was found to be a good means to control the pore shape, size and porosity. The influence of FD on the morphological characteristics was investigated by the scaffold built with a single lay-down pattern (0/90°) and a series of FDs (1.00, 1.25 and 1.50 mm). The measured porosities of the scaffolds fell in the range of 55–72% (Table 12.2) which are in close agreement with each other. This result indicates that the filament distance also plays a significant role to control the porous characteristics of scaffolds. The micro-CT analysis demonstrates that both lay-down pattern and FD

Table 12.2 Influence of variation in extruding pressure on the morphological and mechanical properties (T = 1550C & Speed = 5.0 mm/sec)

Sample	Pressure (bar)	Filament dia (mm)	Pore size (mm)	Porosity (%)	Comp modulus (MPa)	Yield strength (MPa)
P1	4.5	0.41	1.09	67	2	19.3
P2	5.0	0.49	1.01	60	2.8	24.4
P3	5.5	0.64	0.86	51	3.6	35.8

Influence of variation in liquefier temperature on the morphological and mechanical properties (P = 5.0 bars, Speed = 5.0 mm/sec)

Sample	Temperature (°C)	Filament dia (mm)	Pore size (mm)	Porosity (%)	Comp modulus (MPa)	Yield strength (MPa)
T1	150	0.39	1.11	69	1.4	12.7
T2	155	0.49	1.01	60	2.8	24.4
T3	160	0.66	0.84	40	5	48.9

Influence of variation in dispensing speed on the morphological and mechanical properties (T = 155°C & P = 5.0 bars)

Sample	Speed (mm/sec)	Filament dia (mm)	Pore size (mm)	Porosity (%)	Comp modulus (MPa)	Yield strength (MPa)
S1	4.0	0.59	0.91	51	3.4	34.1
S2	5.0	0.49	1.01	60	2.8	24.5
S3	6.0	0.37	1.13	72	1.4	13.8

influence the scaffold's surface to volume ratio. The scaffolds with 2-angle pattern (0/90°) that resulted in highest porosity (72%) gave rise to highest surface to volume ratio (12.05 mm^2/mm^3). On the other hand, scaffolds with 6-angle pattern (0/30°), which produced lowest porosity (62%) resulted in lowest surface to volume ratio (9.02 mm^2/mm^3). Likewise, widest FD (1.50 mm) that produced highest porosity (72%) resulted in highest surface to volume ratio (12.05 mm^2/mm^3) and scaffolds with narrowest FD (1.00 mm) that provided lowest porosity (56%) produced lowest surface to volume ratio (7.21 mm^2/mm^3).

In principle, the ST can be equivalent to the nozzle diameter but in practice, it is set slightly to a lower value to improve layer-to-layer adhesion. For example, in case of nozzle diameter of 500 μm, the ST was set at 348 μm which resulted in good layer-to-layer adhesion (Figs. 12.7 and 12.9). However, the decrease in ST might compromise the scaffold porosity. Therefore, the pore width in x or y direction directly increased with the increase of FD and eventually increased the porosity. The scaffolds with widest filament distance (1.50 mm) produced largest pores (1000±50 μm), hence the structure became most porous (72% porosity). On the other hand, the scaffold with narrowest filament distance (1.00 mm) resulted in smallest pores (500±30 μm) and thus gave rise to the densest structure (55% porosity). This is simply because of the fact that when the filaments are laid far from each other less material is present in the bulk scaffold that makes more empty spaces available. On the other hand, when the filaments are laid closely more material is present and less empty spaces become available, and consequently the scaffold porosity is decreased. This can also be explained analytically by means of the equation of relative density of cellular solid, (square open-pore) given by Gibson & Ashby (1997):

$$\frac{\rho^*}{\rho_s} = 2\frac{t}{l}\left(1 - \frac{1}{2}\frac{t}{l}\right) \tag{12.3}$$

where, ρ^* is the density of cellular solid (scaffold), ρ_s is the density of material from which the cells (i.e. scaffolds) are made, l is the edge length (FD) and t is the edge thickness (road width or filament diameter) as shown in Fig. 12.7.

According to the Eq. (12.3), with a specific filament diameter the relative density (i.e. porosity) of scaffold increases with the increase of FD. The empirical relationship between pore size, porosity and FD is shown in Fig. 12.8. Experimentally, the pore size in x or y direction increased in a consistent and predictable manner with the increase of FD following a linear relationship as stated by Eq. (12.4), and accordingly the porosity also increased in the same manner following the relationship of Eq. (12.5). These empirical relationships now allow developing scaffolds with targeted pore morphology.

$$y = 1000x - 500 \tag{12.4}$$

$$y = 35x + 20.30 \tag{12.5}$$

12 Design, Fabrication and Physical Characterization of Scaffolds

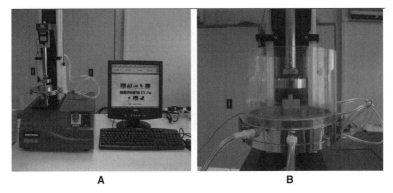

Fig. 12.9a compression test machine (INSTRON) A: Whole set up; B: Chamber simulated physiological conditions (PBS at 37 C)

Therefore, the filament distance also plays a significant role in controlling pore size in x or y direction, and consequently porosity of the scaffold. However, there should be a limit to increase the FD otherwise, the filament bars will slack in-between two junctions. The top and bottom bars even may merge together which will significantly decrease the porosity and ultimately interrupt the interconnectivity of the pores. Unlikely, the FD did not influence the pore height in z-direction. In addition, the variation in nozzle size can also influence the pore size and porosity of the scaffold with specific lay-down pattern and FD. With smaller nozzle size, keeping the pattern and FD constant, larger pores and thus, higher porosity can be obtained.

12.4.3.3 Mechanical Properties

In tissue engineering applications, porous scaffolds must have sufficient mechanical strength to restrain their initial structures after implantation, particularly in the reconstruction of hard, load-bearing tissues, such as bones and cartilages. The biostability of many implants depends on factors such as strength, elasticity, absorption at the material interface and chemical degradation. Therefore, the investigation of compressive properties is of primary importance in determining the suitability of the designed scaffold (Fig. 12.9a/b). Other mechanical properties like, tensile or flexural properties are secondary to the compressive properties for some basic reasons. The specimen specification for tensile and bending test s that requires dog bone and large bar shapes is a limiting factor as it is often time consuming and expensive

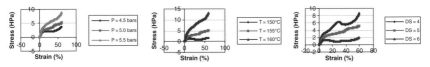

Fig. 12.9b Graphs showing the variation in compressive profiles due to change of process parameters; (a): Pressure, (b): Temperature & (c): Dispensing Speed

to prepare the scaffold samples accordingly. Therefore, the majority of research findings on tissue engineering scaffolds had been focused on their compressive properties when reporting on scaffolds mechanical properties. Most formulations, as reviewed by Gibson & Ashby (1997), found that the mechanical properties of a porous solid depended mainly on its relative density, the properties of the material that made up the pore edges or walls and the anisotropic nature, if any, of the solid.

As practical example, the compression test results of the scaffolds developed by DRBRP technique are described here. This test determined the influence of lay-down-pattern, variation of material nature, change in filament distance and influence of process parameters on the scaffolds' mechanical properties. For each structural configuration, five samples ($6.0 \times 6.0 \times 5.0$ mm) were tested. They were tested using a uniaxial testing system (Instron 4502) and a 1 kN load-cell (Canton, MA, USA) adopting the guidelines for compression testing of acrylic bone cement set in ASTM F451-99a. This is the latest edition of the standard currently used by a number of research groups (Peter et al., 1998; Thomson et al., 1998; Zein et al., 2002) to characterize the mechanical properties of bioresorbable scaffolds of similar geometry. The specimens were compressed in the z-direction of scaffold fabrication process (Fig. 12.9a) at a crosshead speed of 1 mm/min between two steel platens up to a strain level of approximately 60%. The modulus of elasticity called stiffness, E (MPa) was calculated as the slope of initial linear portion of the stress-strain curve neglecting any toe region due to the initial settling of the specimen. Compressive strength at yield σ_y (MPa) was defined as the intersection of the stress–strain curve with the modulus slope at an offset of 1.0% strain. For the mechanical properties measured, a Student's t test was performed in comparing results from two independent sample groups. In all the statistical tests performed, a significance level of 0.05 was used.

The responses of various scaffold designs to compression are demonstrated by plotting stress versus strain of up to 60% as shown in Fig. 12.9b. All the stress–strain curves reflected a typical behavior of a honeycomb-like porous material undergoing deformation (Gibson & Ashby, 1997; Moroni et al., 2006). As strain increased, the 3D pores of the scaffolds were crushed and underwent a densification process. When the rods and struts were crushed, the scaffold became stiffer and the stress level rose quickly. Therefore, stress–strain curves typically followed three distinct regions: (1) a linear–elastic region, (2) a plateau of roughly constant stress and (3) a final region of steeply rising stress as exemplified in Fig.12.9b. When the scaffolds were compressed in z-direction it was the filament junctions of adjacent layers that mainly bore the applied load at the beginning. In this case, the initial linear-elastic deformation involved significant shear deformation of the filament joints. On further compression, the linear-elastic regime was truncated by sliding of filament layers, which also manifested as a plateau of constant stress on the stress-strain curve. The final failure occurred when the filaments of adjacent layers were crushed. To strengthen the scaffold structure when compressed (in z-direction), a large number of filament joints would be expected. The strengthening effect can also be dependent on the bond strength i.e. the perfection of fusion between filaments at their joints, which in turn dependent on the design and process parameters of the fabrication technique.

To investigate the influence of lay-down pattern on mechanical properties, the scaffolds developed with two polymers (PCL and PCL-PEG) and three lay-down patterns (0/30, 0/60 and 0/90) were tested. The filament distance and nozzle size for all the scaffolds were 1.5 mm and 500 µm respectively. The values of compressive stiffness, E (MPa) and 1% offset yield strength, σ_y (MPa) are presented in Table 12.2. In case of both materials (PCL and PCL-PEG), the scaffolds with 0/90 lay-down pattern had highest stiffness and 1% offset yield strength. In comparison, the scaffolds with 0/30 lay-down pattern resulted in lowest stiffness and 1% offset yield strength. Therefore, 0/90 or 2-angle pattern scaffold was the stiffest, whereas 0/30/60/90/120/150 or 6-angle pattern scaffold was the most deformable. This scaling might be attributed to the fact that a smaller angle (e.g., 0/30) pattern has a wider fused zone at filament junctions than the larger angle (e.g., 0/90) pattern that likely act as sliding/sharing plane. When the scaffold is compressed, the filament layers can slide easily from each other in small angle pattern, thus explaining why the scaffold with 0/30 pattern became most deformable. On the other hand, in the 0/90 pattern the upper layer filaments are deposited exactly onto the lower layer filaments in the cross way and the junctions behave like columns when the scaffold is compressed. Therefore, the 0/90 pattern scaffold appeared as the most resistant to compression as confirmed by the statistical analysis.

The influence of FD on mechanical properties was investigated by compressing the scaffolds fabricated with two polymers (PCL and PCL-PEG) and three FDs (1.00, 1.25 and 1.50 mm). The lay-down pattern and the nozzle size for all the scaffolds were of 0/90 and 500 µm respectively. Under compression again all the scaffolds illustrated the typical behavior of honeycomb-like open-pore cellular solids, whereby their response to compressive load is porosity dependent (Gibson & Ashby, 1997; Moroni et al., 2006). The stiffness, E (MPa) and 1% offset yield strength, σ_y (MPa) are plotted as function of FD for both polymers as shown in Figures respectively. The filament distance had a direct influence on the porosity and consequently on mechanical properties of the scaffolds. In case of both materials (PCL and PCL-PEG), the scaffolds with narrowest filament distance (1.00 mm) producing smallest pores (500±10 µm) and lowest porosity (56%) resulted in highest stiffness and yield strength. On the other hand, scaffolds with widest filament distance (1.50 mm) having largest pores (1000±20 µm) and highest porosity (73%) produced lowest stiffness and yield strength. Therefore, the scaffold with narrowest filament distance was found to be the stiffest and that with widest filament distance was the most deformable. This variation in mechanical properties due to change of filament distance can be attributed to the fact that the filament junctions irrespective of material nature mainly resist the deformation when the scaffold is compressed (in z-direction). Under compression, the junctions in the z-direction behave like columns. The scaffold with narrowest filament distance has maximum number of columns in a specific area (e.g. 6 columns in 6mm scaffold sample of 1.0 mm FD; Fig. 12.4a) that resist the deformation most and thus the scaffold becomes stiffest. In contrast, the scaffold with widest filament distance has minimum number of columns in the same area (e.g. 4 columns in 6mm scaffold sample of 1.5mm FD; Fig. 12.4c) that resist the deformation least causing the scaffold most deformable.

To investigate the influence of process parameters (liquefier temperature, extruding pressure & deposition speed) on mechanical characteristics, the scaffolds were fabricated with two polymers (PCL and PCL-PEG) and three different values of each process parameters (Temperatures 80, 90 & 100 °C; Pressures 3, 4 & 5 bars; Speed 240, 300 & 360 mm/min). The lay-down pattern, the nozzle size and FD for all the scaffolds were of 0/90 and 500 μm and 1.5 mm respectively. The process parameters had direct influences on filament thickness (RW), in turn on porosity and consequently on mechanical properties (Fig. 12.9). In case of both materials (PCL and PCL-PEG), increase of liquefier temperature and extruding pressure resulted in thickening of extruded filaments and accordingly, decreased the pore size and porosity (Table 12.2) and hence increased the stiffness and yield strength. Unlikely, increase of deposition speed caused narrowing of filaments and consequently, increased the pore size and porosity and thus decreased the stiffness and yield strength. Therefore, the process parameters play significant role in controlling the mechanical properties of the scaffolds while other design parameters (lay-down pattern, nozzle size, FD) remain the same.

References

Ali E, Hamid M (1998) Densities of poly(ethylene glycol) + water mixtures in the 298.15–328.15 K temperature range. J Chem Eng Data 43:719–721

Ang TH, Sultana FSA, Hutmacher DW, Wong YS, Fuh JYH, M. XM, Loh HT, Burdet E and Teoh SH (2002) Fabrication of 3D chitosan- hydroxyapatite scaffolds using a robotic dispensing system. Materials Science & Engineering: C 20(1–2):35–42

Anna K (1997) Rapid prototyping trends. Rapid Prototyping J 3(4):150–152

Baldwin DF, Shimbo M, Suh NP (1995) The role of gas dissolution and induced crystallization during microcellular polymer processing: A study of poly(ethylene terephthalate) and carbon dioxide systems. J Eng Mater-T ASME 117(1):62–74

Brauker JH, Carr-Brendel VE, Martinson LA, Crudele J, Johnston WD, Johnson RC (1995) Neovascularization of synthetic membranes directed by membrane microarchitecture. J Biomed Mater Res 29(12):1517–1524

Calvert P, Crockett R (1997) Chemical solid free-form fabrication: Making shapes without molds. Chem Mater 9:650–663

Comb JW, Priedeman WR, Turley PW (1994a) Layered manufacturing control parameters and material selection criteria. Manufacturing Science and Engineering Vol. 2, PED-vol. 68–2, ASME

Comb JW et al (1994b) FDM technology process improvements. In: Marcus HL et al (eds) Proc of the Solid Freeform Fabrication Symposium 5:42–49

D'Urso PS, Earwaker WJ, Barker TM, Redmond MJ, Thompson RG, Effeney DJ, Tomlinson FH (2000) Custom cranioplasty using stereolithography and acrylic. Br J Plast Surg 53(3):200–4

Deng X, Zhu Z, Xiong C Zhang L (1997) J Polym Sci Part A: Polym Chem 35:703

Dobrzynski P, Kasperczyk J, Bero M (1999) Macromolecules 32:4735–4737

Endres M et al (2003) Osteogenic induction of human bone marrow derived mesenchymal progenitor cells in novel synthetic polymerhydrogel matrices. Tissue Eng 9:689–702

Feng XD, Song CX, Chen WS (1983), *J. Polym. Sci. Polym. Lett. Edn.* **21**, 593

Gibson LJ, Ashby MF (eds) (1997) Cellular solids: Structure and properties. Cambridge University Press, Cambridge

Harris LD, Kim BS, Mooney DJ (1998) Open biodegradable matrices formed with gas foaming. Journal of Biomedical Materials Research 42:396–402

Healy KE, Whang K, Thomas CH (1998) Method of fabricating emulsion freeze-dried scaffold bodies and resulting products. US Patent 5,723,508

Ho ST, Hutmacher DW (2005) Application of micro CT and computation modeling in bone tissue engineering. Computer-Aided Design 37(11):1151–1161

Hollister SJ (2006) Porous scaffold design for tissue engineering. Nat Mater 4(7):518–524

Hollister SJ et al (2000) An image based approach to design and manufacture craniofacial scaffolds. Int J Oral Maxillofac Surg 29:67–71

Holy CE, Shoichet MS, Davies JE (2000) Engineering 3-D bone tissue in vitro using biodegradable scaffolds: Investigating initial cell-seeding density and culture period. J Biomed Mater Res 51(3):376–382

Hoque ME, Hutmacher DW, Feng W, Li S, Huang M-H, Vert M, Wong YS (2005) Fabrication using a rapid prototyping system and in vitro characterization of PEG-PCL-PLA scaffolds for tissue engineering. J Biomater Sci Polym Ed. 16(12):1595–1610

Huang M-H, Li S, Hutmacher DW, Schantz J-T, Vacanti CA, Braud C, Vert M (2004) Degradation and cell culture studies on block copolymers prepared by ring opening polymerization of ε-caprolactone in the presence of poly(ethylene glycol). J Biomed Mater Res 69A: 417–427

Hutmacher DW (2000a) Polymeric scaffolds in tissue engineering bone and cartilage. Biomaterials 21:2529–2543

Hutmacher DW, Teoh SH, Zein I, Ng KW, Schantz JT, Leahy JC (2000b) Design and Fabrication of a 3D Scaffold for Tissue Engineering Bone. In: Synthetic bioabsorbable polymers for implants; Agrawal CM, Parr JE, Lin ST (eds), STP 1396-EB, p.152

Hutmacher DW (2001a) Scaffold design and fabrication technologies for engineering tissues — state of the art and future perspectives. J Biomater Sci Polym Edn 12(1):107–124

Hutmacher DW, Schantz T, Zein I, Ng KW, Teoh SH, Tan KC (2001b) Mechanical properties and cell cultural response of polycaprolactone scaffolds designed and fabricated via fused deposition modeling. J Biomed Mater Res 55:203

Hutmacher DW, Cool S (2007) Concepts of scaffold-based tissue engineering the rational to use solid free-form fabrication techniques. J Cell Mol Med. 11(4):654–669

Hutmacher DW, Sittinger M, Risbud MV (2004) Scaffold-based tissue engineering: Rationale for computer-aided design and solid free-form fabrication systems. TRENDS in Biotechnology 22(7):354–362

Hutmacher DW, Woodfield T, Dalton PD, Lewis JA (in press) Scaffold design and fabrication. In: Clemens van Blitterswijk (chief editor), Cancedda R, Hubbell J, Lindahl A, Thomsen P, Williams D (eds) Textbook on Tissue Engineering. Elsevier

Koch KU, Biesinger B, Arnholz C, Jansson V, (1998) Creating of bio-compatible, high stress resistant and resorbable implants using multiphase jet solidification technology. In: Time-Compression Technologies, Interactive Computing Europe, CATIA-CADAM Solutions, formation, International Business Machines Corporation – IBM: Time-Compression Technologies '98 Conference London, GB: Rapid News Publications, pp 209–214

Landers R, Hubner U, Schmelzeisen R, Mulhaupt R (2002b) Rapid prototyping of scaffolds derived from thermoreversible hydrogels and tailored for applications in tissue engineering. Biomaterials 23(23):4437–47

Landers R, Pfister A, Hubner U, John H, Schmelzeisen R, Mulhaupt R (2002a) Fabrication of soft tissue engineering scaffolds by means of rapid prototyping techniques. J Mater Sci 37(15):3107–3116

Landers R, Mülhaupt R (2000) Desktop manufacturing of complex objects, prototypes & biomedical scaffolds by means of computer-assisted design combined with computer-guided 3-D plotting of polymers & reactive oligomers. Macromolecular Materials & Engrg 282: 17–21

Lee JW, Hua F, Lee DS, (2001) Thermoreversible gelation of biodegradable poly(ε-caprolactone) and poly(ethylene glycol) multiblock copolymers in aqueous solutions. J. Controlled Release 73: 315–327

Leong KF, Cheah CM, Chua CK (2003) Solid freeform fabrication of 3-D scaffolds for engineering replacement tissues and organs. Biomaterials 24:2363–2378

Leong KF, Phua KK, Chua CK, Du ZH, Teo KO (2001) Fabrication of porous polymeric matrix drug delivery devices using the selective laser sintering technique. Proc Inst Mech Eng H 215:191–201

Li S, Garreau H, Pauvert B, McGrath J, Toniolo A, Vert M (2002) Enzymatic degradation of block copolymers prepared from ε-caprolactone and poly(ethylene glycol). Biomacromolecules 3:525–530

Li SM, Rashkov I, Espartero JL, Manolova N, Vert M (1996) Synthesis, Characterization, and Hydrolytic Degradation of PLA/PEO/PLA Triblock Copolymers with Long Poly(L-lactic acid) Blocks, J. Macromolecules 29(1): 57–62

Lo H, Ponticiello MS, Leong KW (1995) Fabrication of controlled release biodegradable foams by phase separation. Tissue Eng 1:15–28

Longhai P, Zhongli D, Mingxiao D, Xuesi C, Xiabin J (2003) Synthesis and characterization of PCL/PEG/PCL triblock copolymers by using calcium catalyst. Polymer 44:2025–2031

Lu CH, Lin WJ (2002). Permeation of protein from porous poly(ε-caprolactone) films. J Biomed Mater Res 63:220–225

Lu L, Mikos AG(1996) The importance of new processing techniques in tissue engineering. MRS Bulletin 21(11):28–32

Ma PX, Zhang RY (1999b) Synthetic nano-scale fibrous extracellular matrix. J Biomed Mater Res 46(1):60–72

Mano JF, Vaz CM, Mendes SC et al (1999) Dynamic mechanical properties of hydroxyapatite-reinforced and porous starch-based degradable biomaterials. J Mater Sci 10:857

Maquet V, Jerome R (1997) Design of macroporous biodegradable polymer scaffolds for cell transplantation. Mater. Sci. Forum 250:15–42

Middleton JC, Tipton AJ (1998) Synthetic biodegradable polymers as medical devices. Med Plastics Biomater Mag 3:30

Mikos AG, Sarakinos G, Leite SM, Vacanti JP, Langer R (1993b) Laminated 3-D biodegradable foams for use in tissue engineering. Biomaterials 14:323–330

Mikos AG, Sarakinos G, Vacanti JP, Langer RS, Cima LG (1996) Biocompatible polymer membranes and methods of preparation of 3-D membrane structures. US Patent 5,514,378

Mooney DJ, Baldwin DF, Suh NP, Vacanti JP, Langer R (1996) Novel approach to fabricate porous sponges of poly(D,L-lactic-co-glycolic acid) without the use of organic solvents. Biomaterials 17:1417–1422

Moroni L, de Wijn JR, van Blitterswijk CA (2006). 3-D fiber-deposited scaffolds for tissue engineering: Influence of pores geometry and architecture on dynamic mechanical properties. Biomaterials 27(7):974–785

Perrin DE, English JP (1997) Polyglycolide and polylactide. In: Domb AJ, Kost J, Wiseman DM (eds) Handbook of biodegradable polymers. Harwood Academic Publishers, New York, pp 3–27

Perrin DE, English JP (1998) Polyglycolide and polylactide. In: Domb AJ, Kost J, Wiseman DM (eds) Handbook of biodegradable polymers. Harwood Academic Publishers, New York, pp 3–27

Peter SJ, Miller ST, Zhu G, Yasko AW, Mikos AG (1998) In vivo degradation of a poly (propylene fumarate) β-tricalcium phosphate injectable composite scaffold, J Biomed Mater Res 41(1):1–7

Peters MC, Mooney DJ (1997) Synthetic extracellular matrices for cell transplantation. Mater Sci Forum 250:43–52

Pfister A et al (2004) Biofunctional rapid prototyping for tissue engineering applications: 3-D bioplotting versus 3-D printing. J Polym Sci 42:624–638

Pitt CG (1990) poly(ε-caprolactone) and its copolymers. In: Chasin M, Langer R (eds) Biodegradable polymers as drug delivery systems. Marcel Dekker, New York p 71–120

Thomson RC, Yaszemski MJ, Powers JM, Mikos AG (1998) Biomaterials 19:1935

Rai B, Teoh SH, Ho KH, Hutmacher DW, Cao T, Chen F, Yacob K (2004) The effect of rhBMP-2 on canine osteoblasts seeded onto 3-D bioactive polycaprolactone scaffolds. Biomaterials 25(24):5499–506

Rashkov I, Manolova N, Li S, Espartero JL, Vert M (1996) Macromolecules 29:50

Recheis W, Weber GW, Schafer K, Knapp R, Seidler H (1999) Virtual reality and anthropology. Eur J Radiol 31(2):88–96

Sachlos E, Czernuszka JT (2003) Making tissue engineering scaffolds work. Review on the application of solid freeform fabrication technology to the production of tissue engineering scaffolds. European Cells and Materials 5:29–40

Saltzman WM (1999) Delivering tissue regeneration Nature Biotechnol 17:534–535

Sanghera B, Naique S, Papaharilaou Y, Amis A (2001) Preliminary study of rapid prototype medical models. Rapid Prototyping J 7(5):275–284

Shalaby SW, Johnson RA (1994) Synthetic absorbable polyesters. In: Shalaby SW (eds) Biomedical polymers: Designed-to-degrade systems. Hanser Publishers, New York, pp 1–34

Shastri VP, Martin I, Langer R (2000) Macroporous polymer foams by hydrocarbon templating. Proc Natl Acad Sci USA, 97(5):1970–1975

Sherwood JK et al (2002) A 3-D osteochondral composite scaffold for articular cartilage repair. Biomaterials 23:4739–4751

Thomson RC, Yaszemski MJ, Powers JM, Mikos AG (1995a) Fabrication of biodegradable polymer scaffolds to engineer trabecular bone. J Biomater Sci Polym Edn 7(1):23–38

Thomson RC, Shung AK, Yaszemski MJ, Mikos AG (2000) Polymer scaffold processing. In: Lanza RP, Langer R, Vacanti JP (eds) Principles of tissue engineering, 2nd edn. Academic Press, San Diego Chapter 21, 251–262

Vanezi P, Vanezis M, McCombe G, Niblett T (2000) Facial reconstruction using 3-D computer graphics. Forensic Sci Int, 108(2):81–95

Vats A, Tolley NS, Polak JM, Gough JE (2003) Clin Otolaryngol 28:165

Vert M, Li S, Spenlehauer G, Guerin P (1992) Bioresorbability and biocompatibility of aliphatic polyesters. J Mater Sci: Mater in Med 3:432–446

Wang F, Shor L, Darling A, Khalil S, Sun W, Güçeri S, Lau A (2004) Precision extruding deposition and characterization of cellular poly-ε-caprolactone tissue scaffolds. Rapid Prototyping J 10(1):420–429

Wang N, Butler JP, Ingber DE (1993) Mechanotransduction across the cell surface and through the cytoskeleton. Science 260:1124–1127

Whang K, Thomas CK, Nuber G, Healy KE (1995) A novel method to fabricate bioabsorbable scaffolds. Polymer 36(4):837–842

Whang K, Tsai DC, Nam EK, Aitken M, Sprague SM, Patel PK, Healy KE (1998) Ectopic bone formation via rhBMP-2 delivery from porous bioabsorbable polymer scaffolds. J Biomed Mater Res 42:491–499

Widmer MS, Mikos AG (1998) Fabrication of biodegradable polymer scaffolds for tissue engineering. In: Patrick Jr CW, Mikos AG, McIntire LV (eds) Frontiers in tissue engineering. Elsevier Sciences, New York pp 107–120

Woodfield TBF, Malda J, De Wijn J, Péters F, Riesle J, Van Blitterswijk CA (2004) Design of porous scaffolds for cartilage tissue engineering using a 3-D fiber-deposition technique. Biomaterials 25:4149–4161

Xiong Z et al (2001) The fabrication of porous poly [L-lactic acid] scaffolds for bone tissue engineering via precise extrusion. Scr Mater 45:773–779

Yang SF, Leong KF, Du ZH, Chua CK (2001) The design of scaffolds for use in tissue engineering: Part 1-traditional factors. Tissue Eng 7(6):679–89

Yuan ML, Wang YH, Li XH, Xiong CD, Deng XM (2000) Polymerization of Lactides and Lactones. 10. Synthesis, Characterization, and Application of Amino-Terminated Poly(ethylene glycol)-co-poly(ε-caprolactone) Block Copolymer, Macromolecules 33(5):1613–1617

Zein IW, Hutmacher DW, Tan KC, Teoh SH (2002) Fused deposition modeling of novel scaffold architecture for tissue engineering application. Biomaterials 23:1169–1185

Zhou YF, Chou AM, Li ZM, Hutmacher DW, Sae-Lim V, Lim TM (2007) Combined marrow stromal cell sheet techniques and high strength biodegradable composite scaffolds for engineered functional bone grafts. Biomaterials 28(5):814–824 Feb 2007

Zollikofer CPE, De Leon MSP (1995) Tools for rapid prototyping in biosciences. IEEE Comput Graph, 15(6):48–55

Index

α-L-guluronic (G blocks) 168-169, 273
β-D-mannuronic (M blocks) 168
β-Tricalcium phosphate (β-TCP) 135, 164, 178

Abaqus 93, 101–102
Acrylonitrile butadiene styrene 12, 181, 277
Adhesion 90, 110, 132, 144, 160, 167, 193, 286
Agar 163, 192, 195
Agarose 146
Alginate 132, 138, 146–147, 154, 168–169, 192, 195
Aliphatic polyurethanes 195, 277
Alkaline phosphatise 137
Allergic reaction 3, 266
Allogenic 151
Allograft 177
Alloys 132, 176, 179, 187
Alumina 158, 192, 198
Angiogenesis 125, 152, 211, 229, 266
Angiography 50, 65
Anisotropic 269, 289
Antibiotic 3
Apatite 178, 192, 197, 201
Articular cartilage 161, 278
Augmented reality 22
Autograft 176, 177
Autologous 151
   Bone 4
   Bone graft 1
   Implant 2
Automated fabrication *see Rapid prototyping*
Axis Aligned Bounding Box (AABB) 31

Ballistic particle manufacturing (BPM) 175, 189–193, 196, 198, 200, 202–203
Bioabsorbable 131
Bioactive molecules 135, 137, 143
Biobuild™ software 9
Bioceramic 154, 196, 202
Biochemical properties 271, 277, 280

Biocompatibility 16, 71, 132, 134, 137, 142–143, 147, 153–154, 195–196, 199, 200, 266, 273, 278–279
Biocompatible 1, 2, 4, 146, 158–159, 170, 194–195, 198, 202, 211, 241, 260
   Glass 178
Biocomposite 69, 82, 83
Biodegradability 71, 132, 153, 273
Biodegradable 193, 202, 211, 280
   Copolymer 158
   Material 82, 154, 170, 273
   Polymer 203, 272, 277
   Polyurethane 277
   Resin 158
   Scaffold 266
   Synthetic polymer 137, 265
Bioengineering 91, 265
Bioglass 178
Biomechanical properties 265
Biomimetic 90, 93, 161, 169
Biomolecules 211–212, 218, 266
Bioplotter 163–164, 187, 192, 193, 195, 276, 277, 280
Bioplotting 163, 187, 195, 277
Biopolymer 154, 183, 189, 193–195, 201–202, 281
   Degradation 122
Bioresorbable 194, 196, 197, 273
   Ceramics 197
   Co-polymer 273
   Implants 20
   Material 16
   Polymer 16, 158, 260, 271, 273
   Thermoplastic 241
   Scaffold 273, 274, 289
Biostereometrics 45
Biosynthetic 132
   Material 132
   Scaffolds 146

Biphasic calcium phosphate 178
Blood 61
   Flow 51, 61
   Velocity 61
   Vessel 52, 70, 125, 211
BMP growth factor 162
BMP-7 159
Bone
   Cells 4, 145, 193, 199
   Graft 1, 3, 177, 178, 179, 198
   Marrow 200, 274
   Matrix 93
   Morphogenic protein (BMP) 153, 159, 179
   Regeneration 126, 149, 260, 269
   Surgery simulation 22, 24–26, 30–33, 40
   Surgery simulator 23, 27, 38
Bovine articular chondrocytes 278
Break-out quality metric (BQM) 246–247, 249, 254

CAD/CAM 4, 9, 16, 213, 243
CAD
   Software 6, 8, 9, 11, 13, 16, 181, 199
   Methods 15, 116
   Modelling 73
Calcium phosphates 178
Carbon dioxide (CO_2) 134–136, 155, 167, 183, 186, 241
Cartilage 134, 144, 147, 161, 229, 271, 278, 288
   Formation 126
   Repair 159
   Scaffold 126
   Tissue 144
      Engineering 194, 276
CAT *see Computed axial tomography*
Ceramics 132, 154, 158, 178, 181, 182, 183, 189, 196, 197, 198, 211,
Chemical degradation 289
Chemotactic 125–126
Chemotaxis 266
Chitin 138, 271
Chitosan 154, 164, 192, 202, 203, 204, 271
Chloroform 133, 134, 136, 143
Chondrocytes 126, 134, 137, 144, 146, 147, 164, 165, 278
Citrate 134
CNC *see Computer numerical control*
Collagen
   Glycosaminoglycan 153, 271
   Type I 144
   Type II 144

Collision detection 34, 38–40
Compressive
   Stiffness 160–162, 290
   Strength 35, 202, 289
Computed
   Axial tomography (CAT) *see Computed tomography*
   Tomography (CT) 4, 5, 27, 50, 73, 255
Computer
   Aided
      Design (CAD) 4, 5, 9, 70, 89, 116, 141, 181, 199, 213, 275
      Manufacturing (CAM) 13, 213
      System for tissue scaffolds or casts (CASTS) 69, 71, 73, 75, 77, 79–80, 85
      Tissue engineering (CATE) 69, 89, 91, 92, 108
   Assisted cranioplasty 1, 2, 4, 15, 16
   Models 96, 233, 271
   Numerical control (CNC) 5, 11, 12, 181
   Simulation 21
Conductivity 115, 118–120, 134, 216, 261
Contractile cardiac hybrids 168
Conventional methods 70, 131, 133, 154
Copolymerization 241, 273
Coral derived hydroxyapatite/calcium carbonate composites 177
Cornstarch 167
Cortical bone 35, 177
Cranial implant 2, 4, 12, 13, 15, 16
Cranioplasty 1–4, 9, 13, 15–17
Crosslinking 139, 146
Crystallinity 83, 122, 183, 194, 273
Crystallization temperature 274
CT *see Computed tomography*
Cuboctahedron 96, 102
Cytokines 125, 279
Cytotoxicity 75, 226–227

Data acquisition 5, 6, 27, 156
Data Exchange Format (DXF) 9
Data transfer 6, 9
Degradability 139, 273
Dehydrothermal treatment 139
Deltoidal Icositetrahedron 97
Dental
   Applications 178
   Implants 52
Design matrix 245, 249–250, 252
Desktop robot based rapid prototyping (DRBRP) 276, 280–282, 289
DICOM 6, 8
Differentiation 19, 90, 125, 132, 152–153, 196, 211, 222–224, 228

Index 295

Diffusivity 116, 119–123, 261
Digital
 Light projection (DLP) 189–190
 Radiography Imaging (X-Rays) 50
 Subtraction Angiography (DSA) 50
Dimensional Accuracy quality metric (DQM) 247, 249, 254
Disc scaffolds 82
Distilled water 137, 141, 167
DNA 220, 221
Duraform^{TM} polymide 69, 74, 76–77, 79–80, 82, 85

Elastic modulus 93, 101–103, 105, 109, 123, 258, 258, 269
Electro-Encephalo-Graphy (EEG) 51
Embryonic 222
Endothelial cell 125, 151, 168, 221, 224
Epithelial
 Cells 144, 167, 195
 Membrane 144, 146
Ethanol 138, 164, 167, 196, 202
Evolutionary structural optimization (ESO) 93
Extracellular matrix 125, 132, 143, 146, 147, 211, 214, 274

FDM *see Fused deposition modelling*
FEA *see Finite element analysis*
FEM *see Finite element method*
Fibre
 Bonding 142–143, 274
 Mesh 142, 145
Fibrin 125, 132, 164, 192
Fibroblast 72, 126, 142, 144, 153, 159, 162, 167, 197, 273
Fibrous tissue 16, 143, 145
Fibrovascular 145
Finite element
 Analysis (FEA) 89–91, 100–102
 Method (FEM) 30
First-order erlan probability density function 122
Fluid
 Flow 90, 105, 218–219, 239–240, 270–271
 Pressure 120
 Property 91
 Rheology 219
 Velocity 120, 126
 Viscosity 120
Fluorapatite 154
Food and drug administration 193, 273
Force
 Generation 38
 Modelling 25, 34

Freeform fabrication *see also Solid freeform fabrication*
Freeze drying 137–138, 155, 202–203, 274
Functional Magnetic Resonance Imaging (fMRI) 50–51, 61
Fused deposition
 Modelling (FDM) 5, 11–12, 16, 72, 160, 175, 181–182, 270, 276–277
 Of ceramics (FDC) 182

Gas foaming and particle leaching (GF/PL) 135–138, 274
Gelatine 163, 167, 195
Gelation 138, 168
Genotype 213, 220, 228, 229
Geometric modelling 25–26, 28, 40, 64
Glucose 144
Glycerol 214, 223, 277
Göpferich model 122
Graphic rendering 23–26, 30–31, 40
Growth factors 133, 143, 146, 151, 152, 153, 211, 228, 260, 278, 279

Haptic
 Device 21, 25–26, 34, 37–38
 Feedback 23–24, 36–37, 40
 Rendering 25–27, 33–34, 37–40
Haptotactic 125, 126
HDPE *see high density polyethylene*
Health Service Institution 15
Hexagonal cells 269
Hexahedron 108
Hexamethylene diisocyanate 139
Hexane 134
High density polyethylene (HDPE) 192, 202, 203
Homogeneous material 25
Homogenization theory 91, 94, 118, 120
Homologous 28, 47–48, 239
Honeycomb 269
 Architecture 270
 Pores 160
Honeycomb-like
 Architectures 273
 Open-pore 290
 Porous material 290
 Scaffolds 160
Hormones 279
Human osteosarcoma cells 223–224, 226
Hyaluronic acid 140, 145, 154
Hydrocarbon 134
 Solvent 134
 Templating 274
Hydrogels 146, 147, 154, 168, 211
Hydrolysis 123, 135, 193, 271

Hydrolytic attack 270
Hydrophilic 116, 146, 273
  Poly(lactic-coglycolic acid)/poly(vinyl alcohol) 140
  Polymer 146
Hydrophobic 122, 146, 273
Hydrostatic pressure 135, 218
Hydroxyapatite
  Coated titanium 117
  Scaffolds 71, 135

Icositetrahedron 96
IGS *see Image-guided surgery*
Image fusion 45, 51–52, 59
Image matching 45, 48
Image processing 4–6, 8, 25, 27, 40, 60, 95
Image segmentation 8, 28, 45, 59, 60, 61, 64
Image-based design (IBD) 116–117, 120, 122, 124
Image-guided surgery (IGS) 62
Imaging modalities 45, 50–51, 56, 61, 116, 132
Immune 151, 271
Infrared 157, 158, 194
Initial Graphics Exchange Specification (IGES) 9
Internal morphology 155, 167, 282
Interstitial fluid 93
Interstitial spaces 138
Intestinal epithelial cells 195
Intracellular 220
Isophorone diisocyanate 277
Isotropic 4–5, 8, 91, 95, 97, 101–102

Knee
  Implants 2
  Replacement 34
  Surgery 22–23

Laser
  Engineering net shaping (LENS) 187
  Scanning 241, 243
Layer object manufacturing (LOM) 73, 175, 186
Ligaments 22, 26
Long bones 31, 35

Magnetic resonance
  Angiography (MRA) 50
  Imaging (MRI) 27, 50, 155
  Spectroscopy (MRS) 51
Magneto-Encephalo-Graphy (MEG) 51
Mammalian cells 218, 220–221, 270
Material degradation 122–124, 219
Matlab 250, 253

Mechanical strength 27, 71, 146–147, 194, 196, 201–202, 279, 282
Medical
  Image segmentation 60
  Photogrammetry 45
  Imaging 1, 13, 15, 27, 45–46, 50–51, 58–60, 64, 70–71, 73, 132, 155
Melt moulding molding 90, 140–141, 274
Methylene chloride 133–134, 137, 143
Micro-computed tomography 85, 259, 260, 271
Microstereolithography 157
MicroTec 157
Milling 5, 11–13, 24, 25, 36
Mimics® software 9, 73, 74, 285, 286
Mineralised bone tissue (MBT) 201
Minimally invasive surgery 22
Minipig mandibular condyle 255
Morphological analysis 104
Motion tracking 37, 45, 61–62, 64
MR *see Magnetic resonance imaging*
MRA *see Magnetic resonance imaging*
MRI *see Magnetic resonance imaging*
Multi jet modelling (MJM) 190
Multiphase jet solidification (MJS) 276
Muscle cells 136, 143–144, 162, 167
Musculoskeletal tissue 269

National science foundation 131, 151
Nerves 151, 229, 271
Neurogenesis 211, 229
Nickel-titanium alloy 177
Non-antigenic 279
Non-carcinogenic 279
Non-toxic 153, 181, 202, 279
Non-uniform rational b-spline (NURBS) 13–15, 116
Non-woven 142, 155
Nuclear Magnetic Resonance (NMR) 50
Nucleation 135–136, 139
NURBS *see Non-uniform rational b-spline*

Octahedron-tetrahedron 75, 99
Octree 30, 32–33, 40
  Data 32
  Modelling 30
  Nodes 30, 32
  Representation 30
  Structure 30
Octree-based
  Marching cube 32
  Volume 25, 30
Ocular repair 229
Oligoetherurethanes 277
Oligoethylene oxide 163, 277

Ophthalmology 61
Optical Coherence Tomography (OCT) 50
Orthopaedic
  Surgery 2–3
  Tissue 89
Orthotropic 97
Osteoblasts 16, 71, 137, 143–144, 147, 176, 202
Osteoclasts 176
Osteoconductive 3, 16, 177, 179
Osteocyte 126, 176
Osteogenesis 144,179
Oxygen 115, 146, 152, 270

PCL–HA 160, 277
PCL–PEG 272, 274, 282, 284, 290–291
PCL–TCP 160, 241
PEEK 69, 72, 78–83
PEEK–HA 69, 78–79, 81–82
PEGDMA 195
PEG–PLLA 136
PEGT/BT 161, 193–194, 278
Pentane 134
Permeability 89–99, 108, 115–116, 118–121, 124, 126, 146, 267, 270
PET see Positron emission tomography
Phase separation 136–137, 139–140, 145, 155, 162, 274
Phenotype 213, 219–220, 222, 228–229
Photogrammetry 45–48, 50, 56, 58–64
Photomicrographs 248
Photosensitive 11, 156, 188–189, 198
Physical modelling 26, 40
PLGA/HA 144
PLLA/PLGA 134, 144,
PLLA/TCP 163, 277
Poisson's ratio 93, 102
Poly(butylenes terephthalate-block-oligoethylene oxide) 163
Poly(butylmethacrylatemethylmethacrylate) (P(BMA/MMA) 164
Poly(caprolactone) 163
Poly(D,L-lactic acid-co-glycolic acid) 123
Poly(dioxanone) 154
Poly(DL-lactide) (PDLLA) 193, 273
Poly(ethylene glycol) (PEG) 132, 134, 144, 146, 272–274, 278, 282, 284, 290–291
Poly(ethylene glycol) dimethacrylate (PEGDMA) 187, 192, 195
Poly(ethylene glycol)-terephthalate-poly(butylene terephthalate) (PEGT/PBT) 161, 193, 194, 278
Poly(ethylene oxideterephtalate)-co-poly(butylene terephtalate) (PEOT/PBT) 164, 277
Poly(glycolic acid) (PGA) 154
Poly(hydroxybutyrate) 154
Poly(hydroxybutyrateco-valeriate) 163
Poly(lactic acid) (PLA) 154
Poly(lactide-co-glycolide) 163
Poly(lactides) 163
Poly(L-lactic acid) (PLLA) 72, 133–134, 136, 138–140, 142–145, 162, 167, 192–193, 196, 276
Poly(α-hydroxy ester)s 271
Polyanhydride 154, 271
Polybutylene teraphthalate 154
Polyethylene oxide 154
Polygonal pores 269, 282
Polyhedra 69, 73, 89–91, 96–100, 102–103, 105, 108–109, 111
Polyhedron shapes 71
Polylactide-coglycolide (PLGA) 72, 133, 134, 136, 137, 139, 140, 141, 143, 144, 145, 167, 192, 195, 196, 272
Polymer chain length 122
Polymer-processing techniques 274
Polymethylmethacrylate (PMMA) 1
Polyorthoesters 271
Polyphosphazens 271
Polypropylene (PP) 192, 202
Polyurethane 1, 142, 192, 195, 277
Polyvinylalcohol (PVA) 192, 203
Pore morphology 188
Pore structure 16, 133, 136, 138, 146, 153, 155, 162
Porogen 72, 120, 133–134, 137, 141
Porous
  Bone replacement 175–176, 180–181, 192–193, 195, 204
  Implants 176
  Material 104, 111, 176, 179, 180, 269, 271, 290
  Scaffold 120, 131, 133, 137–138, 140–143, 167, 175, 184, 187, 192, 194, 269, 276, 282
  Structure 117, 136, 162, 171, 204, 267, 269, 286
Positron Emission Tomography (PET) 27, 50, 51, 52
PP-TCP 160
Precise extrusion manufacturing (PEM) 162, 276
Precision extruding deposition (PED) 162, 278
Pro/Engineer 69, 73–75, 85, 199
Progenitor cells 16
Proliferation 70–71, 125–126, 142, 144–145, 152–153, 162, 167, 179, 198, 270, 274
Prostheses 91

qCT *see Quantitative computed tomography*
Quadtree 31, 40
Quantitative computed tomography (qCT) 94
Quartic Bezier Patch 15

Radiography 50
Radiometric 45, 48, 52, 56, 58, 61–62
Rapid micro product development (RMPD) 157
Reactive biosystems 163
Resorption kinetics 267, 269, 277
Reverse engineering 16, 181
Rheological properties 273, 282, 284
Rhombicuboctahedron 96, 98–99
Rhombitruncated cuboctahedrons 108

Scaffold
 Architecture 16, 71, 120, 124, 127, 269, 271, 274, 282
 Assembly 73, 95, 105, 106, 108
 Degradation 125–128, 132
 Design 70–71, 91, 108–109, 115–117, 122–123, 126, 243, 255–256, 260, 265–267, 269–270, 289
 Engineering 105
 Fabrication 69–70, 72, 73, 78, 85, 90, 117, 131, 159, 265, 274–275, 289
 Porosity 133, 267, 285–286, 288
 Performance 116, 122, 127
 Topology 116
Scaffold-based tissue engineering concepts 266
Scanning electron microscope (SEM) 78, 282
Seeding 79, 101–102, 144, 211–212, 265, 277
Semi-synthetic 132
Sensor geometry 45, 48, 50, 58
SFF *see solid freeform fabrication*
Shape deposition manufacturing (SDM) 276
Shear
 Deformation 290
 Energy 35
 Flow 218–219
 Forces 213, 220
 Rates 167
 Strain 126, 218–219
 Stress 218–222
Single-photon emission computed tomography (SPECT) 27, 50
Skin 2, 3, 4, 6, 8, 70, 153, 155, 229, 247, 269, 271
 Depth 215–216
 Graft 229

SLS *see Selective laser sintering*
Sodium
 Alginate 168
 Chloride 133, 136, 154
 Hydroxide 164, 202–203
Solid architectures 97
Solid freeform fabrication (SFF) 89, 94, 155, 233, 241, 284
Solvent-casting particle-leaching (SC/PL) 131, 133–134, 136–137, 143
SPECT *see Single-photon emission computed tomography*
Spring damping force model 37
Stainless steel 132, 187, 276
Starch 185, 271, 277
Starch-based
 Polymer 141
 Powder 167
Stem cell 143–144
Stereolithography (SLA) 5, 11, 13, 15, 95, 157, 158, 175, 187, 275
Stereolithography apparatus (SLA) *see Stereolithography*
Stereoscopic 63, 77
STL file 5, 8, 9, 79, 155, 243
Stochastic 122
Stokes flow 119–120
Stromal cells 200, 274
Support structures 11, 12, 163, 168, 183, 187, 190
Surface rendering 31, 32, 33, 37, 40
Surface roughness 115–116
Surface tension 82, 217, 219, 234–235, 240, 261, 271
Surgical planning 13
Syngeneic 151

Tartrate 133
Teflon mould 134
Therics incorporated 167
Thermal degradation 134
Three-dimensional
 Reconstruction 50, 58, 64, 73, 255, 285
 Printing (3DP) 166, 175, 185, 276
Time Dependent 115–116, 122–125, 127–128, 216
Tissue
 Formation 134, 143–144, 161, 270
 Grafts 229, 266
 Primitives 89, 92, 94, 96, 99, 101, 104, 106–107
 Regeneration 70, 89, 90–91, 115, 118, 125, 127–128, 143, 151–153

Tissue-derived shapes 97
Titanium 3, 177
  Absorption 216
  Alloys 176, 187
  Cranioplasty 3, 13
  Mesh 1, 3
  Mini-plate 3
  Plate 3
  Screws 52
  Sheet 1–3
Titanium-based alloys 132
Topology 116–117
  Optimization 92, 120, 123–124
Total quality metric (TQM) 249–254
Trabecular
  Bone 35, 89, 91, 94–96, 98, 108, 126, 159, 198, 240, 259
  Scaffolds 161–162
Triakistetrahedron 96
Tricalcium phosphate (TCP) 135, 162, 192, 201–202, 241, 276
Truncated
  Dodecahedron 96
  Icosahedron 96, 102
  Octahedron 96, 108
  Tetrahedron 96, 109
Two photon lithography 189

Ultrasonic Imaging 50

Vascular 20, 132, 154, 167, 229
Vascularization 153, 199, 266, 267, 269, 270, 280

Virtual
  Bone surgery 21–23, 25–26, 36–38, 40
  Reality 21, 22, 25, 40
Viscoelastic property 281
Viscosity 29, 120, 137, 164, 190, 234, 235, 236, 240, 244, 261
Viscous behaviour 274
Void fraction quality metric (VQM) 248–249, 254
Volume rendering 24, 29, 31, 33, 37
Von mises stress 93, 111
Voxel-based
  3D model data 156
  Data 155
  Modelling 28
  Representation 28
  Sculpting 25

Water 3, 5, 59, 123, 134, 137, 139–141, 146, 154, 158, 167, 169, 178, 181, 183, 185, 189, 195–196, 199–200, 203
  Diffusion 123
  Diffusivity 122–123
Wound 4, 23, 144, 146, 279
Woven 142

Xenogenic 151
Xenograft 177
X-rays 27, 50, 51

Yield strength 16, 159–160, 258, 269, 287, 290–291
Young's modulus 94, 102, 126

Printed in the United States of America.